Practical Handbook of
MATERIAL FLOW ANALYSIS

Practical Handbook of
MATERIAL FLOW ANALYSIS

Paul H. Brunner and Helmut Rechberger

LEWIS PUBLISHERS

A CRC Press Company
Boca Raton   London   New York   Washington, D.C.

Library of Congress Cataloging-in-Publication Data

Brunner, Paul H., 1946-
    Practical handbook of material flow analysis / by Paul H. Brunner and Helmut Rechberger.
        p. cm. — (Advanced methods in resource and waste management series ; 1)
    Includes bibliographical references and index.
    ISBN 1-5667-0604-1 (alk. paper)
    1. Materials management—Handbooks, manuals, etc. 2. Environmental engineering—Handbooks, manuals, etc. I. Rechberger, Helmut. II. Title. III. Series.

TS161.B78 2003
658.7—dc21   2003055150

This book contains information obtained from authentic and highly regarded sources. Reprinted material is quoted with permission, and sources are indicated. A wide variety of references are listed. Reasonable efforts have been made to publish reliable data and information, but the author and the publisher cannot assume responsibility for the validity of all materials or for the consequences of their use.

Neither this book nor any part may be reproduced or transmitted in any form or by any means, electronic or mechanical, including photocopying, microfilming, and recording, or by any information storage or retrieval system, without prior permission in writing from the publisher.

The consent of CRC Press LLC does not extend to copying for general distribution, for promotion, for creating new works, or for resale. Specific permission must be obtained in writing from CRC Press LLC for such copying.

Direct all inquiries to CRC Press LLC, 2000 N.W. Corporate Blvd., Boca Raton, Florida 33431.

Trademark Notice: Product or corporate names may be trademarks or registered trademarks, and are used only for identification and explanation, without intent to infringe.

Visit the CRC Press Web site at www.crcpress.com

© 2004 by CRC Press LLC
Lewis Publishers is an imprint of CRC Press LLC

No claim to original U.S. Government works
International Standard Book Number 1-5667-0604-1
Library of Congress Card Number 2003055150
Printed in the United States of America 1 2 3 4 5 6 7 8 9 0
Printed on acid-free paper

Dedication

To Sandra and Heidi

Preface

About 40 years ago, Abel Wolman coined the term *metabolism of cities* in an article for *Scientific American*. His pioneering view of a city as a living organism with inputs, stocks, and outputs of materials and energy has since inspired many others. Today, there are numerous studies describing metabolic processes of companies, regions, cities, and nations. While the phenomenology of the anthroposphere is described in several books and papers, there is no widely accepted methodology for applying these concepts. This handbook was written to help in establishing and disseminating a robust, transparent, and useful methodology for investigating the material metabolism of anthropogenic systems.

After many years of using and developing material flow analysis (MFA), we have seen the value of applying this method in various fields such as environmental management, resource management, waste management, and water quality management. We have written this book to share our experience with engineering students, professionals, and a wider audience of decision makers. The aim is to promote MFA and to facilitate the use of MFA in a uniform way so that future engineers have a common method in their toolboxes for solving resource-oriented problems.

The hidden agenda behind the handbook comprises two objectives: resource conservation and environmental protection, otherwise known as "sustainable materials management." We believe that human activities should not destroy or damage natural resources and systems. Future generations must be able to enjoy resources and the environment as we do. We also believe that this goal can be achieved if technology and social sciences are developed further. The case studies presented in this book exemplify the potential of MFA to contribute to sustainable materials management.

This is a handbook directed toward the practitioner. The 14 case studies demonstrate how to apply MFA in practice. The exercises in the "Problem" sections that appear throughout the book serve to deepen comprehension and expertise. The MFA tool has not yet been perfected, and there is much room for further refinement. If the reader finds that the handbook promotes understanding of anthropogenic systems and leads to better design of such systems, then we have accomplished our goals. Since this book is not the final work on the subject, we would appreciate any comments and suggestions you may have. Our main hope is that this handbook encourages application of and discussion about MFA. We look forward to your comments on the Web site www.iwa.tuwien.ac.at/MFA-handbook.htm, where you will also find the solutions to the exercises presented in this handbook.

We are grateful to Oliver Cencic, who wrote Sections 2.3 and 2.4 in Chapter 2, about data uncertainty and MFA software, and who contributed substantially to Case Study 1. The support of the members of the Waste and Resources Management Group at the Vienna University of Technology in editing the final manuscript is greatly acknowledged. Demet Seyhan was instrumental in preparing the case study on phosphorus. Bob Ayres and Michael Ritthoff supplied important comments on Chapter 2, Section 2.5. We are indebted to Inge Hengl, who did all the artwork and expertly managed the genesis and completion of the manuscript. Helmut Rechberger personally thanks Peter Baccini for conceding him the time to work on this handbook. Finally, we are particularly grateful for critical reviews by Bob Dean, Ulrik Lohm, Stephen Moore, and Jakov Vaisman, who evaluated a first draft of this handbook.

Paul H. Brunner
Helmut Rechberger

Authors

Paul H. Brunner, recognized for his outstanding research work in the fields of waste and resource management, is a professor at the Vienna University of Technology Institute for Water Quality and Waste Management, Austria. Together with Peter Baccini from ETH Zurich, he published the groundbreaking book *Metabolism of the Anthroposphere*, presenting a new view of the interactions among human activities, resources, and the environment. His work at the Institute for Water Quality and Waste Management focuses on advanced methods for waste treatment and on methods to assess, evaluate, and design urban systems. For more than 30 years, Dr. Brunner has been engaged in research and teaching in the United States, Europe, and Asia. He travels frequently throughout the world, lecturing on the application of material flow analysis as a tool for improving decision making in resource and waste management.

Helmut Rechberger is currently a research scientist and lecturer in the fields of waste and resource management at ETH Swiss Federal Institute of Technology Department of Resource and Waste Management, Zurich, Switzerland. He began his academic career as a process engineer at the Vienna University of Technology in Austria with a pioneering Ph.D. thesis on the application of the entropy concept in resource management. Since then, he has been engaged in research and teaching in both the United States and Europe and he has become internationally acknowledged for his research work in the fields of resource management and waste-treatment technology. His current research interests are focused on the further development of methods to advance sustainable regional and urban resource management. Dr. Rechberger was recently appointed professor at the Technische Universität Berlin, Germany.

Table of Contents

Chapter 1 Introduction 1
1.1 Objectives and Scope 1
1.2 What Is MFA? 3
1.3 History of MFA 5
    1.3.1 Santorio's Analysis of the Human Metabolism 5
    1.3.2 Leontief's Economic Input–Output Methodology 8
    1.3.3 Analysis of City Metabolism 8
    1.3.4 Regional Material Balances 9
    1.3.5 Metabolism of the Anthroposphere 12
    1.3.6 Problems — Sections 1.1–1.3 13
1.4 Application of MFA 13
    1.4.1 Environmental Management and Engineering 14
    1.4.2 Industrial Ecology 14
    1.4.3 Resource Management 16
    1.4.4 Waste Management 17
    1.4.5 Anthropogenic Metabolism 19
        1.4.5.1 Unprecedented Growth 19
        1.4.5.2 Anthropogenic Flows Surpass Geogenic Flows 22
        1.4.5.3 Linear Urban Material Flows 23
        1.4.5.4 Material Stocks Grow Fast 25
        1.4.5.5 Consumption Emissions Surpass Production
                Emissions 25
        1.4.5.6 Changes in Amount and Composition of Wastes 28
1.5 Objectives of MFA 28
    1.5.1 Problems — Sections 1.4–1.5 29
References 30

Chapter 2 Methodology of MFA 35
2.1 MFA Terms and Definitions 35
    2.1.1 Substance 35
    2.1.2 Good 36
    2.1.3 Material 37
    2.1.4 Process 37
    2.1.5 Flow and Flux 39
    2.1.6 Transfer Coefficient 40
    2.1.7 System and System Boundaries 43
    2.1.8 Activities 44
    2.1.9 Anthroposphere and Metabolism 48

		2.1.10	Material Flow Analysis	49
		2.1.11	Materials Accounting	51
		2.1.12	Problems — Section 2.1	51
2.2	MFA Procedures			53
	2.2.1	Selection of Substances		54
	2.2.2	System Definition in Space and Time		56
	2.2.3	Identification of Relevant Flows, Stocks, and Processes		58
	2.2.4	Determination of Mass Flows, Stocks, and Concentrations		59
	2.2.5	Assessment of Total Material Flows and Stocks		61
	2.2.6	Presentation of Results		63
	2.2.7	Materials Accounting		64
		2.2.7.1	Initial MFA	64
		2.2.7.2	Determination of Key Processes, Flows, and Stocks	65
		2.2.7.3	Routine Assessment	66
	2.2.8	Problems — Section 2.2		67
2.3	Data Uncertainties			69
	2.3.1	Propagation of Uncertainty		69
		2.3.1.1	Gauss's Law	69
		2.3.1.2	Monte Carlo Simulation	72
	2.3.2	Least-Square Data Fitting		75
		2.3.2.1	Geometrical Approach	76
		2.3.2.2	Analytical Approach	78
	2.3.3	Sensitivity Analysis		79
2.4	Software for MFA			80
		2.4.1	General Software Requirements	80
	2.4.2	Special Requirements for Software for MFA		82
	2.4.3	Software Considered		83
	2.4.4	Case Study		83
	2.4.5	Calculation Methods		85
	2.4.6	Microsoft Excel		86
	2.4.7	Umberto		89
		2.4.7.1	Program Description	89
		2.4.7.2	Quickstart with Umberto	89
		2.4.7.3	Potential Problems	111
	2.4.8	GaBi		113
		2.4.8.1	Program Description	113
		2.4.8.2	Quickstart with GaBi	114
		2.4.8.3	Potential Problems	132
	2.4.9	Comparison		132
		2.4.9.1	Trial Versions	132
		2.4.9.2	Manuals and Support	132
		2.4.9.3	Modeling and Performance	132
2.5	Evaluation Methods for MFA Results			133
	2.5.1	Evaluation Methods		135
		2.5.1.1	Material-Intensity per Service-Unit	136
		2.5.1.2	Sustainable Process Index	137

		2.5.1.3	Life-Cycle Assessment	140
		2.5.1.4	Swiss Ecopoints	141
		2.5.1.5	Exergy	142
		2.5.1.6	Cost–Benefit Analysis	145
		2.5.1.7	Anthropogenic vs. Geogenic Flows	147
		2.5.1.8	Statistical Entropy Analysis	149
References				159

Chapter 3		Case Studies		167
3.1	Environmental Management			168
	3.1.1	Case Study 1: Regional Lead Pollution		168
		3.1.1.1	Procedures	169
		3.1.1.2	Results	180
		3.1.1.3	Basic Data for Calculation of Lead Flows and Stocks in the Bunz Valley	184
	3.1.2	Case Study 2: Regional Phosphorous Management		184
		3.1.2.1	Procedures	185
		3.1.2.2	Results	191
	3.1.3	Case Study 3: Nutrient Pollution in Large Watersheds		194
		3.1.3.1	Procedures	195
		3.1.3.2	Results	196
	3.1.4	Case Study 4: MFA as a Support Tool for Environmental Impact Assessment		200
		3.1.4.1	Description of the Power Plant and Its Periphery	202
		3.1.4.2	System Definition	203
		3.1.4.3	Results of Mass Flows and Substance Balances	204
		3.1.4.4	Definition of Regions of Impact	207
		3.1.4.5	Significance of the Coal Mine, Power Plant, and Landfill for the Region	210
		3.1.4.6	Conclusions	213
	3.1.5	Problems — Section 3.1		213
3.2	Resource Conservation			215
	3.2.1	Case Study 5: Nutrient Management		215
		3.2.1.1	Procedures	216
		3.2.1.2	Results	218
	3.2.2	Case Study 6: Copper Management		220
		3.2.2.1	Procedures	222
		3.2.2.2	Results	228
		3.2.2.3	Conclusions	235
	3.2.3	Case Study 7: Construction Wastes Management		235
		3.2.3.1	The "Hole" Problem	236
		3.2.3.2	Use of MFA to Compare Construction Waste-Sorting Technologies	239
		3.2.3.3	Procedures	239
		3.2.3.4	Results	242

		3.2.3.5 Conclusions 250
	3.2.4	Case Study 8: Plastic Waste Management 251
	3.2.5	Problems — Chapter 3.2 254
3.3	Waste Management 256	
	3.3.1	Use of MFA for Waste Analysis 258
		3.3.1.1 Direct Analysis 259
		3.3.1.2 Indirect Analysis 260
	3.3.2	MFA to Support Decisions in Waste Management 269
		3.3.2.1 Case Study 11: ASTRA 269
		3.3.2.2 Case Study 12: PRIZMA 279
		3.3.2.3 Case Study 13: Recycling of Cadmium by MSW Incineration 287
	3.3.3	Problems — Section 3.3 290
3.4	Regional Materials Management 291	
	3.4.1	Case Study 14: Regional Lead Management 292
		3.4.1.1 Overall Flows and Stocks 292
		3.4.1.2 Lead Stock and Implications 293
		3.4.1.3 Lead Flows and Implications 293
		3.4.1.4 Regional Lead Management 294
	3.4.2	Problems — Section 3.4 295
References 296		

Chapter 4	Outlook: Where to Go? 301
4.1	Vision of MFA 301
4.2	Standardization 304
4.3	MFA and Legislation 305
4.4	Resource-Oriented Metabolism of the Anthroposphere 307
	4.4.1 MFA as a Tool for the Design of Products and Processes 307
	4.4.2 MFA as a Tool to Design Anthropogenic Systems 307
	4.4.3 MFA as a Tool to Design the Anthropogenic Metabolism 308
References 309	

Index 311

1 Introduction

The river Jordan is the great source of blessing for the Holy Land. It's the life source for Palestine, for Israel. This river of blessing flows into the lake of Galilee, and anyone who has ever visited there, at any time of year, will remember the banks of that lake as paradise. Then the Jordan flows out of that lake and on, and eventually empties into the Dead Sea. But this body of water is absolutely dead. No fish can live in it. Its shores are parched desert. The difference between these two bodies of water is that the Jordan flows into the lake of Galilee and then out again: The blessing flows in and the blessing flows out. In the Dead Sea, it only flows in and stays there.

David Steindl-Rast and Sharon Lebell,[1] in *Music of Silence*

1.1 OBJECTIVES AND SCOPE

The meeting was long and intense. After all, more than $50 million had been invested in this mechanical waste-treatment plant, and still the objective of the treatment, namely to produce recycled materials of a given quality, could not be reached. Engineers, plant operators, waste-management experts, financiers, and representatives from government were discussing means to improve the plant to reach its goals. A chemical engineer took a piece of paper and asked about the content of mercury, cadmium, and some other hazardous substances in the incoming waste. The waste experts had no problem indicating a range of concentrations. The engineer then asked about the existing standards for the products, namely compost and cellulose fibers. Again, he got the information needed. After a few calculations, he said, "If the plant is to produce a significant amount of recycling material at the desired specifications, it must be able to divert more than 80% of the hazardous substances from the waste received to the residue for landfilling. Does anybody know of a mechanical treatment process capable of such partitioning?" Since none of the experts present was aware of a technology to solve the problem at affordable costs, the financiers and government representatives started to question why such an expensive and state-of-the-art plant could not reach the objective. It was the mayor of the local community, who experienced most problems with the plant because of citizens complaints about odors and compost quality, who said, "It seems obvious: garbage in, garbage out. What else can you expect?"

The purpose of this book is to prevent such debacles. The methods presented will enable the reader to design processes and systems that facilitate careful resource management. The term *resources* in this context stands for materials, energy, the environment, and wastes. Emphasis is placed on the linkage between sources, pathways, and sinks of materials, always observing the law of conservation of matter. The book is a practical handbook, and it is directed toward the practitioner. Hence, many case studies, examples, and problems are included. Readers who take advantage of these exercises will soon become well acquainted with the techniques needed for successful application of material flow analysis (MFA).

In addition to serving as a practical handbook, this volume also contains directions for readers who are interested in sustainable resource management. The authors share the opinion that the benefits of progress in production economy and technology will be maximized through long-term protection of the environment and the judicious use of resources in a careful, nondissipative way. In this book, they provide evidence that current management of the anthroposphere may result in serious long-term burdens and that changes are needed to improve opportunities for current and future generations. Indeed, some encouraging changes are already taking place that have proved to be feasible, and they will eventually improve the quality of life. The authors are convinced that if decisions for changes are based on MFA, among other criteria, they will yield even better results. They recognize the need for balance not only for technical systems, but also for social systems. The opening quotation by Steindl-Rast and Lebell links these two systems in a compelling way, demonstrating the value of the MFA approach for any field. The current book, however, has been written for a technical readership. Since the authors are engineers and chemists, social science issues are sometimes mentioned in this volume but are never discussed to the necessary depth.

This book is directed toward engineers in the fields of resource management, environmental management, and waste management. Professionals active in the design of new goods, processes, and systems will profit from MFA-based approaches because they facilitate the inclusion of environmental and resource considerations into the design process. The potential audience comprises private companies and consulting engineers operating in the fields mentioned above, government authorities, and educational institutions at the graduate and postgraduate level. In particular, the book is designed as a textbook for engineering students who are looking for a comprehensive and in-depth education in the field of MFA. It is strongly recommended that the students focus on the case studies and problems, which illustrate how MFA is applied in practice and how to interpret and use the results. At the end of relevant chapter sections, a related problem section allows readers to exercise newly acquired knowledge, to learn applications of MFA, to gain experience, and to check their ability to solve MFA problems.

The book is structured into four chapters. Chapter 1 provides a short overview of MFA followed by a discussion of the history, objectives, and application range of MFA. Chapter 2 is the most important from a methodological point of view; it explains comprehensively the terms, definitions, and procedures of MFA, and much of the discussion focuses on the use of software suited for MFA. Chapter 3 provides case studies that illustrate the application range of MFA. The case studies demonstrate that (1) the concept of MFA goes beyond simple input and output balances of single processes and (2) analysis of the flows and stocks of a complex real-world system is a challenging and often an interdisciplinary task. The book ends with Chapter 4 and a short outlook on potential future developments. Literature references are given at the end of each chapter. Problem sections appear as subchapters where appropriate.

Introduction

1.2 WHAT IS MFA?

Material flow analysis (MFA) is a systematic assessment of the flows and stocks of materials within a system defined in space and time. It connects the sources, the pathways, and the intermediate and final sinks of a material. Because of the law of the conservation of matter, the results of an MFA can be controlled by a simple material balance comparing all inputs, stocks, and outputs of a process. It is this distinct characteristic of MFA that makes the method attractive as a decision-support tool in resource management, waste management, and environmental management.

An MFA delivers a complete and consistent set of information about all flows and stocks of a particular material within a system. Through balancing inputs and outputs, the flows of wastes and environmental loadings become visible, and their sources can be identified. The depletion or accumulation of material stocks is identified early enough either to take countermeasures or to promote further buildup and future utilization. Moreover, minor changes that are too small to be measured in short time scales but that could slowly lead to long-term damage also become evident.

Anthropogenic systems consist of more than material flows and stocks (Figure 1.1). Energy, space, information, and socioeconomic issues must also be included if the anthroposphere is to be managed in a responsible way. MFA can be performed without considering these aspects, but in most cases, these other factors are needed to interpret and make use of the MFA results. Thus, MFA is frequently coupled with the analysis of energy, economy, urban planning, and the like.

A common language is needed for the investigation into anthropogenic systems. Such commonality facilitates comparison of results from different MFAs in a transparent and reproducible way. In this handbook, terms and procedures to analyze, describe, and model material flow systems are defined, enabling a comprehensive, reproducible, and transparent account of all flows and stocks of materials within a system. The methodology, presented in greater detail in Chapter 2, is well suited as a base to build MFA software tools (see Section 2.4).

The term *material* stands for both substances and goods. In chemistry, a *substance* is defined as a single type of matter consisting of uniform units.[2] If the units are atoms, the substance is called an element, such as carbon or iron; if they are molecules, it is called a chemical compound, such as carbon dioxide or iron chloride. *Goods* are substances or mixtures of substances that have economic values assigned by markets. The value can be positive (car, fuel, wood) or negative (municipal solid

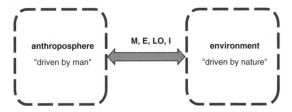

FIGURE 1.1 The two systems "anthroposphere" and "environment" exchange flows of materials (M), energy (E), living organisms (LO), and information (I).

waste, sewage sludge). In economic terms, the word *goods* is more broadly defined to include *immaterial* goods such as energy (e.g., electricity), services, or information. In MFA terminology, however, the term *goods* stands for material goods only. Nevertheless, the link between goods as defined by MFA and other goods as used by economists can be important when MFA is applied, for example, for decisions regarding resource conservation.

A *process* is defined as a transport, transformation, or storage of materials. The transport process can be a natural process, such as the movement of dissolved phosphorous in a river, or it can be man made, such as the flow of gas in a pipeline or waste collection. The same applies to transformations (e.g., oxidation of carbon to carbon dioxide by natural forest fires vs. man-made heating systems) and storages (e.g., natural sedimentation vs. man-made landfilling).

Stocks are defined as material reservoirs (mass) within the analyzed system, and they have the physical unit of kilograms. A stock is part of a process comprising the mass that is stored within the process. Stocks are essential characteristics of a system's metabolism. For steady-state conditions (input equals output), the mean residence time of a material in the stock can be calculated by dividing the material mass in the stock by the material flow in or out of the stock. Stocks can stay constant, or they can increase (accumulation of materials) or decrease (depletion of materials) in size.

Processes are linked by *flows* (mass per time) or *fluxes* (mass per time and cross section) of materials. Flows/fluxes across systems boundaries are called *imports* or *exports*. Flows/fluxes of materials entering a process are named *inputs*, while those exiting are called *outputs*.

A *system* comprises a set of material flows, stocks, and processes within a defined boundary. The smallest possible system consists of just a single process. Examples of common systems for investigations by MFA are: a region, a municipal incinerator, a private household, a factory, a farm, etc. The *system boundary* is defined in space and time. It can consist of geographical borders (region) or virtual limits (e.g., private households, including processes serving the private household such as transportation, waste collection, and sewer system). When the system boundary in time is chosen, criteria such as objectives, data availability, appropriate balancing period, residence time of materials within stocks, and others have to be taken into account. This is discussed further in Chapter 2, Section 2.1.7.

In addition to the basic terms necessary to analyze material flows and stocks, the notion of *activity* is useful when evaluating and designing new anthropogenic processes and systems. An activity comprises a set of systems consisting of flows, stocks, and processes of the many materials that are necessary to fulfill a particular basic human need, such as to nourish, to reside, or to transport and communicate. Analyzing material flows associated with a certain activity allows early recognition of problems such as future environmental loadings and resource depletions. One of the main questions for the future development of mankind will be "Which sets of processes, flows, and stocks of goods, substances, and energy will enable long-term, efficient, and sustainable feeding of the increasing global population?" Equally important is the question, "How to satisfy the transportation needs of an advanced global population without compromising the future resources of mankind?" When

Introduction

alternative scenarios are developed for an activity, MFA can help to identify major changes in material flows. Thus, MFA is a tool to evaluate existing systems for food production, transportation, and other basic human needs, as well as to support the design of new, more efficient systems.

1.3 HISTORY OF MFA

Long before MFA became a tool for managing resources, wastes, and the environment, the mass-balance principle has been applied in such diverse fields as medicine, chemistry, economics, engineering, and life sciences. The basic principle of any MFA — the conservation of matter, or input equals output — was first postulated by Greek philosophers more than 2000 years ago. The French chemist Antoine Lavoisier (1743–1794) provided experimental evidence that the total mass of matter cannot be changed by chemical processes: "Neither man made experiments nor natural changes can create matter, thus it is a principle that in every process the amount of matter does not change."[3]

In the 20th century, MFA concepts have emerged in various fields of study at different times. Before the term *MFA* had been invented, and before its comprehensive methodology had been developed, many researchers used the law of conservation of matter to balance processes. In process and chemical engineering, it was common practice to analyze and balance inputs and outputs of chemical reactions. In the economics field, Leontief introduced input–output tables in the 1930s,[4,5] thus laying the base for widespread application of input–output methods to solve economic problems. The first studies in the fields of resource conservation and environmental management appeared in the 1970s. The two original areas of application were (1) the metabolism of cities and (2) the analysis of pollutant pathways in regions such as watersheds or urban areas. In the following decades, MFA became a widespread tool in many fields, including process control, waste and wastewater treatment, agricultural nutrient management, water-quality management, resource conservation and recovery, product design, life cycle assessment (LCA), and others.

1.3.1 SANTORIO'S ANALYSIS OF THE HUMAN METABOLISM

One of the first reports about an analysis of material flows was prepared in the 17th century by Santorio Santorio (1561–1636), also called S. Sanctuarius.[6] The similarity of the conclusions regarding the analysis of the anthropogenic metabolism between Santorio and modern authors is astonishing. Santorio was a doctor of medicine practicing in Padua and a lecturer at the University in Venice. His main interest was to understand the human metabolism. He developed the first method to balance inputs and outputs of a person (Figure 1.2 A,B,C). Santorio measured the weight of the person, of the food and beverages he ate, and of the excretions he gave off. The result of his investigation was disappointing and surprising at the same time: he could not close the mass balance. However, he found that the visible material output of a person was less than half what the person actually takes in. He suspected that some yet unknown insensible perspiration left the body at night. Thus, he wrapped the person during the night's sleep in a hide. But the little amount of sweat he

FIGURE 1.2 The experimental setup of Santorio Santorio (1561–1636) to analyze the material metabolism of a person. A: The person is sitting on a chair attached to a scale. The weight of the food and the person are measured. B: Despite the fact that all human excreta were collected and weighed, input A and output B do not balance. What is missing? C: Santorio's measurements could not confirm his hypothesis that an unknown fluidum leaves the body during the night, but they proved that more than half of the input mass leaves the body by an unknown pathway.

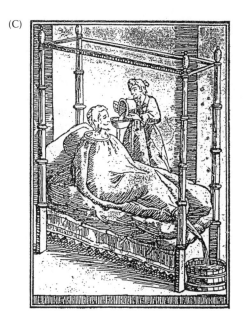

FIGURE 1.2 Continued.

collected by this procedure did not account for the large missing fraction. It was not known yet in the 16th century that the volume of air a person breathes has a certain mass. It should be remembered that Santorio was living a century before Lavoisier investigated oxidation processes and proved the existence of oxygen. Hence, Santorio did not care for the air a person was breathing, neither the intake of fresh air nor the emission of spent air. In 1614, Santorio published a book about his metabolic studies, *De Medicina Statica Aphorismi*, that made him widely known as the "father of the science of human metabolism." He concluded that "a doctor, who carries the responsibility for the health of his patients, and who considers only the visible processes of eating end excreting, and not the invisible processes resulting in the loss of insensible perspiration, will only mislead his patients and never cure them from their disease." Considering that Santorio was certainly the first and probably the only one who performed such metabolic studies at that time, this statement may have secured him the visit of many patients.

The experiments of Santorio allow conclusions similar to recent MFA studies. First, it is still impossible to evaluate and optimize anthropogenic systems, i.e., to cure the patient, without knowing the material flows and stocks — the metabolism of the system. Second, it is difficult to balance a process or system if basic information about this process or system, such as major input and output goods, is missing. Third, it happens quite often that inputs and outputs of a process or system do not balance, pointing to new research questions. And fourth, analytical tools are often not appropriate or precise enough to measure changes in material balances with the necessary accuracy.

1.3.2 LEONTIEF'S ECONOMIC INPUT–OUTPUT METHODOLOGY

Wassily W. Leontief (1906–1999) was an American economist of Russian origin. His research was focused on the interdependence of anthropogenic production systems. He searched for analytical tools to investigate the economic transactions between the various sectors of an economy. One of his major achievements, for which he was awarded the 1973 Nobel Prize in economic sciences, was the development of the input-output method in the 1930s.[4,5] At the core of the method are so-called input-output tables. These tables provide a method for systematically quantifying the mutual interrelationships among the various sectors of a complex economic system. They connect goods, production processes, deliveries, and demand in a stationary as well as in a dynamic way. The production system is described as a network of flows of goods (provisions) between the various production sectors. Input-output analysis of economic sectors has become a widespread tool in economic policy making. It proved to be highly useful for forecasting and planning in market economies as well as in centrally planned economies, and it was often applied to analyze the sudden and large changes in economies due to restructuring.

In order to investigate the effect of production systems on the environment, the original method was later expanded to include production emissions and wastes. More recently, the input-output method has been incorporated into LCA to establish the economic input–output LCA method.[7] This expansion provides a means of assessing the relative emissions and resource consumption of different types of goods, services, and industries. The advantage of using input-output methodology for LCA is the vast amount of available information in the form of input-output tables for many economies. This information can be used for LCA as well.

1.3.3 ANALYSIS OF CITY METABOLISM

Santorio analyzed the physiological, "inner" metabolism of humans, but this is only a minor part of the modern anthropogenic turnover of materials. The "outer" metabolism, consisting of the use and consumption of goods not necessary from a physiological point of view, has grown much larger than the inner metabolism. Hence, in places with a high concentration of population and wealth such as modern urban areas, large amounts of materials, energy, and space are consumed. Today, most cities are rapidly growing in population and size, and they comprise a large and growing stock of materials.

The first author to use the term *metabolism of cities* was Abel Wolman[8] in 1965. He used available U.S. data on consumption and production of goods to establish per capita input and output flows for a hypothetical American city of 1 million inhabitants. He linked the large amounts of wastes that are generated in a city to its inputs. The complex urban metabolism has also fascinated other authors, who have developed more specific methods to quantify the urban turnover of energy and materials and investigated the effects of the large flows on resource depletion and the environment. Two prominent examples are the studies of Brussels by Duvigneaud and Denayeyer-De Smet[9] and of Hong Kong by Newcombe et al.[10]

In 1975, Duvigneaud and Denayeyer-De Smet analyzed the city of Brussels using natural ecosystems as an analogy. They assessed the total imports and exports of goods such as fuel, construction materials, food, water, wastes, sewage, emissions, etc. in and out of the city and established an energy balance. The authors concluded that Brussels was highly dependent on its hinterland, with the city importing all its energy from external sources. Since solar energy theoretically available within the city equalled Brussels's entire energy demand, this dependency could be reduced by shifting from fossil fuels to solar energy. Water produced within the city by precipitation is not utilized; all drinking water is imported. Materials such as construction materials and food are not recycled after their use and are exported as wastes. The linear flows of energy and materials result in high pollution loads that deteriorate the quality of the water, air, and soil of the city and its surroundings. The authors point out the necessity of changing the structures of cities in a way that improves the utilization of energy and materials, creates material cycles, and reduces losses to the environment. Similar to Santorio's observation regarding the relationship of metabolism to the health of a person, the authors conclude that efforts to ensure the continual welfare of a city must be guided by knowledge of the city's metabolism. They recognize that only an interdisciplinary approach will succeed in analyzing, defining, and implementing the necessary measures of change.

At about the same time, in the beginning of the 1970s, Newcombe and colleagues started their investigation into the metabolism of Hong Kong. This Asian city was experiencing a rapid transition period of high population growth and intense economic development due to its privileged position at the interface between Western trade and Eastern production and manufacturing. Hong Kong was an ideal case for metabolic studies, since city limits coincide more or less with state boundaries. Thus, in contrast to the Brussels study, economic data from state statistics were available for an accurate assessment of the import and export goods of Hong Kong. Also, Hong Kong differs from Brussels in its high population density and its lower per capita income and material throughput. The authors found that the material and energy used for Hong Kong's infrastructure was about one order of magnitude smaller than in more highly developed cities. They concluded that a worldwide increase in material consumption to the level of modern cities would require very large amounts of materials and energy and would have negative impacts on global resources and the environment. They also stated that in order to find sustainable solutions for the future development of cities, it is necessary to know the urban metabolism. Thus, it is important to be able to measure the flows of goods, materials, and energy through urban systems. In 1997, König[11] revisited the metabolism of Hong Kong, showing the impact of the large increase in material turnover on the city and its surroundings.

1.3.4 REGIONAL MATERIAL BALANCES

At the end of the 1960s, the first studies on heavy metal accumulation in regions were initiated. In order to identify and quantify the sources of metals, methods such as pathway analysis and material balancing were developed and applied in regional studies. One of the groundbreaking regional material flow studies was reported by Huntzicker et al.[12] In 1972, these authors established a balance for automobile-emitted lead in the Los Angeles basin that was revisited by Lankey et al.[13] in 1998. The authors developed a material-balance method that was based on the measurement of atmospheric particle size distributions, atmospheric lead concentrations, and surface deposition fluxes. The studies show the important sources, pathways, and sinks for lead in the Los Angeles basin, and the results point to potential environmental problems and solutions. The authors state that their method did not allow them to elucidate every detail of substance dispersion in the environment. Yet the mass-balance approach identified all of the important pathways and provided a basis for quantification of flows. They concluded that the "material balance-flow pathway approach" was, in general, a powerful tool for assessing the environmental impact of a pollutant. The tool is also well suited to evaluate environmental management decisions such as the reduction of lead in automobile fuels. The authors show that lead inputs and outputs were roughly in agreement and that the reduction of lead in automobile fuel significantly reduced the overall input of lead in the Los Angeles basin. Their MFA method identified resuspended road dust as an important secondary source of lead that is expected to decrease slowly over time.

Another pioneering study was undertaken by Ayres et al.[14] in the early 1980s. These authors analyzed the sources, pathways, and sinks of major pollutants in the Hudson-Raritan basin for a period of 100 years from 1885 to 1985. They chose heavy metals (Ag, As, Cd, Cr, Cu, Hg, Pb, and Zn), pesticides (dichlorodiphenyltrichloroethane [DDT], tetrachlorodiphenylethane [TDE], aldrin, benzene hexachloride [BHC]/lindane, chlordane, and others), and "other critical pollutants" (polychlorinated biphenyl [PCB], polyaromatic hydrocarbons [PAH], N, P, total organic carbon [TOC], etc.) as objects of their investigation. One of the major motivations for this project was to explore the long-term effects of anthropogenic activities on aquatic environments, in particular on fish populations in the Hudson River Bay. The authors chose the following procedure:

- Definition of systems boundaries
- Establishment of a model linking sources, pathways, and sinks for each of the selected pollutants
- Historical reconstruction of all major flows in the system
- Adoption of data either from other regions or deduced from environmental transport models in cases where major substance flows could not be reconstructed from historical records
- Validation of the model by comparing measured concentrations in the river basin with results calculated by the model

The authors used the same MFA methodology to balance single processes like coal combustion, as well as complex combinations of processes such as consumption processes in the Hudson River basin. Despite the scarcity of data for the historical reconstruction, the authors were able to establish satisfying agreement between their model and values measured in the environment of the basin.

Ayres et al. used the results of their study to identify and discriminate the main sources and sinks for each pollutant. They were able to distinguish the importance of point sources and nonpoint sources and of production and consumption processes,

Introduction

which they identified as prevailing sources for many pollutants. Their model predicted changes in the environment due to alterations in population, land use, regulations, etc. In the following decade, Ayres expanded his studies from single substances to more comprehensive systems, looking finally at the entire "industrial metabolism."[15] He used MFA methodology to study complex material flows and cycles in industrial systems. He aimed at designing a more efficient "industrial metabolism" by improving technological systems, by inducing long-term planning based on resource conservation and environmental protection, and by producing less waste and recycling more materials.

In the 1980s and 1990s, the papers by Huntzicker et al. and by Ayres et al. were followed by many studies on substance flows in regions such as river basins, nations, and on the global scale. (The term *region* is used in this chapter to describe a geographical area on the Earth's surface that can range from a small size of a few square kilometers to a large size such as a continent or even the globe. It is thus not used as defined in regional sciences, e.g., for urban planning.)

Rauhut[16] was among the first to publish national substance inventories similar to those of the U.S. Bureau of Mines.[17] His studies were detailed enough to serve as a base for policy decisions regarding heavy metals, such as managing and regulating cadmium (Cd) as a step toward environmental protection.

Van der Voet, Kleijn, Huppes, Udo de Haes, and others of the Centre of Environmental Science, Leiden University, in the Netherlands prepared several reports on the flows and stock of substances within the European economy and environment. They recognized the power of MFA as a decision support tool for environmental management and waste management. Their comprehensive MFA studies included chlorine (Cl),[18] Cd[19] and other heavy metals, and nutrients. Based on their experience with problems of transnational data acquisition, they advocated an internationally standardized use of MFA.

Stigliani and associates from the International Institute for Applied Systems Analysis (IIASA) in Laxenburg, Austria, assessed the flows of pollutants in the Rhine River basin using MFA methodology.[20] They identified the main sources and sinks of selected heavy metals and drew conclusions regarding the future management of pollutants in river basins. At the same institute, Ayres et al. applied the materials-balance principle for selected chemicals. Their report[21] concentrated on four widely used inorganic chemicals (Cl, Br, S, and N) and provided a better knowledge of their environmental implications. This work also presented a new understanding on how societies produce, process, consume, and dispose of materials, and it linked these activities to resource conservation and environmental change. Ayres et al. achieved this by embedding MFA in the concept of industrial metabolism.

On a global scale, studies by geochemists such as Lantzy and McKenzie[22] have been important for the understanding of large-scale geogenic and anthropogenic metal cycles of substances between the lithosphere, hydrosphere, and atmosphere. In his books, Nriagu has investigated the sources, fate, and behavior of substances such as arsenic,[23] vanadium, mercury, cadmium, and many others in the context of the protection of humans and the environment. He presents a critical assessment of the chemistry and toxic effects of these substances as well as comprehensive information on their local and global flows.

All of the regional and global studies mentioned previously are based on material flow analysis. They have been selected to demonstrate the wide range of application of MFA and because they are recognized as pioneering studies in their research areas. The reader will certainly encounter additional regional studies by other authors active in the field that are not cited here.

1.3.5 Metabolism of the Anthroposphere

Baccini, Brunner, and Bader[24,25] extended material flow analysis by defining a systematic and comprehensive methodology and by introducing the concepts of "activity" and "metabolism of the anthroposphere" (see Sections 2.1.8 and 2.1.9). Their main goals were (1) to develop methods to analyze, evaluate, and control metabolic processes in man-made systems and (2) to apply these methods to improve resource utilization and environmental protection on a regional level. Engaged in solving waste-management problems, they recognized that the so-called filter strategy at the back end of the materials chain is often of limited efficiency. It is more cost-effective to focus on the total substance flow and not just on the waste stream. Their integrated approach is directed toward (1) the turnover of materials and energy, (2) activities and structures, and (3) the interdependency of these aspects in regions. In the project SYNOIKOS, Baccini, collaborating with Oswald[26] and a group of architects, combined physiological approaches with structural approaches to analyze, redefine, and restructure urban regions. This project shows the full power of the combination of MFA with other disciplines to design new, more efficient, and sustainable anthropogenic systems.

Lohm, originally an entomologist studying the metabolism of ants, and Bergbäck[27] were also among the first to use the notion "metabolism of the anthroposphere" to study metabolic processes by MFA. In their pioneering study of the metabolism of Stockholm, they focused on the *stock* of materials and substances in private households and the corresponding infrastructure. They identified the very large reservoir of potentially valuable substances, such as copper and lead, within the city. Lohm and Bergbäck drew attention to urban systems in an effort to prevent environmental pollution by the emissions of stocks and to conserve and use the valuable substances hidden and hibernating in the city.

Fischer Kowalsky et al.[28] employed a similar set of tools and expanded the methodology by adopting approaches used in the social sciences. They coined the term *colonization* to describe the management of nature by human societies and investigated the transition from early agricultural societies to today's enhanced metabolism. Wackernagel et al.[29] developed a method to measure the ecological footprints of regions, a method that is based, in part, on MFA. They concluded that regions in affluent societies use a very large "hinterland" for their supply and disposal, and they suggest the need to compare and ultimately reduce the ecological footprints of regions. They argue that our concept of progress must be redefined. The "progress" observed in most of today's societies does not translate to an increase in the general welfare, if measured properly, and thus there is a need to change the direction of development.[30]

In other works also based on MFA, both Schmidt-Bleek[31] and von Weizsäcker[32] from the Wuppertal Institute for Climate and Energy concluded that, considering

Introduction

environmental loadings and resource conservation, the turnover of materials in modern economies is much too high. In order to achieve sustainable development, they recommend a reduction of material flows by a factor of four to ten. Bringezu,[33] from the same group in Wuppertal, established a platform for the discussion of materials-accounting methodology, called "Conaccount," that was joined by many European research groups. The Wuppertal Institute also started to collect and compare information about national material flows in several countries from Europe and Asia and in the U.S. In *The Weight of Nations*, Matthews and colleagues document and compare the material outflows from five industrial economies (Austria, Germany, Japan, Netherlands, and the U.S.).[34] They developed physical indicators of material flows that complement national economic indicators, such as gross domestic product (GDP). In 2000, a group of researchers from the Center for Industrial Ecology at Yale University launched a several-years-long project to establish material balances for copper and zinc on national, continental, and global levels (the stocks and flows project [STAF]).[35]

1.3.6 PROBLEMS — SECTIONS 1.1–1.3

Problem 1.1: (a) MFA is based on a major principle of physics. Name it and describe its content. (b) What is the main benefit of the principle for MFA?

Problem 1.2: Divide the following 14 materials into two categories of "substances" and "goods": cadmium, polyvinyl chloride, molecular nitrogen (N_2), melamine, wood, drinking water, personal computer, steel, iron, copper, brass, separately collected wastepaper, glucose, and copper ore.

Problem 1.3: Estimate roughly your total personal daily material turnover (not including the "ecological rucksack"). Which category is dominant on a mass basis: (a) solid materials and fossil fuels, (b) aqueous materials, or (c) gaseous materials?

Problem 1.4: The flux of zinc into the city of Stockholm is about 2.7 kg/capita/year; the output flux amounts to ca. 1.0 kg/capita/year; and the stock is about 40 kg/capita.[36] Calculate the time until the stock of zinc will double, assuming stationary conditions.

The solutions to the problems are given on the Web site www.iwa.tuwien.ac.at/MFA-handbook.

1.4 APPLICATION OF MFA

The historical development summarized in Section 1.3 shows that MFA has been applied as a basic tool in such diverse fields as economics, environmental management, resource management, and waste management. The most important application areas of MFA are discussed in the following sections. Reasons are given to explain why MFA is an indispensable tool for these applications. The potential uses and the limits of MFA in each field are outlined. The methodological differences between MFA and other similar approaches such as pathway analysis, input-output analysis, and LCA are discussed in Chapter 2.

1.4.1 ENVIRONMENTAL MANAGEMENT AND ENGINEERING

The environment is a complex system comprising living organisms, energy, matter, space, and information. The human species, like all other species, has used the environment for production and disposal. We produce food and shelter — drawing on soil, water, and air — and in return we gave back wastes such as feces, exhaust air, and debris. Environmental engineering has been described as (1) the study of the fate, transport, and effects of substances in the natural and engineered environments and (2) the design and realization of options for treatment and prevention of pollution.[37] The objectives of management and engineering measures are to ensure that (1) substance flows and concentrations in water, air, and soil are kept at a level that allows the genuine functioning of natural systems and (2) the associated costs can be carried by the actors involved.

MFA is used in a variety of environmental-engineering and management applications, including environmental-impact statements, remediation of hazardous-waste sites, design of air-pollution control strategies, nutrient management in watersheds, planning of soil-monitoring programs, and sewage-sludge management. All of these tasks require a thorough understanding of the flows and stocks of materials within and between the environment and the anthroposphere. Without such knowledge, the relevant measures might not be focused on priority sources and pathways, and thus they could be inefficient and costly.

MFA is also important in management and engineering because it provides transparency. This is especially important for environmental-impact statements. Emission values alone do not allow cross-checking when a change in boundary conditions (e.g., change in input or process design) is appropriate to meet regulations. However, if the material balances and transfer coefficients of the relevant processes are known, the results of varying conditions can be cross-checked.

There are clear limits to the application of MFA in the fields of environmental engineering and management. MFA alone is not a sufficient tool to assess or support engineering or management measures. Nevertheless, MFA is an indispensable first step and a necessary base for every such task, and it should be followed by an evaluation or design step, as described in Chapter 2, Section 2.5.

1.4.2 INDUSTRIAL ECOLOGY

Although earlier traces can be tracked down, the concept of industrial ecology evolved in the early 1990s.[38,39] So far, there is no generally accepted definition of industrial ecology.[40] Jelinski et al. define it as a concept in which an industrial system is viewed not in isolation from its surrounding systems but in concert with them. Industrial ecology seeks to optimize the total materials cycle from virgin material to finished material, to component, to product, to waste product, and to ultimate disposal.[41] While industrial metabolism, as defined, e.g., by Ayres[42] and by Ayres and Simonis,[43] explores the material and energy flows through the industrial economy, industrial ecology goes farther. Similar to what is known about natural ecosystems, the approach strives to develop methods to restructure the economy into a sustainable system. The industrial system is seen as a kind of special biological

Introduction

ecosystem or as an analogue of the natural system.[44] Other pioneers of industrial ecology define it even as the science of sustainability.[45] In this context, it is worth mentioning that sustainability has not only an ecological but also social and economic dimensions. These aspects are addressed by Allenby[46] in the newly founded *Journal of Industrial Ecology*, headquartered at Yale's School of Forestry and Environmental Studies.

Clearly, the variety of topics and approaches demonstrates the breadth of the field that industrial ecology attempts to span. This fact is also used to criticize the approach as being vague and mired in its own ambiguity and weakness. The legitimacy of the analogy between industrial and ecological ecosystems is also questioned.[47] However, there are several basic design principles in industrial ecology (adapted after Ehrenfeld[48]) that suggest the utilization of MFA:

1. Controlling pathways for materials use and industrial processes
2. Creating loop-closing industrial practices
3. Dematerializing industrial output
4. Systematizing patterns of energy use
5. Balancing industrial input and output to natural ecosystem capacity

The following applications illustrate the role of MFA in industrial ecology. First, a better understanding of industrial metabolism requires a description of the most relevant material flows through the industrial economy. This encompasses the selection of relevant materials on the "goods level" (e.g., energy carriers, mineral construction materials, steel, fertilizers) and on the "substance level" (e.g., carbon, iron, aluminum, nitrogen, phosphorus, cadmium). The system boundaries must be defined in such a way that the pathways of materials are covered from the cradle (exploitation) to the grave (final sink for the material). The results of an MFA reveal the most important processes during the life cycle of a material, detect relevant stocks of the material in the economy and the environment, show the losses to the environment and the final sinks, and track down internal recycling loops. Additionally, MFA can be used to compare options on the process level and at the system level.

Second, the concept's imminent call for closed loops (realized, for example, in the form of a cluster of companies, a so-called industrial symbiosis) requires information about the composition of wastes to become feedstock again and about the characteristics of the technological processes involved. In particular, the implementation of industrial loops requires controls by appropriate MFA, since loops have the potential of accumulating pollutants in goods and stocks. The fact that waste is recycled or reused is not yet a guarantee for a positive result. Two negative examples are the use of contaminated fly ash in cement production or the reuse of animal protein causing bovine spongiform encephalopathy (BSE), known as "mad cow" disease.

A third objective in industrial ecology is dematerialization. This can be achieved by providing functions or services rather than products. Again, MFA can be used to check whether a dematerialization concept (e.g., the paperless office) succeeds in practice. Other ways of dematerialization are to prolong the lifetime of products or to produce lighter goods.

Up to now, most applications of MFA have served to investigate the industrial metabolism for selected materials such as heavy metals, important economic goods, or nutrients.[34,49–56] The city of Kalundborg, Denmark, is frequently mentioned as an example of an "industrial ecosystem" in the industrial ecology literature. Materials (fly ash, sulfur, sludges, and yeast slurry) and energy (steam, heat) are exchanged between firms and factories within a radius of about 3 km.[57] Using waste heat for district heating and other purposes (e.g., cooling) has long been recognized as good industrial practice (known as power-heat coupling). The comparatively few material flow links between the actors in Kalundborg show that the concept of (apparently) closed loops is difficult to accomplish in reality. Materials balancing is seen as a major tool to support industrial ecosystems.

1.4.3 Resource Management

There are two kinds of resources: first, natural resources such as minerals, water, air, soil, information, land, and biomass (including plants, animals, and humans), and second, human-induced resources such as the anthroposphere as a whole, including materials, energy, information (e.g., "cultural heritage," knowledge in science and technology, art, ways of life), and manpower. The human-induced or so-called anthropogenic resources are located in (1) private households, agriculture, industry, trade, commerce, administration, education, health care, defense, and security systems and (2) infrastructure and networks for supply, transportation and communication, and disposal. Given the large-scale exploitation of mines and ores, many natural resources are massively transformed into anthropogenic resources (see Section 1.4.5). Thus, the growing stocks of the anthroposphere will become increasingly more important as a resource in the future.

Resource management comprises the analysis, planning and allocation, exploitation, and upgrading of resources. MFA is of prime importance for analysis and planning. It is the basis for modeling resource consumption as well as changes in stocks, and therefore it is important in forecasting the scarcity of resources. MFA is helpful in identifying the accumulation and depletion of materials in natural and anthropogenic environments. Without it, it is impossible to identify the shift of material stocks from "natural" reserves to "anthropogenic" accumulations. In addition, if MFA is performed in a uniform way at the front and back end of the anthropogenic system, it is instrumental in linking resources management to environmental and waste management. It shows the need for final sinks and for recycling measures, and it is helpful in designing strategies for recycling and disposal.

Balancing all inflows and outflows of a given stock yields information on the time period until the stock reaches a critical state of depletion or accumulation. This could be the slow exhaustion of available phosphorous in agricultural soil due to the lack of appropriate fertilizer, or it could be the unnoticed buildup of valuable metals in a landfill of incinerator ash and electroplating sludge. It is difficult to estimate the change in the stock by direct measurement, especially for stocks with a high variability in composition and slow changes in time. In such cases, it is more accurate and cost effective to calculate critical time scales (the time when a limiting or reference value is reached) by comparing the difference between input and output to the stock from its flow balance. Direct measurement requires extensive sampling programs with much analysis, and the heterogeneity of the flows produces large standard deviations of the mean values. Thus, it takes large differences between mean values until a change becomes statistically significant. A slow change in a heterogeneous material can be proved on statistical grounds only over long measuring periods. As a result, MFA is better suited and more cost-effective than continuous soil monitoring in early recognition of changes in resource quality, such as harmful accumulations in the soil. For more information, see Obrist et al.[58] and Chapter 3, Section 3.1.1.

The use (and preferably conservation) of resources to manufacture a particular good or to render a specific service is often investigated by LCA. The result of an LCA includes the amount of emissions and the resources consumed. Since MFA is the first step of every LCA, MFA is also a base for resource conservation.

The quality and price of a resource usually depends on the substance's concentration. Thus, it is important to know whether a natural or anthropogenic process concentrates or dilutes a given substance. MFA is instrumental in the application of such evaluation tools as statistical entropy analysis (see Chapter 2, Section 2.5.1.8 and Chapter 3, Section 3.2.2.), which is used to compare the potential of processes and systems to accumulate or dilute valuable or hazardous substances.

1.4.4 Waste Management

Waste management takes place at the interface between the anthroposphere and the environment. The definition and objectives of waste management have changed over time and are still changing. The first signs of organized waste management appeared when people started to collect garbage and remove it from their immediate living areas. This was an important step regarding hygiene and helped to prevent epidemics. These practices were improved over the centuries. However, dramatic changes in the quantity and composition of wastes during the 20th century caused new problems. First, the emissions of the dumping sites (landfills) polluted groundwater and produced greenhouse gases. Second, landfill space became scarce in densely populated areas. Even the concept of sanitary landfilling could not solve these problems in the long run. Today, waste management is an integrated concept of different practices and treatment options comprising prevention and collection strategies; separation steps for producing recyclables or for subsequent processing using biological, physical, chemical, and thermal treatment technologies; and different landfill types. People now have the opportunity (or, in some places, the duty) to separate paper, glass, metals, biodegradables, plastics, hazardous wastes, and other materials into individual fractions.

The goals of modern waste management are to:

- Protect human health and the environment
- Conserve resources such as materials, energy, and space
- Treat wastes before disposal so that they do not need aftercare when finally stored in landfills

The last goal is also known as the final-storage concept and is part of the precautionary principle: the wastes of today's generation must not impose an economic or ecological burden to future generations. Similar goals can be found in many instances of modern waste-management legislation,[59–63] which were written to comply with the requirements for sustainable development. The aforementioned goals make it clear that the focus in waste management is not necessarily on goods (paper, plastics, etc.) or on functions of materials (e.g., packaging). The focus is on the nature of the substances.

Hazardous substances threaten human health. The threat occurs when municipal solid waste (MSW) is burned in poorly equipped furnaces and volatile heavy metals escape into air. It is not the *good* leachate of a landfill that imposes danger to the groundwater. The danger resides in the cocktail of hazardous *substances* in the leachate of the landfill. The fact that a material has been used for packaging is irrelevant for recycling. What is important is its elemental composition, which determines whether it is appropriate for recycling. Hence, advanced waste-management procedures are implemented to control and direct the disposition of substances at the interface between the anthroposphere and the environment to achieve the following two goals.

1. Materials that can be recycled without inducing high costs or negative substance flows should be recycled and reused. Negative flows can appear as emissions or by-products during the recycling process. The recycling process itself can also lead to enrichment of pollutants in goods and reservoirs. For example, the recycling process can increase heavy metal contents in recycled plastics, or it can lead to accumulation of metals in the soil when sewage sludge is applied to agricultural fields.
2. Nonrecyclables should be treated to prevent the flow of hazardous substances to the environment. A tailor-made final sink — defined as a reservoir where a substance resides for a long period of time (>10,000 years) without having a negative impact on the environment — should be assigned for each substance.

MFA is a valuable tool in substance management because it can cost-efficiently determine the elemental composition of wastes exactly (see Chapter 3, Section 3.3.1). This information is crucial if the goal is to assign a waste stream to the best-suited recycling/treatment technology and to plan and design new waste-treatment facilities. For example, mixed plastic wastes that cannot be recycled for process reasons can be used as a secondary fuel in industrial boilers as long as their content of heavy metals and other contaminants is not too high (see Chapter 3, Sections 3.3.2.1 and 3.3.2.2).

MFA is also helpful in investigating the substance management of recycling/treatment facilities. For instance, substance control by an incinerator is different from substance control by a mechanical–biological treatment facility. Such information is a prerequisite for the design of a sustainable waste-management system. Nordrhein-Westfalen (Germany) is the first region that requires MFA by legislation as a standard tool in waste-management planning.[64]

Introduction

Finally, MFA can contribute to the design of better products that are more easily recycled or treated once they become obsolete and turn into "waste." These practices are known as design for recycling, design for disposal, or design for environment.

An MFA-based total material balance shows whether given goals have been achieved. An MFA balance also identifies the processes and flows that have the highest potential for improvements.

Waste management is an integral part of the economy. Some experts who have experience with MFA suggest that waste management should be replaced by materials and resource management. They assert that controlling the material flows through the total economy is more efficient than the current practice of separating management of wastes from the management of production supply and consumption.

1.4.5 ANTHROPOGENIC METABOLISM

Baccini and Brunner[24] applied MFA to analyze, evaluate, and optimize some of the key processes and goods of the "metabolism of the anthroposphere." In a more recent study, Brunner and Rechberger[65] systematically summarized relevant phenomena of the anthropogenic metabolism. The following examples illustrate the power of MFA to identify key issues for resource management, environmental management, and waste management.

1.4.5.1 Unprecedented Growth

In prehistoric times, the total anthropogenic metabolism (input, output, and stock of materials and the energy needed to satisfy all human needs for provisions, housing, transportation, etc.) was nearly identical to human physiological metabolism. It was mainly determined by the need for food, for air to breathe, and for shelter. For modern man, the material turnover is 10 to 20 times greater (Figure 1.3). The fraction that is used today for food and breathing is comparatively small. More important is the turnover for other activities, such as to clean, to reside, and to transport and communicate (Table 1.1). These activities require thousands of goods and substances that were of no metabolic significance in prehistoric times.

The consumption of goods has increased over the past two centuries, and there are no clear signs yet that this will change. Figure 1.4 displays the growth in ordinary materials such as paper, plastic, and tires. Figure 1.5 shows the increase in construction materials in the U.S. for a period of 100 years. Growth of material flows is closely associated with economic growth. Economic progress is defined in a way that causes an increase in material turnover. It is important to develop new economic models that decouple economic growth from material growth, thus promoting long-term welfare without a constant increase in resource consumption.

The need for a new economic model becomes even more evident when one considers the growth rates of potentially hazardous substances such as heavy metals or persistent and toxic organic substances. On the level of substances, the increase in consumption is by far greater than one order of magnitude. For instance, the global anthropogenic lead flow (Figure 1.6) increased in the last few thousand years by about 10^6, i.e., by six orders of magnitude. Material growth is not just an issue

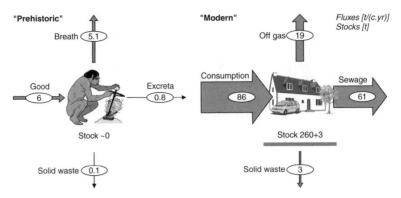

FIGURE 1.3 The material turnover of primitive man in his "private household" was about an order of magnitude smaller than today's consumption of goods. Note that the figure includes direct material flows only. Materials (and wastes) turned over outside of households to manufacture the goods consumed in households are larger than 100 t/(c.yr). (From Brunner, P.H. and Rechberger, H., Anthropogenic metabolism and environmental legacies, in *Encyclopedia of Global Environmental Change*, Vol. 3, Munn, T., Ed., John Wiley & Sons, West Sussex, U.K., 2001. With permission.)

TABLE 1.1
Material Flows and Stocks for Selected Activities of Modern Man

Activity	Input, t/(c.yr)	Output, t/(c.yr)			Stock, t/Capita
		Sewage	Off Gas	Solid Residues	
To nourish[a]	5.7	0.9	4.7	0.1	<0.1
To clean[b]	60	60	0	0.02	0.1
To reside[c]	10	0	7.6	1	100 + 1
To transport[d]	10	0	6	1.6	160 + 2
Total	86	61	19	2.7	260 + 3

Note: The most outstanding and unprecedented feature of today's economies is the very large stock of material that has accumulated in private households.[65]

[a] Includes all flows of goods associated with the consumption of food within private households, e.g., food, water for cooking, etc.
[b] Comprises water, chemicals, and equipment needed for laundry, dishwashing, personal hygiene, toilet, etc.
[c] Consists of the buildings, furniture, appliances, etc. needed for living.
[d] Accounts for materials (cars, trains, fuel, air, etc.) used for the transport of persons, goods, energy, and information to accommodate the needs of private households (including materials for road construction).

Introduction

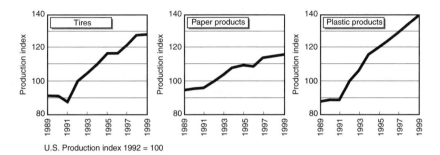

U.S. Production index 1992 = 100

FIGURE 1.4 Growth in consumer product consumption. The examples of paper, plastics, and tires consumed in the U.S. show the high growth rate of material flows from 1989 to 1999.[68] (Reprinted from Storck, W.J. et al., *Chem. Eng. News*, 78, 76, June 26, 2000, p. 16, American Chemical Society.)

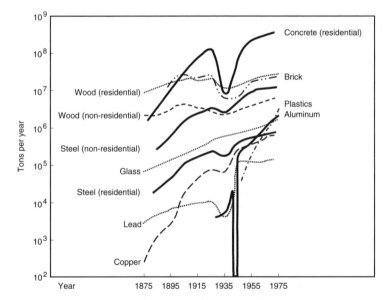

FIGURE 1.5 Growth of construction materials used in the U.S. from 1875 to 1975.[69] (Adapted from Wilson, D., Recycling of demolition wastes, in *Concise Encyclopedia of Building and Construction Materials*, Moavenzadeh, F., Ed., Pergamon/Elsevier, Oxford, 1990. With permission from Elsevier.) Around 1900, wood was surpassed by concrete as the main construction material. Plastics and aluminum were among the fastest growing construction materials in the 1970s. Changes in the consumption of construction materials alter the construction stock, too, and hence are important for future waste management; large amounts of plastics and metals such as aluminum and copper will have to be handled safely and efficiently.

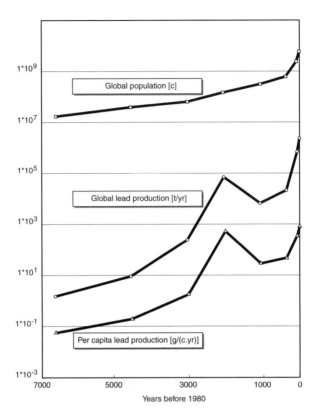

FIGURE 1.6 Increase in global production of lead. Due to tremendous progress in mining technology and economic development, lead mining increased from 0.1 g/capita/year to 1000 g/capita/year. Taking into account population growth, which increased at a smaller rate than the per capita lead production, the total amount of lead produced from mining increased from 1 ton per year to more than 3 million tons per year (lead data from Settle and Patterson[70]). (From Brunner, P.H. and Rechberger, H., Anthropogenic metabolism and environmental legacies, in *Encyclopedia of Global Environmental Change*, Vol. 3, Munn, T., Ed., John Wiley & Sons, West Sussex, U.K., 2001. With permission.)

from a quantitative point of view; it is also of qualitative concern. If total material flows have increased by one to two orders of magnitude and certain hazardous substances by more than five orders of magnitude, then the stock as well as the output of the anthropogenic system will be highly enriched with such substances. Hence, in future, stocks and wastes will have to be managed with greater care to avoid harmful accumulations of these substances. It becomes clear that, from an environmental point of view, the growth of substance flows is more important than the growth rate of mass goods.

1.4.5.2 Anthropogenic Flows Surpass Geogenic Flows

A consequence of the constant increase in material exploitation due to the large per capita consumption is that the anthropogenic flow of certain substances already

Introduction

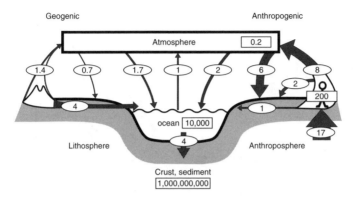

FIGURE 1.7 Global flows (kt/year) and stocks (kt) of cadmium in the 1980s. Man-made flows (right-hand side) surpass natural flows (left-hand side). Comparatively high atmospheric emissions lead to a significant accumulation of anthropogenic cadmium in the soil. The stock of cadmium in the anthroposphere grows by 3% per year. In the future, this stock needs to be managed carefully if negative effects on the environment are to be avoided.[72] (From Brunner, P.H. and Baccini, P., Neue Zürcher Zeitung, Beilage Forsch. *Technik* 70, 65, 1981. With permission.)

surpasses the geogenic (or "natural") flow of such substances. While this is quite clear for those organic chemicals that are exclusively man made (such as polyvinylchloride [PVC]), it is less obvious for metals like Cd (Figure 1.7). For this metal, the man-made flow from the lithosphere to the anthroposphere is about three to four times larger than the geogenic flows by erosion, weathering, leaching, and volcanoes. As a consequence, the Cd concentrations will increase in some environmental compartments such as the soil. The Cd stock in the anthroposphere represents an unwanted legacy for future generations, which will have to bear the high cost of managing this stock very carefully to avoid hazards associated with this toxic metal.

The global phenomenon of anthropogenic flows surpassing natural flows has been demonstrated on a national scale, too. Bergbäck[71] compared the rates for weathering and erosion with those of anthropogenic emissions in Sweden. He found that emission rates of lead, chromium, and cadmium are exceeding natural rates. Similar results have been observed in other countries with advanced economies. This demonstrates that modern economies with a large material turnover slowly change the concentrations of substances in environmental compartments (Figure 1.8).

1.4.5.3 Linear Urban Material Flows

Figure 1.9 displays the material flows through Vienna. Like in any other city, the flows are mainly linear. To change from a linear flow system to a cycling system requires more than new technologies. It means a complete change in lifestyle, value system, priorities, as well as large-scale changes in technology and the economy. At present, such a change is neither in sight nor is there convincing evidence that it is needed.[67]

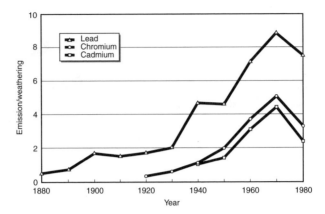

FIGURE 1.8 Rates of anthropogenic emissions of lead, chromium, and cadmium compared with weathering rates in Sweden. (From Bergbäck, B., Industrial Metabolism: The Emerging Landscape of Heavy Metal Emission in Sweden, Ph.D. thesis, Linköping University, Sweden, 1992. With permission.)

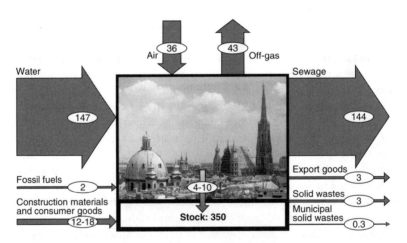

FIGURE 1.9 In the 1990s, about 200 tons of materials were consumed per capita per year in the city of Vienna, and somewhat less than this amount left the city in a linear way. The small difference of 4 to 10 t/(c.yr) accumulates, resulting in a continuous growth of the stock of 350 t/capita that doubles in 50 to 100 years. Basically, the modern anthropogenic metabolism is a linear-throughput reactor, with less than 1% of materials being involved in regional cycles. Note the differences from Figure 1.3 and Table 1.1 (material turnover in private households): the numbers for Vienna also include material turnovers outside private households in industry, the service sector, public service and administration, etc. (From Brunner, P.H. and Rechberger, H., Anthropogenic metabolism and environmental legacies, in *Encyclopedia of Global Environmental Change*, Vol. 3, Munn, T., Ed., John Wiley & Sons, West Sussex, U.K., 2001. With permission.)

Introduction

1.4.5.4 Material Stocks Grow Fast

Because material inputs in growing regional economies are, as a rule, larger than outputs, most modern regions accumulate a large material stock within their boundaries. Exceptions are regions that exploit and export resources on a large scale, such as coal, metal ores, sand, gravel and stones, or timber. There are many stocks in the anthroposphere: mining residues left behind as tailings; material stocks of industry, trade, and agriculture; urban stock of private households and of the infrastructure for transport, communication, administration, education, etc.; and the comparatively small but growing stock of wastes in landfills. The difference in stock accumulation between prehistoric (<1 ton) and modern humans is striking. Present-day material stocks amount to about 200 to 400 tons per urban citizen. This stock has to be managed and maintained. Today's decisions about the renewal and maintenance of the urban stocks, such as buildings and networks for energy supply, transport, and communication, have far-reaching consequences. The residence time of materials in the stock can range up to more than 100 years. This means that once a material is built into the stock, it will not show up quickly in the output of the stock, namely in waste management.

For the management of resources and the environment, the urban stock is important for several reasons:

1. It is a growing reservoir of valuable resources and holds a tremendous potential for future recycling.
2. It represents a mostly unknown source, whose importance is not yet in focus, which awaits assessment with respect to its significance as an economic resource and as a threat to the environment.
3. It is a potential long-term source of large flows of pollutants to the environment. Urban stocks contain more hazardous materials than so-called hazardous-waste landfills, which are at present a focus of environmental protection.

Planners and engineers are asked to design new urban systems. In the future, the location and amount of materials in a city's stock should be known. Materials should be incorporated into the stock in a way that allows easy reuse and environmental control. Economies are challenged to maintain high growth rates, building up ever larger stocks, and setting aside sufficient resources to maintain and renew this stock properly over long periods of time.

1.4.5.5 Consumption Emissions Surpass Production Emissions

In service-oriented societies, most production emissions are decreasing, while some consumer emissions, for example CO_2 and heavy metals, are increasing. This is due to the tremendous efforts that have been undertaken in many countries to reduce industrial emissions. On the other hand, inputs into the consumption process still grow, and thus consumer outputs such as emission loads and waste amounts are rising, too. Anderberg and colleagues[66] indicate that for heavy metals in Sweden,

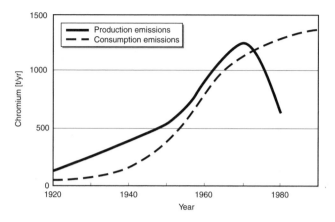

FIGURE 1.10 Consumption emissions vs. production emissions for chromium in Sweden. Production emissions decreased due to pollution-abatement measures as well as decreasing activities of heavy industry, and consumer emissions increased due to rising demand in consumer goods and dissipative applications of materials. (From Anderberg, S. et al., *Ambio*, 18, 4, 216, 1989. With permission.)

industrial activities were the most important emission sources up to the 1970s. After that period, nonpoint consumer-oriented emissions became dominant (Figure 1.10).

An actual example, which is based on MFA, is given in Figure 1.11. A galvanizing company optimized its production process so that most of its zinc residues and wastes were either internally or externally reused. From the point of view of the production process, the company has protected the environment in the best possible way. An MFA reveals that now the largest zinc flow to the environment is caused by the corrosion of the zinc coating on the product: during the lifetime of the galvanized surface, much of the coating (zinc) is lost to the environment. Thus, despite pollution-prevention measures during the electroplating process, as much as 85% of the zinc may end up in the environment. Since the corrosion process is slow and the residence time of metals on surfaces can be rather long, the metal flows to the environment last for years to decades. The legacies of today's protected surfaces are tomorrow's environmental loadings. At present, the significance of such emissions has not been assessed and evaluated. There are no methods available yet to measure and appraise these sources of pollutants to the environment. For state-of-the-art surface coating, it is not the production waste that causes the largest losses to the environment; it is the product itself that is the important source of environmental loadings.

This new situation is due to, on the one hand, advanced technology and legislation in the field of industrial environmental protection. On the other hand, the high and still growing rate of consumption has led to large stocks discharging significant amounts of materials in a hardly noticeable way into the water, air, and soil. Examples of such consumer-related emission sources are the weathering of surfaces of buildings (zinc, copper, iron, etc.), the corrosion and erosion (chassis, tires, brakes) of vehicles and infrastructure for traffic, the emission of carbon dioxide and other greenhouse gases due to space heating or cooling and transportation, and the growing

Introduction

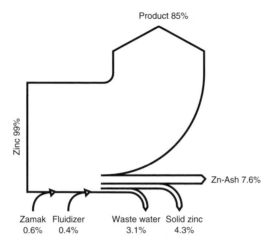

FIGURE 1.11 Zinc flow through an advanced electroplating plant: 85% of the zinc used as surface coating for corrosion protection is sold with the product. Nearly all of the 15% zinc not sold is recycled. The emissions to the environment during production are very small. During the lifetime use of the Zn-coated product, most of the zinc is likely to be dispersed in the environment due to corrosion. Hence, despite the effort to protect the environment by pollution-prevention measures in the factory (cleaner production), most zinc ends up in the environment. (From Brunner, P.H. and Rechberger, H., Anthropogenic metabolism and environmental legacies, in *Encyclopedia of Global Environmental Change*, Vol. 3, Munn, T., Ed., John Wiley & Sons, West Sussex, U.K., 2001. With permission.)

problem of nitrogen overload. Cities are hot spots of such nonpoint sources. The new situation requires a shift in pollution abatement. In order to be efficient, environmental management must concentrate on the priority sources. Thus, in the future, the output of the consumer has to be in the focus of the product designer and environmental regulator. Since consumer behavior and lifestyle determine the material throughput through the anthroposphere, it is important to explore means to influence this behavior.

It is more difficult to prevent consumer-related emissions than it is to stop industrial pollution:

1. It is a much more demanding challenge for a company to change or discontinue a product than to add an air pollution-reducing device to its production facility.
2. The number of consumer sources is many orders of magnitude larger than that of production facilities.
3. The individual emission may be very small, and only the multiplication by the large number of sources may cause a hazard for the environment. Thus, consumer awareness for the overall relevance of his/her small contribution must be created first in order to be able to tackle the problem.
4. In general, emissions of industrial sources can be reduced with a much higher efficiency than widespread small-scale emissions from households or cars.

1.4.5.6 Changes in Amount and Composition of Wastes

Due to the large growth in consumption, the amount and composition of wastes are changing and will continue to vary in the future, too. For materials with long residence times, such as construction materials, a huge stock is being built up before wastes become noticeable at the back end of the life cycle. Waste management inherits most of the materials in the stock. Due to the increasing complexity of the goods produced, wastes comprise rising amounts of new materials, many of them being composed of mixtures that cannot be taken apart by physical methods. The long-lived stock needs to be protected from degradation by microorganisms, ultraviolet light, temperature, weathering, erosion, etc. Thus, it contains more hazardous substances (heavy metals and persistent organic materials as stabilizers and additives) than the short-lived consumer products that are recycled today (packaging materials, newsprint, glass). Therefore, in the future a large amount of hazardous substances will have to be removed from the wastes of the long-lived stock if materials of the stock are to be recycled safely.

1.5 OBJECTIVES OF MFA

MFA is an appropriate tool to investigate the flows and stocks of any material-based system. It gives insight into the behavior of the system, and when combined with other disciplines such as energy-flow analysis, economic analysis, and consumer-oriented analysis, it facilitates the control of an anthropogenic system. The objectives of MFA are to:

1. Delineate a system of material flows and stocks by well-defined, uniform terms
2. Reduce the complexity of the system as far as possible while still guaranteeing a basis for sound decision making
3. Assess the relevant flows and stocks in quantitative terms, thereby applying the balance principle and revealing sensitivities and uncertainties
4. Present results about flows and stocks of a system in a reproducible, understandable, and transparent way
5. Use the results as a basis for managing resources, the environment, and wastes, in particular for:
    a. Early recognition of potentially harmful or beneficial accumulations and depletions of stocks, as well as for timely prediction of future environmental loadings
    b. The setting of priorities regarding measures for environmental protection, resource conservation, and waste management (what is most important; what comes first?)
    c. The design of goods, processes, and systems that promote environmental protection, resource conservation, and waste management (green design, ecodesign, design for recycling, design for disposal, etc.)

1.5.1 PROBLEMS — SECTIONS 1.4–1.5

Problem 1.5: Assuming you want to repeat Santorio's experiment, which of the three approaches do you prefer for measuring the input of the test person? (a) Weigh the mass of food the person eats, or (b) determine the difference of the mass of the person before and after eating the meal by weighing the person twice, or (c) both. Take into consideration that different scales are usually used to weigh 1 kg and 100 kg.

Problem 1.6: In order to conserve resources, "Factor 4" and "Factor 10" have been proposed:[31,32] the total turnover of materials should be decreased by 75 and 90% respectively. (a) How much can you reduce the flow of materials through your household by existing technologies and at affordable costs? Take Table 1.1 as a reference and discuss for each activity the potential to reduce material flows by Factor 4. (b) What do you recognize as the main advantage and disadvantage of the "Factor X" philosophy?

Problem 1.7: The global flows and stocks of cadmium in the 1980s are presented in Figure 1.7. Two decades later, an (unpublished) review in 2003 reveals the following changes in the flows and stocks of the anthroposphere: cadmium exploitation has slightly increased and amounts to 18 to 20 kt/year. The anthropogenic stock has more than doubled (+300 kt). Due to environmental protection, emissions to air and water have both decreased by about 80%. Data about the residence time of goods containing cadmium suggest that about 20 kt/year of cadmium are changing into goods that are no longer in use; they are either discarded as wastes or "hibernating" in the anthroposphere. The amount of cadmium transferred to landfills is not known, it may range between 2 and 20 kt/year. (a) Discuss the effect of reduced atmospheric cadmium emissions on the soil. (b) In view of the increase in anthropogenic stock from 200 to 500 kt of cadmium during the last 20 years, assess the cadmium flow to landfills. (c) Around the year 2000, NiCd batteries are responsible for about 75% of all cadmium used. What measures do you suggest to control problems that may arise from the use and disposal of cadmium batteries?

Problem 1.8: Prepare a quantitative list of solid wastes that you produce in your household (in kg/capita/year). Start with an inventory of major inputs of goods in your household. Use Table 1.1, your solution to Problem 1.3, and your own observations. Include all wastes from materials such as your home, car, water, food, etc., and arrange them in groups according to their collection (e.g., MSW, construction wastes, hazardous wastes).

Problem 1.9: Read the latest two issues of the *Journal of Industrial Ecology* and look for recent applications of MFA. Identify the goals of the MFA, the procedures, and the results, and discuss whether the conclusions and implications have been appropriately visualized by MFA. Evaluate whether MFA was the only possible approach to reach the objectives, or whether other methods would have allowed the same conclusions.

The solutions to the problems are given on the Web site www.iwa.tuwien.ac.at/ MFA-handbook.

REFERENCES

1. Steindl-Rast, D. and Lebell, S., *Music of Silence*, Seastone, Berkeley, CA, 2002.
2. Atkins, P.W. and Beran, J.A., *General Chemistry*, Scientific American Inc., New York, 1992, p. 5.
3. Vidal, B., *Histoire de la Chemie*, Presses Universitaire de France, Paris, 1985, p. 123.
4. Leontief, W., *The Structure of the American Economy, 1919–1929*, repr. International Arts and Sciences Press (now M.E. Sharpe), White Plains, N.Y., 1977.
5. Leontief, W., *Input-Output Economics*, Oxford University Press, New York, 1966.
6. Major, R.H., Santorio Santorio, *Ann. Medical Hist.*, 10, 369,1938.
7. Hendrickson, C., Horvath, A., Joshi, S., and Lave, L.B., Economic input-output models for environmental life cycle analysis, *Environ. Sci. Technol.*, 32 184A, 1998.
8. Wolman, A., The metabolism of cities, *Sci. Am.,* 213, 179, 1965.
9. Duvigneaud, P. and Denayeyer-De Smet, S., L'Ecosystème Urbs, in *L'Ecosystème Urbain Bruxellois, in Productivité Biologique en Belgique*, Duvigneaud, P. and Kestemont, P., Eds., Traveaux de la Section Belge du Programme Biologique International, Bruxelles, 1975, pp. 581–597.
10. Newcombe, K., Kalma, I.D., and Aston, A.R., The metabolism of a city: the case of Hong Kong, *Ambio*, 7, 3, 1978.
11. König, A., The Urban Metabolism of Hong Kong: Rates, Trends, Limits and Environmental Impacts, paper presented at POLMET (Pollution in the Metropolitan and Urban Environment) Conference, Hong Kong, November 1997.
12. Huntzicker, J.J., Friedlander, S.K., and Davidson, C.I., Material balance for automobile-emitted lead in the Los Angeles Basin, *Environ. Sci. Technol.*, 9, 448, 1975.
13. Lankey, R.L., Davidson, C.I., and McMichael, F.C., Mass balance for lead in the California South Coast Air Basin: an update, *Environ. Res., Sect. A*, 78, 86, 1998.
14. Ayres, R.U., Ayres, L.W., McCurley, J., Small, M.J., Tarr, J.A., and Ridgery, R.C., *A Historical Reconstruction of Major Pollutant Levels in the Hudson-Raritan Basin: 1880–1980,* Variflex Corp., Pittsburgh, 1985.
15. Ayres, R.U., *Industrial Metabolism Technology and Environment*, National Academy Press, Washington, DC, 1989.
16. Rauhut, A. and D. Balzer, Verbrauch und Verblieb von Cadmium in der Bundesrepublik Deutschland im Jahr 1973, *Metall.,* 30, 269, 1976.
17. U.S. Bureau of Mines, *Minerals Yearbook 1932–1994*, U.S. Department of the Interior (formerly U.S. Bureau of Mines), Washington, DC, 1933–1996.
18. Kleijn, R., Van der Voet, E., and Udo de Haes, H.A., Controlling substance flows: the case of chlorine, *Environ. Manage.*, 8, 523, 1994.
19. Voet, van der, E., Egmond, L., Kleijn, R., and Huppes, G., Cadmium in the European community: a policy-oriented analysis, *Waste Manage. Res.*, 12, 507, 1994.
20. Stigliani, W.M., Jaffé, P.R., and Anderberg, S., Heavy metal pollution in the Rhine Basin, *Environ. Sci. Technol.*, 27, 786, 1993.
21. Ayres, R.U., Norberg-Bohm, V., Prince, J., Stigliani, W.M., and Yanowitz, J., *Industrial Metabolism, the Environment, and Application of Materials-Balance Principles for Selected Chemicals*, International Institute for Applied System Analysis (IIASA), Laxenburg, Austria, 1989.

22. Lantzy, R.J. and McKenzie, F.T., Atmospheric trace metals: global cycles and assessment of manis impact, *Geochim. Cosmochim. Acta*, 43, 511, 1979.
23. Nriagu, J.O., *Arsenic in the Environment*, Wiley, West Sussex, U.K., 1994.
24. Baccini, P. and Brunner, P.H., *Metabolism of the Anthroposphere*, Springer, New York, 1991.
25. Baccini, P. and Bader, H.P., *Regionaler Stoffhaushalt — Erfassung, Bewertung, Steuerung*, Spektrum Akademischer Verlag GmbH, Heidelberg, 1996.
26. Baccini, P. and Oswald, F., *Netzstadt — Transdisziplinäre Methoden zum Umbau urbaner Systeme*, Vdf Hochschulverlag AG an der ETH, Zürich, 1998.
27. Lohm, U., Anderberg, S., and Bergbäck, B., Industrial metabolism at the national level: a case-study on chromium and lead pollution in Sweden, 1880–1980, in *Industrial Metabolism — Restructuring for Sustainable Development*, Ayres, R.U. and Simonis, U.E., Eds., The United Nations University, Tokyo, 1994.
28. Fischer-Kowalski, M., Haberl, H., Hüttler, W., Payer, H., Schandl, H., Winiwarter, V., and Zangerl-Weisz, H., *Gesellschaftlicher Stoffwechsel und Kolonisierung von Natur: Ein Versuch in Sozialer Ökologie*, Gordon and Breach Fakultas, Amsterdam, 1997.
29. Wackernagel, M., Monfreda, C., and Deumling, D., *Ecological Footprint of Nations* (November 2002 update), Redefining Progress, Oakland, 2002; www.redefiningprogress.org/publications/ef1999.pdf.
30. Wackernagel, M. and Rees, W., *Our Ecological Footprint — Reducing Human Impact on the Earth*, New Society Publishers, Philadelphia, 1996.
31. Schmidt-Bleek, F., *Wieviel Umwelt braucht der Mensch? Faktor 10 — das Maß für ökologisches Wirtschaften*, DTV, Munich, 1997.
32. von Weizsäcker, E.U., Lovins, A.B., and Lovins, L.H., *Faktor vier. Doppelter Wohlstand — halbierter Verbrauch*, Droemer Knaur, Munich, 1997.
33. Bringezu, S. et al., *The ConAccount Agenda: The Concerted Action on Material Flow Analysis and its Agenda for Research and Development*, Wuppertal Institute, Wuppertal, Germany, 1998.
34. Matthews, E., *The Weight of Nations: Material Outflows from Industrial Economies*, World Resource Institute, Washington, DC, 2000.
35. Graedel, T.E., The contemporary European copper cycle: introduction, *Ecol. Econ.*, 42, 2002, 5.
36. Bergbäck, B., Johansson, K., and Mohlander, U., Urban metal flows — a case study of Stockholm, *Water Air Soil Pollut. Focus*, 1, 3, 2001.
37. Valsaraj, K.T., *Elements of Environmental Engineering*, CRC Press, Boca Raton, FL, 2000.
38. Erkman, S., Industrial ecology: an historical view, *J. Cleaner Prod.*, 5, 1, 1997.
39. Desrochers, P., Market processes and the closing of "industrial loops": a historical reappraisal, *J. Ind. Ecol.*, 4, 29, 2000.
40. O'Rourke, D., Connelly, L., and Koshland, C.P., Industrial ecology: a critical review, *Int. J. Environ. Pollut.*, 6, 1996, 89.
41. Jelinski, L.W. et al., Industrial ecology: concepts and approaches, *Proc. Natl. Acad. Sci. U.S.A.*, 89, 793, 1992.
42. Ayres, R.U., Toxic heavy metals: materials cycle optimization, *Proc. Natl. Acad. Sci. U.S.A.*, 89, 815, 1992.
43. Ayres, R.U. and Simonis, U.E., Eds., *Industrial Metabolism: Restructuring for Sustainable Development*, United Nations University Press, Tokyo, 1994.
44. Frosch, R.A. and Gallopoulos, N.E., Strategies for manufacturing, *Sci. Am.*, 261, No. 3, 94, 1989.

45. Graedel, T.E. and Allenby, B.R., *Industrial Ecology*, 2nd ed., Prentice Hall, Upper Saddle River, NJ, 2001, chap. 2.
46. Allenby, B., Culture and industrial ecology, *J. Ind. Ecol.*, 3, 2, 1999.
47. Commoner, B., The relation between industrial and ecological systems, *J. Cleaner Prod.*, 5, 125, 1997.
48. Ehrenfeld, J.R., Industrial ecology: a framework for product and process design, *J. Cleaner Prod.*, 5, 87, 1997.
49. Gorter, J., Zinc balance for the Netherlands, in *Material Flow Accounting: Experience of Statistical Institutes in Europe*, Statistical Office of the European Communities, Brussels, 2000, p. 205.
50. Jasinski, S.M., The materials flow of mercury in the United States, *Resour. Conserv. Recycling*, 15, 145, 1995.
51. Kleijn, R., Tukker, A., and van der Voet, E., Chlorine in the Netherlands, part I, and an overview, *J. Ind. Ecol.*, 1, 95, 1997.
52. Landner, L. and Lindeström, L., *Zinc in Society and in the Environment*, Swedish Environmental Research Group, Kil, Sweden, 1998.
53. Landner, L. and Lindeström, L., *Copper in Society and in the Environment*, 2nd ed., Swedish Environmental Research Group, Vaesteras, Sweden, 1999.
54. Llewellyn, T.O., Cadmium Materials Flow, Information Circular 9380, U.S. Bureau of Mines, Washington, DC, 1994.
55. Kinzig, A.P. and Socolow, R.H., Human impacts on the nitrogen cycle, *Phys. Today*, 47, 24, 1994.
56. Thomas, V. and Spiro, T., Emissions and Exposures to Metals: Cadmium and Lead, in *Industrial Ecology and Global Change*, Socolow, R., Andrews, C., Berkhout, F., and Thomas, V., Eds., Cambridge University Press, Cambridge, U.K., 1994.
57. Chertow, M.R., Industrial symbiosis: literature and taxonomy, *Annu. Rev. Energy Environ.*, 25, 313, 2000.
58. Obrist, J., von Steiger, B., Schulin, R., Schärer, F., and Baccini, P., Regionale Früherkennung der Schwermetall- und Phosphorbelastung von Landwirtschaftsböden mit der Stoffbuchhaltung "Proterra," *Landwirtsch. Schweiz.*, 6, 513, 1993.
59. Resource Conservation and Recovery Act, U.S. Code Collection, Title 42, Chap. 82, Solid Waste Disposal, Subchap. I, General Provisions, sec. 6902, 1976; www4.law.cornell.edu/uscode/42/6902.html.
60. Federal Ministry for Environment, Youth, and Family, Österreichisches Abfallwirtschaftsgesetz, BGBl, No. 325/1990, Federal Ministry for Environment, Youth, and Family, Vienna, 1990.
61. Bundesministerium für Umwelt und Reaktorsicherheit, Gesetz zur Förderung der Kreislaufwirtschaft und Sicherung der umweltverträglichen Beseitigung von Abfällen KrW-/AbfG - Kreislaufwirtschafts- und Abfallgesetz, Juristisches Informationssystem für die Bundesrepublik Deutschland, Berlin, 1994.
62. Eidgenössische Kommission für Abfallwirtschaft, Leitbild für die Schweizerische Abfallwirtschaft, Schriftenreihe Umweltschutz No. 51, Bundesamt für Umwelt, Wald und Landschaft, Bern, 1986.
63. Council Directive 75/442/EEC on Waste (so-called EC-Waste Framework Directive), Official Journal of the European Union, Brussels, Belgium, 1975.
64. Die Stoffflussanalyse im immissionsschutzrechtlichen Genehmigungsverfahren, Erlass des Ministeriums für Umwelt und Naturschutz, Landwirtschaft und Verbraucherschutz des Landes, Nordrhein Westfahlen, Düsseldorf, Germany, 2000.

65. Brunner, P.H. and Rechberger, H., Anthropogenic Metabolism and Environmental Legacies, in *Encyclopedia of Global Environmental Change*, Vol. 3, Munn, T., Ed., John Wiley and Sons, West Sussex, U.K., 2001, pp. 54–77.
66. Anderberg, S., Bergbäck, B., and Lohm, U., Flow and distribution of chromium in the Swedish environment: a new approach to studying environmental pollution, *Ambio*, 18, 4, 216, 1989.
67. Becker-Boost, E. and Fiala, E., *Wachstum ohne Grenzen*, Springer, New York, 2001.
68. Storck, W.J. et al., Facts and figures for the chemical industry: markets: housing rose again, *Chem. Eng. News*, 78, 76, 2000.
69. Wilson, D., Recycling of demolition wastes, in *Concise Encyclopedia of Building and Construction Materials*, Moavenzadeh, F., Ed., Pergamon/Elsevier, Oxford, 1990, pp. 517–518.
70. Settle, D.M. and Patterson, C.C., Lead in albacore: a guide to lead pollution in Americans, *Science*, 207, 1167, 1980.
71. Bergbäck, B., Industrial Metabolism: The Emerging Landscape of Heavy Metal Emission in Sweden, Ph.D. thesis, Linköping University, Sweden, 1992.
72. Brunner, P.H. and Baccini, P., Neue Zürcher Zeitung, Beilage Forsch. *Technik* 70, 65, 1981.

2 Methodology of MFA

This chapter introduces and describes the terminology of material flow analysis (MFA). The exact definition of *terms* and *procedures* is a prerequisite for generating reproducible and transparent results and for facilitating communication among users of MFA. The definitions as developed by Baccini and Brunner[1] during the 1980s are employed. As described in Chapter 1, Section 1.3, there have been several parallel developments of MFA, resulting in diverse usage and meanings for single terms (for example, the term "substance" has been applied to substances as well as goods) and in other cases, the application of several names for the same object. Needless to say, the presence of such a Babel within the global MFA community complicates the application and usage of MFA, especially for the new user. Differences appear not only in the employment of single terms but also on a higher conceptual and methodological level. In this chapter, these aspects are addressed and discussed whenever appropriate. The authors' intention is to contribute to a generalized terminology in MFA.

2.1 MFA TERMS AND DEFINITIONS

2.1.1 SUBSTANCE

Looking up *substance* in a dictionary may yield the following definitions relevant to the MFA field: the physical matter of which a thing consists; material;[2] matter of particular or definite chemical constitution;[3] or: matter; material of specified, especially complex, constitution.[4] For *material* one may find: of matter; of substance;[2] the elements, constituents, or substances of which something is composed or can be made;[3] and the substance or substances out of which a thing is or may be constructed.[4] All of these definitions seem convincing, but they are not precise enough for our purposes. As a matter of fact, it would be difficult to decide based on the above definitions whether "wood" should be addressed as a "substance" or "material." In everyday language both terms are used more or less synonymously. Therefore, MFA relies on the term *substance* as defined by chemical science. *A substance is any (chemical) element or compound composed of uniform units. All substances are characterized by a unique and identical constitution and are thus homogeneous.*[5] Using this definition makes clear that "wood" is not a substance. It is composed of many different substances such as cellulose, hydrogen, oxygen, and many others. Thus, a substance consists of identical units only. If these units are atoms, one speaks about a (chemical) element such as carbon (C), nitrogen (N), or cadmium (Cd). In cases where the units represent molecules such as carbon dioxide (CO_2), ammonium (NH_3), or cadmium chloride ($CdCl_2$) they are designated as (chemical) compounds. *In MFA, chemical elements and compounds both are correctly addressed as substances.*

Note that some authors use an additional and complementary definition for substance as follows: substance is the umbrella term for matter that differs only in size and shape but not in other specific properties such as color, density, electric conductivity, solubility, etc.[6] This means that a substance has to have a shape. According to this definition, it is not correct to identify Cd in a piece of polyvinyl chloride (PVC) as a substance. As soon as the substance Cd is incorporated and dispersed in the PVC matrix, forming a new heterogeneous mixture of homogeneous substances, Cd loses its shape and therefore the status of a substance. In MFA this definition is not useful, since MFA tracks the fate of substances through a system regardless of the physical and chemical form. Hence, the MFA definition of substances excludes shape. Thus it is correct to refer to Cd and other substances in a window frame made of stabilized PVC.

For MFA, substances are designated as *conservative* when they are not destroyed/transformed in a process. This applies to all elements and certain compounds that remain stable under process conditions. For example, cadmium can be transformed to cadmium chloride in a combustion process. Hence, it is not possible to establish a closed balance for cadmium chloride for the process. Cadmium chloride has to be addressed as a nonconservative substance, while the element cadmium can be designated as conservative because it does not disappear. In chemistry, the definition of conservative is somewhat different and is attributed to substances that are nonreactive. Also, in the environmental sciences, the term *conservative* characterizes substances that are persistent in the environment, having characteristics that are resistant to ordinary biological or biochemical degradation.

Substances are important for environmental and resources management. Many MFAs are carried out to determine flows of potentially hazardous substances to the environment and to find out more about the fate of these substances in environmental compartments such as water bodies and soils. Other MFA studies are commenced to understand better the flows and stocks of a resource in a system. Often these are substances such as heavy metals (Cu, Zn) or nutrients (N, P). Material flows in our industrial society can pose a problem because of both their composition (substances) and their quantity. The importance of discussing substances *and* materials is further explained in Section 2.1.10.

2.1.2 Good

The term *good* describes merchandise and wares. It is mostly used as a plural ("goods"). The opposite of a "good" is not a "bad," meaning that waste materials are also "goods."[7] The term is not used as an adjective but as a noun. *Goods are defined as economic entities of matter with a positive or negative economic value. Goods are made up of one or several substances.* Hence, a correct designation for "wood" is found: it is a "good." Examples for goods are drinking water (includes, besides H_2O, calcium and other trace elements and is therefore not a substance), mineral ores, concrete, TV sets, automobiles, garbage, and sewage sludge. All of these materials are valued and financially rated by the economy. Their turnover in

Methodology of MFA

terms of economic value and, less often, mass is usually reported in all kinds of statistics (by governments, national or multinational organizations and associations, companies, etc.). Hence, production figures (e.g., t/year) of most goods are more or less well known. Such information is a prerequisite for establishing material balances.

Sometimes the words *product*, *merchandise*, or *commodity* are used synonymously for "goods." The advantage of these synonyms is that they are common words. Their drawback is that *product* usually designates the output but not the input of a process or reaction. *Merchandise* and *commodity* are usually used to describe goods with a positive economic value and are rarely applied to negatively valued goods like garbage and sewage sludge. There are only a few goods that have no economic value, such as air or rainfall.

2.1.3 Material

The term *material* has already been explained in Section 2.1.1. While daily language does not distinguish between substance and material, in MFA *material* serves as an umbrella term for *both* substances and goods. So carbon as well as wood can be addressed as a material.

2.1.4 Process

A *process* is defined as the transformation, transport, or storage of materials.[1] Materials are transformed throughout the entire industrial economy on different levels. Transformation takes place in primary production processes such as in the mining and metal industry, where metals are extracted from mineral ores. Consumption processes such as private households transform goods into wastes and emissions. Examples of transformation processes are:

- The human body, where food, water, and air are transformed mainly into CO_2, urine, and feces
- The entirety of private households in a defined region, where countless inputs are converted to sewage, wastes, emissions, and some useful outputs
- Wastewater treatment plants, where sewage is transformed to purified wastewater and sewage sludge
- A car, where fossil fuels are oxidized to CO_2 at a rate of 0.003 l/sec
- A power station with a fuel transformation rate of 60 l/sec (turnover 20,000 × that of a car)
- The metabolism of a city, where many inputs are converted to corresponding outputs
- The total agriculture of a collective of states, such as the U.S. or all countries of a river catchment area, transforming inorganic nitrogen to organic nitrogen, etc.

Transformation of materials is not restricted to anthropogenic processes but is even more relevant for natural systems. Examples are forests that transform carbon

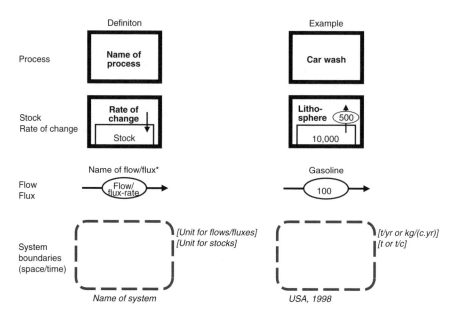

FIGURE 2.1 Main symbols used in MFA diagrams. (*If space is limited, ovals around flow/flux rates can be omitted.)

dioxide into oxygen and biomass, and soils where biomass residues are transformed into humus and CO_2.

The transport of goods, persons, energy, and information can be described as a process, too. During a transportation process, the goods moved are not transformed but relocated over a certain distance. The transportation process covers all material flows that are needed to carry out transport and all wastes and emissions that result from the transport. Transformation and transport processes are both symbolized by rectangular boxes (see Figure 2.1). Usually, processes are defined as black box processes, meaning that processes within the box are not taken into account. Only the inputs and the outputs are of interest. If this is not the case, the process must be divided into two or more subprocesses. This facilitates investigation and analysis of the functions of the overall process in greater detail (see Figure 2.2).

There is one exception to the "black box" approach: the third type of process is used to describe the quantity of materials within a process. The total amount of materials stored in a process is designated as the "stock of materials" in the process. Both the mass of the stock as well as the rate of change of the stock per unit time (accumulation or depletion of materials) are important parameters for describing a process. A "final sink" is a process where materials have very long residence times (t_R >1000 year).

Examples of storage processes are private households storing goods like furniture, electrical appliances, etc. for some years, or landfills where human society stores most of its discards over centuries. Natural storage processes are the atmosphere and the oceans for carbon dioxide or soils for heavy metals. A smaller box within the "process" box symbolizes the "stock" of a process (see Figure 2.1).

Methodology of MFA

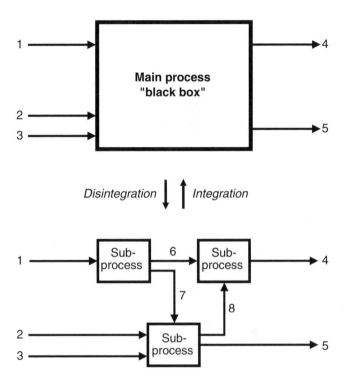

FIGURE 2.2 Opening up a black box by subdividing a single process into several subprocesses provides additional information about the black box.

2.1.5 FLOW AND FLUX

For MFA, the terms *flows* and *fluxes* have been often used in a random way. A *flow* is defined as a "mass flow rate." This is the ratio of mass per time that flows through a conductor, e.g., a water pipe. The physical unit of a flow might be given in units of kg/sec or t/year. A *flux* is defined as a flow per "cross section." For the water pipe, this means that the flow is related to the cross section of the pipe. The flux might then be given in units of kg/(sec·m^2). Note that flux and flow for a system (the water pipe) have different units. The flux can be considered as a *specific* flow. Integration of the flux over the cross section yields the total flow (see Figure 2.3).

In MFA, commonly used cross sections are a person, the surface area of the system, or an entity such as a private household or an enterprise. Some authors also consider the system itself as a cross section and prefer the term *fluxes* for all material exchanges between processes. For instance, the total flow of oil into a region is then designated as a flux. The cross section is the *region* of study. The same flow per capita is also designated as a flux. The cross section in this case is one inhabitant of the region.

In this handbook, a slightly different definition is used. The actual exchange of materials determined for a system is designated as a flow. The system itself is *not* considered to be a cross section. Only specific flows that are related to a cross section

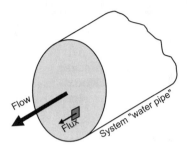

FIGURE 2.3 Total flow and specific flow (flux) of water through a pipe.

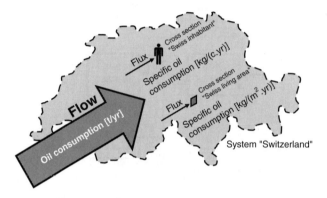

FIGURE 2.4 Flows and fluxes for the system "Switzerland." Fluxes are specific flows and are related to a cross section.

are designated as fluxes (see Figure 2.4). Table 2.1 gives some examples for flows, cross sections, and associated fluxes.

The advantage of using fluxes is that they can be easily compared among different processes and systems, since fluxes are specific values (such as density, heat capacity). Usually fluxes are kept in mind as reference values for a "quick and dirty" assessment of anthropogenic systems, processes, or material flows: an affluent society generates 500 kg/capita/year of municipal solid waste (MSW); an economical car consumes no more than 5 l/100 km of gasoline.

An *arrow* symbolizes flows and fluxes (see Figure 2.1). For each flow and flux a "process of origin" and a "process of destination" has to be defined. The process of origin of a flow or flux that *enters* the system is called an "import process." Flows and fluxes that *leave* the system lead to "export processes." By definition, both import and export processes are located outside of the system and therefore are not balanced or taken into account further (see Figure 2.7). The flow of a good is indicated with $\dot{m}$ and the flow of a substance with $\dot{X}$.

2.1.6 Transfer Coefficient

Transfer coefficients (TC) describe the partitioning of a substance in a process. They are defined as follows and as illustrated in Figure 2.5.

TABLE 2.1
Examples of Flows and Fluxes

	System	Cross Section	Numerical Value of Cross Section	Flow	Flux
Paper consumption	Switzerland	Swiss population	7.3 million	1.8 million t paper/year[a]	246 kg paper/ (c·yr)
Waste treatment	MSW incinerator	grate	50 m²	15 t MSW/h	300 kg MSW/(m²·h)
Emission of SO_2	Switzerland	area of state	42,000 km²	30,000 t SO_2/ year[a]	0.7 g SO_2/ (m²·year)
Total deposition of nitrogen	Vienna	area of city	415 km²	1,400 t N/year[129]	3.4 g N/ (m²·year)

[a] In 2000.

FIGURE 2.5 The transfer of substance X into output flow 2 is given as $TC_2 = \dot{X}_{O,2} \Big/ \sum_i \dot{X}_{I,i}$.

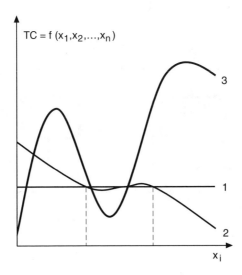

FIGURE 2.6 Different kinds of transfer coefficients: TC_1 is independent of parameter x_i; TC_2 can be treated as a constant within a certain range of x_i; TC_3 is highly sensitive to variation in parameter x_i.

$$TC_i = \frac{\dot{X}_{O,i}}{\sum_{i=1}^{k_I} \dot{X}_{I,i}} \quad (2.1)$$

$$\sum_{i=1}^{k_O} TC_i = 1 \quad (2.2)$$

where
k_I = number of input flows
k_O = number of output flows

Transfer coefficients are defined for each output good of a process. Multiplied by 100, the transfer coefficient gives the percentage of the total throughput of a substance that is transferred into a specific output good (designated as partitioning). It is a substance-specific value. Furthermore, transfer coefficients are technology-specific values and stand for the characteristics of a process. Transfer coefficients are not necessarily constant. They can depend on many variables such as process conditions (e.g., temperature, pressure) as well as input composition. Sometimes transfer coefficients can be regarded as constant within a certain range (see Figure 2.6). This makes them useful for sensitivity analysis of the investigated system and for scenario analysis.

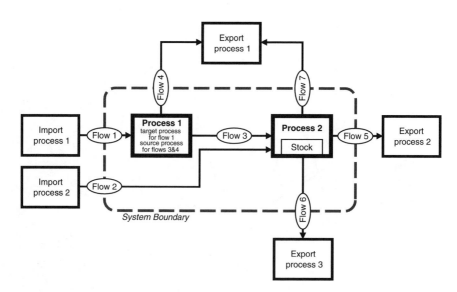

FIGURE 2.7 Exemplary MFA system illustrating selected terms. Ovals of import and export flows are placed directly on the system boundary to facilitate visual understanding of the system.

2.1.7 SYSTEM AND SYSTEM BOUNDARIES

The *system* is the actual object of an MFA investigation. A system is defined by a group of elements, the interaction between these elements, and the boundaries between these and other elements in space and time. It is a group of physical components connected or related in such a manner as to form and/or act as an entire unit.[8] An open system interacts with its surroundings; it has either material or energy imports and exports or both. A closed system is conceived as a system with complete isolation, preventing material and energy flows across the system boundary. In MFA, the physical components are processes, and the connections/relations are given by the flows that link the processes. A single process or a combination of several processes can represent a system. The actual definition of the system is a decisive and demanding task. Poor results of MFA can often be traced back to an unsuitable system definition. The system boundaries are defined in time and space (temporal and spatial system boundary).

The temporal boundary depends on the kind of system inspected and the given problem. It is the time span over which the system is investigated and balanced. Theoretically it can range from 1 sec for a combustion process to 1000 years for landfills. Commonly applied temporal boundaries are 1 h, 1 day, or 1 year. For anthropogenic systems such as an enterprise, a city, or a nation, periods of 1 year are chosen for reasons of data availability. Because financial accounting and other reporting is typically done on an annual base, information that covers periods of 1 year is usually easier to retrieve than data for shorter or longer time periods.

The spatial system boundary is usually fixed by the geographical area in which the processes are located. This can be the premises of a company, a town or city, a region such as the Long Island Sound or the Danube River watershed, a country such as the U.S., a continent such as Europe, or the whole planet. Abstract areas also can serve for a boundary definition when MFA is applied to a specific part of the economy, e.g., the waste-management system of a county, the average private household, or a virtually assembled region within a defined economy. For most studies, it is necessary to define the spatial system boundary also for the third dimension (vertically). Above the Earth's surface, usually the first 500 m of the atmosphere are considered. Within this layer (so-called planetary boundary layer), the main exchange of air (and pollutants) between regions takes place, and there is negligible material exchange with the air compartment above. Under the Earth's surface, the system boundary has to be extended far enough that the groundwater flows are included. A rectangle with broken lines and rounded corners symbolizes the spatial system boundary (see Figure 2.1 and Figure 2.7).

For material balances of larger systems such as nations or continents, the term *cycle* is frequently used.[9–12] This term stems from the grand biochemical cycles of carbon, oxygen, nitrogen, hydrogen, sulfur, and phosphorus that drive the biosphere. While these natural cycles indeed show cyclic behavior, anthropogenic systems hardly do (e.g., see Chapter 1, Section 1.4.5.3 on urban material flows and Chapter 3, Section 3.2.2.1 on copper flows). Therefore, applying the term *cycle* to anthropogenic systems such as urban areas or national economies could be misleading.

2.1.8 ACTIVITIES

Regardless of a community's social, cultural, technical, or economic development, there is a set of basic human needs such as to eat, to breath, to reside, to communicate, to transport, and others.[1] The main goal of a sustainable economy is to satisfy these needs best at the least cost. An activity is defined as comprising all relevant processes, flows, and stocks of goods and substances that are necessary to carry out and maintain a certain human need.* The purpose of defining activities is to facilitate the analysis of a given way of fulfilling human needs, to evaluate the constraints and optimization potentials, and to suggest strategies and measures for optimizing the way the needs are satisfied.

The most important activities can be defined as follows. (Note that this list is not complete. Additional activities such as "leisure," "health and sports," etc. can be defined, if necessary, to analyze and solve a particular resource-oriented problem.):[1]

To nourish: This activity comprises all processes, goods, and substances used to produce, process, distribute, and consume solid and liquid food. "To nourish" starts with agricultural production (e.g., the goods "seed corn," "water," "air," "soil," "fertilizer" and the process "crop raising," e.g., beans), food production (e.g., the process "cannery," the good "canned beans"), distribution (e.g., the process "grocery"), consumption (e.g., the processes "storage," "preparation," and "consumption" of canned beans in private households), and ends with the release of off-gases (breath), feces and urine, and solid wastes (can, leftover beans) to the atmosphere and the waste- and wastewater-treatment systems. These systems already belong to the activity "to clean," as discussed in the following paragraph. Figure 2.8 shows the main processes for the activity.

To clean: In anthropogenic processes, "wanted" materials are often separated from "unwanted" materials. When sugar is produced from sugar cane, sucrose is separated from cellulose and impurities. In dry cleaning, dirt is removed from the surface of clothes by organic solvents such as perchloroethylene. People need to remove dirt and sweat from their body surfaces. Also, they need to remove materials not useful for their metabolism and wastes from their body, such as carbon dioxide in breath, salts in urine, or undigested biomass in feces. Since many of these processes are called "cleaning," the separation of valuable from useless materials has been defined as the activity "to clean." It is an essential activity for human beings, since it is necessary for everybody to keep material input and output in a balance. Examples for cleaning processes on the individual level are "private and commercial laundry" (see Figure 2.9), "dishwashing," and "housecleaning"; on the industrial level there is "refinery," "metal purification," and "flue-gas treatment"; and on the community level there is "sewage and

---

* Human beings have also nonmaterial needs, such as love, security, intellectual exchange, social recognition, etc., that cannot be quantified and are therefore not tangible for an MFA approach.

Methodology of MFA

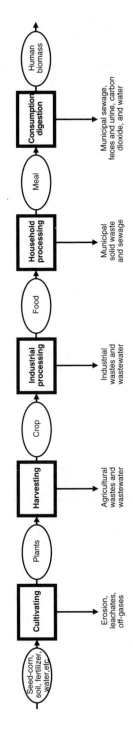

FIGURE 2.8 Process chain of main processes associated with the activity "to nourish." Required goods to operate the processes (energy, machinery, tools, etc.) are omitted.[1] (From Baccini, P. and Brunner, P.H., *Metabolism of the Anthroposphere*, Springer, New York, 1991. With permission.)

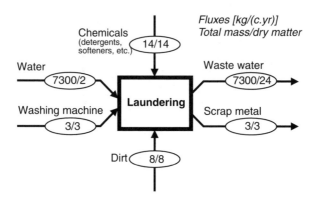

FIGURE 2.9 Mass flows of goods for the activity "to clean" through the process "laundry," which is a subprocess of the process "household." About 100 kg of water are needed to separate 1 kg of dirt from textiles (kg/capita/year; left side of figure: total mass; right side of figure: dry matter).[1] (From Baccini, P. and Brunner, P.H., *Metabolism of the Anthroposphere*, Springer, New York, 1991. With permission.)

waste treatment" and "public cleaning." "To clean" is also a very important activity for public health.

To reside and work: This activity comprises all processes that are necessary to build, operate, and maintain residential units and working facilities. Important processes are "building construction," "operation and maintenance of buildings," "machine construction," "operation and maintenance of machinery," "manufacture of furniture and household appliances," "manufacture of clothing and leisure appliances," and "consumption." Table 2.2 gives an example for subprocesses and related goods for the process "building construction." Figure 2.10 outlines a flowchart for the activity "to reside and work."

The functions and services that are expected from a building are manifold. One is that it should provide an agreeable temperature inside. This can be realized by different heating and cooling systems, different types of wall construction, and the use of different materials for better insulation. But other approaches are also possible to fulfill the service "agreeable body

TABLE 2.2
Subprocesses and Their Input and Output Goods for the Process "Building Construction"

Process	Subprocess	Inputs	Outputs
Building construction	concrete production, steel and metal production, quarry, lumber mill, energy supply	gravel, sand, stone, limestone, marl, metal ores, wood, fuels, water, air	buildings, construction and demolition waste, wastewater, off-gas

Methodology of MFA

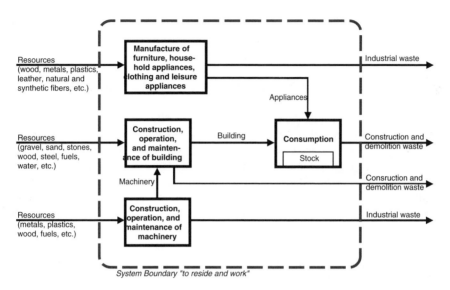

FIGURE 2.10 Relevant processes and goods for the activity "to reside and work." Only solid wastes are indicated.[1]

temperature" during the cool season. Besides measures for the outer skin (the wall), a combination of reduced heating and wearing a pullover (insulation of the inner skin) can also fulfill the task. All three approaches (heating, insulation, clothes) result in different materials and energy consumption.

To transport and communicate: This activity comprises all processes that have been developed to transport energy, materials, persons, and information. The processes range from "road construction, operation, and maintenance of networks and transport vehicles" to "administration" and "production and transport of printed and electronic media." Rapid technological progress causes swift changes in processes and goods that are associated with this activity. Today's networks are established by roads, railway tracks, air tracks, cables, satellites, radio, etc. Hence, there are many ways of transmitting information over longer distances. This can be done by the transport of persons, by the transport of information carriers (printed paper, compact disc, magnetic recording, etc.), or transmission via cable, fiber optics, radio, etc. This activity can be used to decide which way of transmitting information is less resource consuming. Transport systems and strategies for goods and persons can also be compared by applying the activity approach to evaluate different transportation modes (e.g., gasoline- vs. hydrogen-powered engines) as well as different management modes (individual car vs. using the car pool of a service provider).

Goods and processes can be part of several activities. For instance, an automobile is used to drive to work and to do business, such as a taxi. It belongs to the activity "to transport and communicate" as well as to the activity "to reside and work" because it is an essential part of both. It is important to note that there are no strict rules regarding the linkage of processes to activities. It is done according to the utilitarian principle: if it serves the purpose to associate a process with a certain activity, it should be done.

2.1.9 ANTHROPOSPHERE AND METABOLISM

The human sphere of life — a complex system of energy, material, and information flows in space — is called the "anthroposphere." It is part of planet Earth and contains all processes that are driven by mankind. The anthroposphere can be seen as a living organism. In analogy to the physical processes in plants, animals, lakes, or forests, the "metabolism" of the anthroposphere includes the uptake, transport, and storage of all substances; the total biochemical transformation within the organism; and the quantity and quality of all off-products such as flue gas, sewage, and wastes.[1] The complementary part to the anthroposphere is designated as the environment. The anthroposphere interacts with the environment via extraction of resources (air, water, and minerals) and emission of off-products and wastes. The anthroposphere can be defined as part of any region where human activities take place.

As a first step, the anthroposphere can be divided into the four compartments (main processes in Figure 2.11):

1. Agriculture
2. Industry, trade, and commerce
3. Private households (consumption)
4. Waste management

All along the anthropogenic process chain there is exchange of materials and energy with the environment, which comprises four compartments: atmosphere, hydrosphere, pedosphere, and lithosphere. Often, terms such as *water*, *air*, and *soil* (for pedosphere) are used synonymously. Some authors use the terms *technosphere* or *biosphere* instead of *anthroposphere*.

Sometimes the interface between anthroposphere and environment is not clear. For example, a soil that is used by humans can be regarded as a part of the anthroposphere as well as of the environment. Hence, the definition of the anthroposphere is somewhat subjective. Some authors claim that all soils belong to the anthroposphere because anthropogenic trace substances have been detected in all soils on Earth; there are no soils anymore that are in a natural, uninfluenced state, and thus they do not belong to the environment anymore. Other authors include only those soils in the anthroposphere that are actively managed by mankind. In MFA practice, such allocation problems are of little relevance. They should be acknowledged but not overemphasized.

The term *metabolism* is used in many different fields and contexts. Initially it was used to name the turnover, which is the uptake, internal transformation, and emission of energy and matter within living bodies (organisms).[13] Later on, due to the compelling analogy between man-made systems and biological organisms, the term *metabolism* was applied to anthropogenic as well as geogenic (natural) processes and systems.[1,9] In MFA, metabolism of a system stands for the transfer,

Methodology of MFA

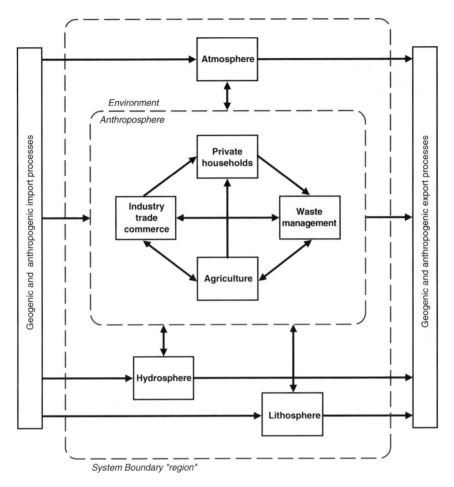

FIGURE 2.11 A region where anthropogenic activity takes place can be divided into the subsystems "anthroposphere" and "environment." Exchange of materials takes place within the subsystems as well as between the subsystems and processes outside the region.

storage, and transformation of materials within the system and the exchange of materials with its environment. Metabolism is applied to anthropogenic as well as geogenic (natural) processes and systems. Patzel and Baccini[13] elaborate the pros and cons of the term with regard to literal and metaphoric usage as compared with *household* and *physiology* and give their preference to *physiology*. It is likely that both terms, *metabolism* and *physiology*, will be used synonymously in future.

2.1.10 MATERIAL FLOW ANALYSIS

Material flow analysis (MFA) is a method to describe, investigate, and evaluate the metabolism of anthropogenic and geogenic systems. MFA defines terms and procedures to establish *material balances* of systems. A complete description of how to perform an MFA is given in Section 2.2. As described before, MFA includes balances

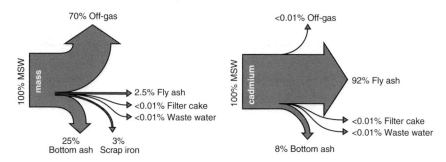

FIGURE 2.12 Mass balances for goods and cadmium of an MSW incinerator.[17] Each balance alone provides limited information. The ensemble of all balances yields a comprehensive understanding of the material flows in an MSW incinerator. Hence, both balances of goods and substances are necessary.

of both goods and substances. In Chapter 1, Section 1.4.4. and in Section 2.1.1 in this chapter, it has been explained why it is often important to base decisions on substance balances and not on balances of goods alone.

One of the first steps in MFA is to establish a *mass balance of goods* for the system of study. This is the base for the determination of balances of *selected substances*. The reason for making the couple "goods/substances" so central in MFA is that a substance balance without the underlying mass balance of goods does not provide the full information.

First, for each substance flow in a system, one has to know the goods flow on which it is based. This is important because often the flows of goods can be controlled directly, but the flows of substances can only be influenced indirectly. For example, application of sewage sludge can cause accumulation of heavy metals in agricultural soil. The flow of heavy metals to soil can be indirectly controlled through an action directed toward "sewage sludge," which is a good.

Second, it is important to know the concentration of a substance in a good. For example, if a hazardous-substance flow has to be reduced, one must know the concentration of the substance in the good. The higher the concentration, the smaller the quantity of the good must be in order to achieve an acceptable flow to the environment. This quantity is decisive when developing an appropriate strategy for reduction or prevention.

Third, establishing mass balances for goods *and* substances helps to detect sources of error that otherwise could not be found with mass balances for either goods or substances alone. Figure 2.12 shows mass balances of goods and cadmium for an MSW incinerator. The information given by the two graphs combined is more comprehensive and allows one to draw more conclusions than would be possible using only a single figure on goods or substances.

In addition to *MFA*, the term *substance flow analysis* (SFA) is also used occasionally. Some authors do not distinguish between goods, materials, and substances in a rigid way as described in this chapter. Therefore, SFA is sometimes used as another name for what is here defined as MFA. At other times, it refers to mass balances of one or several substances.[10,12] Graedel and Allenby[11] differentiate

Methodology of MFA

between elemental flow analysis (EFA), molecular flow analysis (MoFA), SFA, and MFA. In cases where stocks are considered, EFA mutates to elemental stock and flow analysis (ESFA), and so on. The differentiation is not only based on the element, molecule, substance, or material of study but, rather, on the objective and scope of the analysis. A zinc balance for a system can be the result of an EFA, SFA, or MFA, depending on whether the emphasis is on the atom (EFA), on the different speciation of zinc as compounds (SFA), or on the reservoirs* (i.e., processes) among which zinc is exchanged (MFA). Other schools use the term MFA to designate exclusively mass balances of goods without considering substances. These studies are frequently termed as bulk-MFA and the term *material* is used synonymously for goods and not for substances. Such studies at the level of goods do not include environmental impacts, in contrast to SFA, which does. Bulk-MFA studies have been shown to provide valuable information and to create awareness about the turnover of goods in developed economies.

In the following sections, the term *MFA* stands for the investigation and quantification of the metabolism of anthropogenic and geogenic systems. This includes the definition and selection of system boundaries as well as relevant goods, substances, and processes. In MFA, mass balances of both goods and substances are essential.

2.1.11 MATERIALS ACCOUNTING

Materials accounting (sometimes referred to as materials bookkeeping) is the routine updating of the results of an MFA by a few measurements of the key flows and stocks only. Materials accounting yields time series and facilitates detection of trends. It is instrumental for the control of measures in resource management, environmental management, and waste management. The base for material accounting is an MFA, which serves in understanding the system of study and in identifying the appropriate parameters (flows, stocks, concentrations) for the routine analysis. The art of materials accounting is to find those key parameters that yield maximum information at minimum cost. Materials accounting can be applied to all sizes of systems, from single companies to national economies.

2.1.12 PROBLEMS — SECTION 2.1

Problem 2.1: Consider the following flowchart and detect nine errors (see Figure 2.13).

Problem 2.2: The following list comprises physical units for flows, fluxes, and stocks. Assign the units to the three categories and define some more units: t/year, l/(m²·sec), kg/h, mg/capita, m/s, Tg, kg/capita/year, μg/(m²·year), kg/m³, l/day.

---

* Some authors prefer the term *reservoir* instead of *process*. Note that *reservoir* is a narrower term and does not consider all characteristics that can be attributed to a process. While the term *process* implies that something is changed in its quality (feedstock becomes product and waste), a reservoir mainly changes its quantity (mineral deposit, groundwater, landfill). So *reservoir* is merely a synonym for stock, and *process* is the more extensive term comprising change in quality and quantity (location being part of quality).

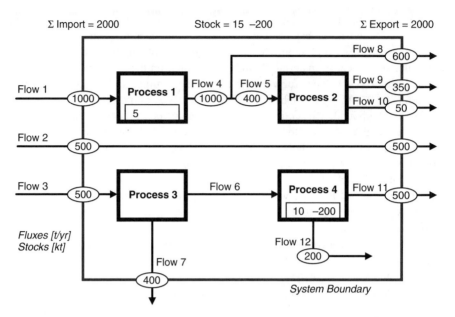

FIGURE 2.13 MFA flowchart with multiple errors.

Problem 2.3: (a) Assess your average daily water consumption. First, list the processes of your daily life that consume water. Second, try to assess the specific water use of these processes. Third, add up your estimates to a final result in l/capita/day. Use ranges when estimates are uncertain. (b) Some data can be easily found on the World Wide Web. Compare your estimates with the result of an Internet search.

Problem 2.4: MSW is separated by a mechanical sorting plant into a combustible fraction (30%) and another fraction (70%) that is digested by a subsequent biochemical process. Approximately half of the latter fraction is transformed to biogas during the anaerobic treatment. (a) Draw the system as a quantitative flowchart. (b) The concentrations of Hg are: MSW input, 1.5 mg/kg; combustible fraction, 1.0 mg/kg; biogas, 0.005 mg/kg; digestion product, 3.4 mg/kg. Calculate all mass flows and transfer coefficients for Hg.

Problem 2.5: Allocate the following substances to one of three categories of element concentrations in the Earth's crust: matrix elements (>10%), trace elements (<1%), or elements in between (1 to 10%): Au, Cr, O, Na, Cu, Ca, Hg, Sb, Ag, Mg, Mn, Zn, Ti, V, Fe, Si, Rb, Cl, Al, As, K, Pb, Se, H, C, Tl, and Cd.

Problem 2.6: Assign the following goods and processes to activities (to nourish, to clean, to reside and work, to transport and communicate): brick, cement kiln, electrostatic precipitator, chicken farm, fertilizer, fuel oil, gasoline, jeans, mobile phone, paper, shovel excavator, sal ammoniac, truck, vacuum cleaner. Note that multiple assignments of terms to activities are possible.

Methodology of MFA

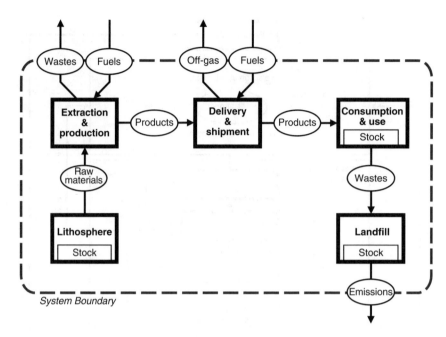

FIGURE 2.14 Flowchart comprising different types of processes.

Problem 2.7: What types of processes (transformation, transport, storage) make up the metabolic system shown in Figure 2.14?

Problem 2.8: Which of the following processes are anthropogenic? Which are geogenic? — agriculture, agricultural soil, atmosphere, compost production, crop production, forest management, landfill, pedosphere, planetary boundary layer, river, volcano, zoo.

The solutions to the problems are given on the Web site, www.iwa.tuwien.ac.at/MFA-handbook.

2.2 MFA PROCEDURES

An MFA consists of several steps that are discussed in stages in the following chapters. In general, an MFA begins with the definition of the problem and of adequate goals. Then relevant substances and appropriate system boundaries, processes, and goods are selected. Next, mass flows of goods and substance concentrations in these flows are assessed. Substance flows and stocks are calculated, and uncertainties considered. The results are presented in an appropriate way to visualize conclusions and to facilitate implementation of goal-oriented decisions. It is important to note that these procedures must not be executed in a strictly consecutive way. The procedures have to be optimized iteratively. The selections and provisions that are taken during the course of the MFA have to be checked continuously. If necessary, they must be adapted to accommodate the objectives of the project. In general, it is best to start with rough estimations and provisional data, and then to constantly

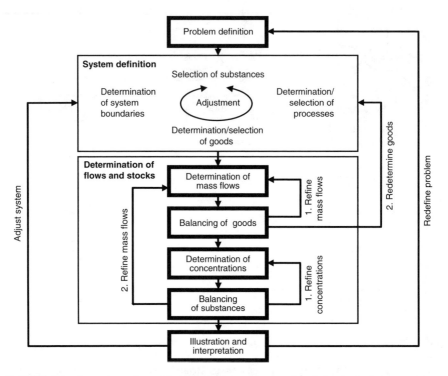

FIGURE 2.15 Procedures for MFA. MFA is an iterative process. Selections of elements (i.e., substances, processes, goods, boundaries) and accuracy of data have to be checked with regard to the objectives of the investigation.

refine and improve the system and data until the required certainty of data quality has been achieved (see Figure 2.15).

2.2.1 SELECTION OF SUBSTANCES

There are various approaches to choosing substances relevant for an MFA. On the one hand, they depend on the purpose of the MFA, and on the other hand, they depend on the kind of system on which the MFA is based.

First, legislation such as the Clean Air Act or standards such as building, materials quality, and safety codes provide listings of relevant substances that are regulated. This approach benefits from existing knowledge and ensures that all substances that have already been selected by the respective authorities are considered for inclusion in the MFA.

Second, the relevance of substances in the important flows of goods has to be evaluated. At the beginning of a study, a practicable rule to determine those flows is as follows. Group all import and export flows of goods of the system into solids, liquids, and gases. For each group, select as many flows as needed to cover at least 90% of the total mass flow of the group. This yields a set of important flows of goods (i) for the system. An elegant way to determine indicator substances (j) is to establish the ratio of substance concentrations in the selected flows of goods (c_{ij}) to geogenic

Methodology of MFA

TABLE 2.3
Selection of Substances for MFA by Goods/EC Ratio[a]

Substance	Earth's Crust (EC),[130] mg/kg	Printed Circuit Boards (PCB)[131]		Coal (C)[132]		Mixed Plastic Waste (MPW)[133]	
		mg/kg	Ratio[a] PCB/EC	mg/kg	Ratio[a] C/EC	mg/kg	Ratio[a] MPW/EC
Ag	0.07	640	**9100**	—	—	—	—
As	1.8	—	—	10	**5.6**	1.3	0.7
Au	0.003	570	**190,000**	0.01	3.3	—	—
Cd	0.15	395	**2600**	1	7	73	**490**
Cl	130	—	—	1000	8	—	—
Cr	100	—	—	20	0.2	48	0.5
Cu	50	143,000	**2900**	20	0.4	220	4
Hg	0.02	9	450	0.15	**8**	1.3	**65**
Mn	1000	—	—	40	0.04	17	0.02
Ni	75	11,000	150	22	0.3	10	0.1
Pb	13	22,000	**1700**	20	1.1	390	**30**
Sb	0.2	4,500	**22,500**	2	**10**	21	**110**
Se	0.05	—	—	3	**60**	6.7	**130**
Sn	2.5	20,000	**8000**	8	3	4	2
Zn	70	4,000	57	35	0.5	550	8

[a] The ratio of "element concentration in selected goods" to "element concentration in the Earth's crust" (EC) helps to identify substances that may be of environmental relevance or of importance as a resource. Candidates for MFA study are in boldface.

reference materials (c_{geog}). For solid material flows, the average concentrations in the Earth's crust ($c_{geog} \equiv c_{EC}$) can be taken as a reference. Similarly, for liquid material flows, the average concentration in geogenic water bodies ($c_{geog} \equiv c_{hydro}$) can be taken as a reference. For gaseous material flows, the average concentration of the atmosphere ($c_{geog} \equiv c_{atmo}$) can be taken as a reference. Substances with a ratio of c_{ij}/c_{geog} > 10 are candidates to be selected for the study (see Table 2.3). The proportions among the substance-specific ratios for a good also have to be considered. If all or most of the ratios are <10, just select those substances with the highest ratios. Note that this rule is only a tool that will assist in the selection process. The result has to be checked for consistency and reasonability during the course of the MFA. These two approaches can be used when the system of study is determined already by the scope of the study. For example, these approaches could be used when the task of the project is to determine the metabolism of a manufacturing firm or a special wastewater-treatment technology in order to find potentials for optimization.

The third case is not a real approach but rather stems from the fact that many MFA studies are carried out to determine a system's metabolism of one or several substances for resource and/or environmental impact aspects. In such cases, the selection of substances is part of the project definition (e.g., project "copper household of Europe"). Also, when activities are investigated, some substances such as C, N, and P are determined for the activity "to nourish."

As mentioned before, a main objective of MFA is to reduce as much as possible the number of parameters that have to be taken into account. To achieve this goal, so-called indicator elements are selected. This is an important means of achieving maximum information with minimum effort. An indicator element represents a group of substances. It shows a characteristic physical, biochemical, and/or chemical behavior that is a specific property of all members of the group. Indicator substances can therefore be used to predict the behavior of other substances. For combustion and high-temperature processes, for instance, substances can be separated into "atmophile" and "lithophile" elements.* Atmophile elements have a lower boiling point and thus are transferred to the off-gas. Examples of atmophile elements are Cd, Zn, Sb, Tl, and Pb. Thus, Cd serves well as an indicator for this group. Lithophile elements and their compounds have a high boiling point and therefore tend to remain in the bottom ash or slag. Lithophile elements include Ti, V, Cr, Fe, Co, and Ni. Fe is representative for this group. Note that the boiling point of the element is not the only decisive factor in whether a substance tends to behave as an atmophile or lithophile in a process. The boiling points of the occurring species (chlorides, oxides, silicates, etc.) are also important. It is obvious that indicator elements have to be chosen in accordance with the system and processes analyzed. The above indicator metals are appropriate when thermal processing of MSW is investigated. In biochemical processing, where evaporation is not a path to the environment for most metals except mercury, other indicators are likely to be more appropriate.

The selection of substances depends on the scope, the grade of precision, and the resources (financial and human) that are available for the MFA study. Experience with MFA shows that many anthropogenic and natural systems can be roughly characterized by a comparatively small number of substances, such as about five to ten elements. Table 2.4 gives a list of frequently used indicator elements in MFA.

2.2.2 SYSTEM DEFINITION IN SPACE AND TIME

The spatial system boundary is usually determined by the scope of the project (carbon balance of a community, MFA of a power plant, etc.). It coincides often with the politically defined region, the estate of a company, or a hydrologically defined region such as the catchment area of a river. Often, the only possibility is to define system boundaries as administrative regions, such as nations, states, or cities, because information is systematically collected on these levels. One advantage of choosing these kinds of system boundaries is that the political and administrative stakeholders are within the regional boundary. Thus, measures based on the results of MFA can be implemented more easily in such administratively defined regions.

In general, any system should be chosen to be as small and consistent as possible while still being broad enough to include all necessary processes and material flows. Consider the following example. From the point of view of resource conservation as well as environmental protection, nitrogen flows and stocks in a city are important.

---

* Geochemistry groups elements as follows:[130] siderophile elements are those elements that are concentrated in the Earth's iron core; chalcophile elements tend to combine with sulfur in sulfide minerals; lithophile elements generally occur in or with silicates; atmophile elements are prominent in air and other natural gases.

TABLE 2.4
Examples of Indicator Elements in MFA

Indicator Element	Symbol	Relevant Activity[a]	Properties
Carbon (organic)	C_{org}	N, R&W, C, T&C	Carrier of chemical energy Carrier of nutrients Main matrix element in many toxic compounds
Nitrogen	N	N, C, T&C	As NO_3^-, an essential nutrient As NO_x, a potential air pollutant As NH_4^+ and NO_2^-, toxic for fish Eutrophication of aquatic ecosystems
Fluorine	F	C	As F^-, a strong inorganic ligand In incineration, HF is formed, a strong acid
Phosphorous	P	N, C	As PO_4^{3-}, an essential nutrient Eutrophication of aquatic ecosystems
Chlorine	Cl	N, C	Forms as Cl^-, soluble salts Forms stable, sometimes toxic chloro-organic compounds (e.g., PCB and dioxins)
Iron	Fe	R&W, T&C	Forms as Fe^{3+}, poorly soluble oxides, and hydroxides As metal, easily oxidized under atmospheric conditions (H_2O and CO_2) Recycling of metallic iron is economical Resource in construction
Copper	Cu	R&W, C, T&C	Forms as Cu^{2+}, stable complexes with organic ligands As metal, an important electrical conductor Even in small concentrations, toxic for unicellular organisms Resource in construction
Zinc	Zn	R&W, C	Forms as Zn^{2+}, soluble salts Important anticorrosive and rubber additive Atmophile Resource in construction
Cadmium	Cd	C, T&C	Stabilizer for PVC, pigment, anticorrosive Constituent of rechargeable batteries Toxic for humans and animals, less so for plants Atmophile
Mercury	Hg	C	Forms metal-organic and toxic compounds under reduced conditions Atmophile and liquid metal
Lead	Pb	R&W, C, T&C	Forms as Pb^{2+}, stable complexes with natural organic ligands Lithophile element Was and still is an important additive in gasoline

[a] N, to nourish; R&W, to reside and work; C, to clean; T&C, to transport and communicate.

Source: From Baccini, P. and Brunner, P.H., *Metabolism of the Anthroposphere,* Springer, New York, 1991. With permission.

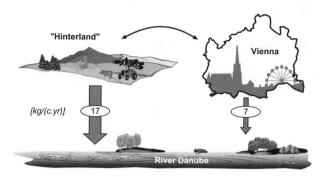

FIGURE 2.16 Nitrogen flux in kg/(capita·year) to the river Danube from the city of Vienna and the corresponding hinterland.[124]

The people in the city must be fed, and the water and air leaving the city must not be polluted. The production and processing of food to nourish the people in the city results in emissions in the agricultural region that supplies the food, the so-called hinterland. Agricultural losses from farming as well as production wastes from sugar refineries, canneries, frozen food processing, etc. yield nutrient flows that are either directly or indirectly led to the river surface waters. Obviously, the choice of the system boundary is important. If the system includes the city only, the major emission flow is not observed (see Figure 2.16). If both the hinterland and the city are included within the system boundary, it becomes apparent that the flow of nutrients to the river is more than twice as large as measured in the city alone, and that the city is responsible for large emissions in the hinterland, too. Hence, it is crucial to include the hinterland within the scope of study. The most appropriate solution for the definition of the system boundary has to be individually established for each case study and objective.

Temporal system boundaries are comparatively easier to determine. This is especially the case when average flows and stocks over a longer period of time are of interest. In this case, the time span of investigation, which is identical with the temporal system boundary, has to be extended long enough to outweigh the momentary unsteadiness of the system. For many anthropogenic systems, this is the case for time periods of investigation of one year. If higher resolutions in time are preferred instead of averages, shorter time periods may be more appropriate for system boundaries in time. Short time periods allow detection of short-term anomalies and nonlinear flows.

2.2.3 IDENTIFICATION OF RELEVANT FLOWS, STOCKS, AND PROCESSES

After selecting substances and defining system boundaries, a first rough balance of goods is carried out for the system. Information about material flows is taken from the literature or other sources such as company and national reports. Sometimes, data have to be assessed, e.g., by contacting experts or visiting government agencies. At this stage, material flows <1% of the total system throughput are neglected.

Methodology of MFA

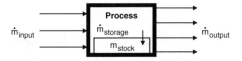

FIGURE 2.17 Consideration of a process's stock.

Nevertheless, these small flows can make a significant contribution to one or more substance balances that are established later. Therefore, when in a subsequent step the substance flows are investigated in greater detail, it is important to check whether these small neglected flows are relevant in view of the objectives of the investigation.

The number of processes necessary to describe the system depends on the objectives of the study and on the complexity of the system. Generally, processes can be subdivided into subprocesses, and vice versa, any number of processes can be merged into a single process (see Figure 2.2). In most cases, systems comprising more than 15 processes (exclusive of import and export processes) turn out to be clumsy and unnecessarily complex. Remember that one of the main purposes of MFA is to develop simple and reliable models to picture reality. For mathematical formulations, let the number of flows in a system be k, the number of processes be p, and the number of substances be n. According to the mass-balance principle, the mass of all inputs into a process equals the mass of all outputs of this process plus a storage term that considers accumulation or depletion of materials in the process (Equation 2.3 and Figure 2.17; the symbol • designates a flow or flux).

$$\sum_{k_I} \dot{m}_{input} = \sum_{k_O} \dot{m}_{output} + \dot{m}_{storage} \qquad (2.3)$$

If inputs and outputs do not balance, one or several flows are either missing or they have been determined erroneously. The mass–balance principle applies to systems as well as processes. A true material balance of a process or system is only achieved if all input and output flows are known, and if either $\dot{m}_{storage} = 0$ or $\dot{m}_{storage}$ can be measured. In practice $\dot{m}_{storage}$ is often calculated by the difference between inputs and outputs. Table 2.5 shows how data can be managed during the course of an MFA.

2.2.4 Determination of Mass Flows, Stocks, and Concentrations

Information about mass flows is usually taken from databases or measured directly or indirectly on site. Regional, national, and international administrative bodies such as bureaus of statistics, industrial associations, professional societies, and consumer organizations can be good sources of specific data, sometimes including time series. Such information often comprises figures on production, consumption, and sales of commodities and goods.[14,15] Data on waste flows, emissions of pollutants, and substance concentrations in air, water, and soil are systematically collected by national and international environmental protection agencies. Papers in scientific

TABLE 2.5
Development of a Data Spreadsheet in the Course of an MFA[a]

Goods	Flow Rate, t/year	Concentration of Substance $\underline{S_1}, \underline{S_2}, \underline{S_3}, \cdots, \underline{S_n}$, unit					Substance Flow Rate $\underline{S_1}, \underline{S_2}, \underline{S_3}, \cdots, \underline{S_n}$, unit					
$\underline{G_1}$	$\underline{\dot{m}_1}$	•	•	•	$\cdots$	•	•	•	•	•	$\cdots$	•
$\underline{G_2}$	$\underline{\dot{m}_2}$	•	•	•	$\cdots$	•	•	•	•	•	$\cdots$	•
$\underline{G_3}$	$\underline{\dot{m}_3}$	•	•	•	$\cdots$	•	•	•	•	•	$\cdots$	•
$\vdots$	$\vdots$	$\vdots$	$\vdots$	$\vdots$	$\cdots$	$\vdots$	$\vdots$	$\vdots$	$\vdots$	$\vdots$		$\vdots$
$\underline{G_k}$	$\underline{\dot{m}_k}$	•	•	•	$\cdots$	•	•	•	•	•	$\cdots$	•

Note: Entries are underlined (selected goods and substances, flow rates of goods). Additionally required data are marked with •. G = name of good; S = name of substance.

[a] See also Table 2.6 and Table 2.7.

journals, proceedings, and books are also a rich source of information. The search for and the collection, evaluation, and handling of data are core tasks in MFA. The more experience MFA experts have, the less time they need and the better results they get when acquiring data for an MFA.

Some material flows are assessed based on assumptions, cross comparisons between similar systems, or so-called proxy data. "Proxies" are figures that help in approximating or estimating the actual data of interest. An example of proxy data is given as follows. The objective of a study is to determine the total loss of zinc by the wear of tires from passenger cars in the U.S. In Sweden, Landner and Lindeström[16] assessed the loss per car as 0.032 kg Zn/year. Assuming that this figure is also valid for the U.S.,* the total loss of zinc in the U.S. can be determined by multiplying the Swedish figure for Zn loss by the active U.S. vehicle fleet of 140 million passenger cars, yielding approximately 4500 t/year. The proxy datum in this case is the specific loss per car of 0.032 kg Zn/year determined for Sweden. It is always necessary to check whether a proxy is eligible to be transferred from one system to another.

Depending on the financial resources available for an MFA, mass flows of goods and substance concentrations can be actually measured. If large systems or extended time periods are to be balanced, this can be quite costly. Therefore, flows, stocks, and concentrations are preferentially measured in smaller systems such as a wastewater-treatment plant, a company, a farm, or a single private household. Such field studies require intensive and timely planning of the measuring procedure and campaign. Schachermayer et al.[17] and Morf et al.[18,19] carried out such measurements for MSW incinerators. They developed a plan to minimize sampling locations, number and quantities of samples, and time of sampling. Such elaborate sampling plans are instrumental in keeping project costs low while producing rigorous and reproducible

---

* This is probably an underestimate because people in the U.S. drive longer distances than people in Sweden do. It depends on the objective of the assessment whether this is a sufficiently exact determination.

TABLE 2.6
Data Spreadsheet after Determination of Substance Concentrations[a]

Goods	Flow Rate, t/year	Concentration of Substance $S_1, S_2, S_3, \ldots, S_n$, mg/kg					Substance Flow Rate $S_1, S_2, S_3, \ldots, S_n$, unit				
G_1	$\dot{m}_1$	c_{11}	c_{12}	c_{13}	$\ldots$	c_{1n}	•	•	•	$\ldots$	•
G_2	$\dot{m}_2$	c_{21}	c_{22}	c_{23}	$\ldots$	c_{2n}	•	•	•	$\ldots$	•
G_3	$\dot{m}_3$	c_{31}	c_{32}	c_{33}	$\ldots$	c_{3n}	•	•	•	$\ldots$	•
$\vdots$	$\vdots$	$\vdots$	$\vdots$	$\vdots$	$\ldots$	$\vdots$	$\vdots$	$\vdots$	$\vdots$	$\ldots$	$\vdots$
G_k	$\dot{m}_k$	c_{k1}	c_{k2}	c_{k3}	$\ldots$	c_{kn}	•	•	•	$\ldots$	•

Note: New entries are underlined. Additionally required data are marked with •. G = name of good; S = name of substance.

[a] See also Table 2.5 and Table 2.7.

results. Nevertheless, it is very rare that a balance between inputs and outputs of a measured system yields an error less than 10% of the total flow. In principle, flows and concentrations can also be measured directly in larger systems, such as regions or entire watersheds. In any case, it is necessary to assess whether the required accuracy can be achieved by using available data or whether additional data have to be collected. Table 2.6 shows the progress for the data table after mass flows and concentrations have been measured or determined.

2.2.5 ASSESSMENT OF TOTAL MATERIAL FLOWS AND STOCKS

The substance flows ($\dot{X}$) that are induced by the flows of goods can be directly calculated from the mass flows of the goods ($\dot{m}$) and the substance concentrations (c) in these goods.

$$\dot{X}_{ij} = \dot{m}_i \cdot c_{ij} \qquad (2.4)$$

where
  $i = 1, \ldots, k$ as the index for goods
  $j = 1, \ldots, n$ as the index for substances

Again, the mass-balance principle applies for each substance in each process in the entire system. As for goods, the reason for a failure to balance can be missing flows; this source of error is eliminated when establishing the balance at the level of goods (see Section 2.2.3.). Another reason could be errors in substance concentrations. Balance differences between input and output of 10% are common and are usually not significant for the conclusions. Section 2.3 shows how balances can be optimized and how data gaps and uncertainty can be handled using mathematical

TABLE 2.7
Completed Data Spreadsheet[a]

Goods	Flow Rate, t/year	Concentration of Substance $S_1, S_2, S_3, \cdots, S_n$, mg/kg					Substance Flow Rate $S_1, S_2, S_3, \cdots, S_n$, kg/year				
G_1	$\dot{m}_1$	c_{11}	c_{12}	c_{13}	$\cdots$	c_{1n}	$\underline{\dot{X}}_{11}$	$\underline{\dot{X}}_{12}$	$\underline{\dot{X}}_{13}$	$\cdots$	$\underline{\dot{X}}_{1n}$
G_2	$\dot{m}_2$	c_{21}	c_{22}	c_{23}	$\cdots$	c_{2n}	$\underline{\dot{X}}_{21}$	$\underline{\dot{X}}_{22}$	$\underline{\dot{X}}_{23}$	$\cdots$	$\underline{\dot{X}}_{2n}$
G_3	$\dot{m}_3$	c_{31}	c_{32}	c_{33}	$\cdots$	c_{3n}	$\underline{\dot{X}}_{31}$	$\underline{\dot{X}}_{32}$	$\underline{\dot{X}}_{33}$	$\cdots$	$\underline{\dot{X}}_{3n}$
$\vdots$	$\vdots$	$\vdots$	$\vdots$	$\vdots$	$\cdots$	$\vdots$	$\vdots$	$\vdots$	$\vdots$	$\cdots$	$\vdots$
G_k	$\dot{m}_k$	c_{k1}	c_{k2}	c_{k3}	$\cdots$	c_{kn}	$\underline{\dot{X}}_{k1}$	$\underline{\dot{X}}_{k2}$	$\underline{\dot{X}}_{k3}$	$\cdots$	$\underline{\dot{X}}_{kn}$

Note: Final entries (substance flow rates) are underlined. G = name of good; S = name of substance.

[a] See also Table 2.5 and Table 2.6.

and statistical tools. Table 2.7 shows the complete data table after the final step of adding substance flow rates.

There are two ways of assessing the amount of materials in stocks. First, the total mass of the stock can be determined either by direct measurement of the mass or by assessing the volume and the density of the stock. This approach is often used when stocks do not change significantly for long time periods ($\dot{m}_{storage} / m_{stock} < 0.01$). This is the case for stocks, e.g., in natural processes such as soils or large lakes. The second approach is applied to fast-changing stocks ($\dot{m}_{storage} / m_{stock} > 0.05$). Then the magnitude of the stock can be calculated by the difference between inputs and outputs over an appropriate time span ($t_0 - t$). The size of the stock at t_0 has to be known (see Figure 2.18). Usually $\dot{m}_{input}$ and $\dot{m}_{output}$ are functions of time. Applying Equation 2.5, the stock (m_{stock}) can be calculated for any time t.

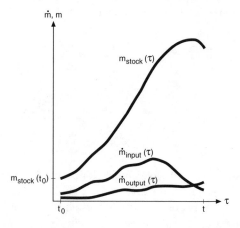

FIGURE 2.18 The stock of a nonsteady-state process. Functions $\dot{m}_{input}$ and $\dot{m}_{output}$ have to be known to calculate m_{stock}.

Methodology of MFA

$$m_{stock}(t) = \int_{t_0}^{t} \dot{m}_{input}(\tau)d\tau - \int_{t_0}^{t} \dot{m}_{output}(\tau)d\tau + m_{stock}(t_0) \quad (2.5)$$

For rough assessments, $\dot{m}_{input}$ and $\dot{m}_{output}$ are often assumed to be time independent. Fast-changing stocks are characteristic of anthropogenic activities. Examples are landfills, metals in urban settlements, plastic materials and electric appliances in private households, and nutrients in agricultural soils.

2.2.6 Presentation of Results

It is important to present the results of an MFA in an appropriate way. The relevant results of the study have to be condensed into a clear message that can be presented in an easily comprehensible manner. The main goal of the presentation is to stage this message to make it clear, understandable, reproducible, and trustworthy. It is important to keep in mind that there are two crucial audiences for an MFA study. On the one hand, there are the technical experts in MFA, LCA, and environmental and resource management; and on the other hand, there are the stakeholders who are not familiar with MFA terminology and procedures, and who probably have no technical or scientific background. It is often this second, decision-oriented audience that is more important because it controls the policy- and decision-making processes. Hence, the results of an MFA have to be presented first in a comprehensive technical report and then in an executive summary.

For the technical report, the following content is suggested. The first page consists of an abstract that does not contain any jargon or technical terms and that can be understood by the general public. The abstract is followed by a table of contents. Next, goals and objectives of the study are given in detail, and the state of knowledge is summarized, including the relevant literature and previous studies by other authors. In the following chapter, a full account of the procedures is given: methodology used, system definition, data acquisition and evaluation, handling of uncertainties. It is necessary to state the source of all data used. Reproducibility and transparency of the procedures is emphasized. In the chapter on results and discussions, the numerical results are presented as well as the conclusions and consequences for the field of study, such as additional research needs or actions to be taken by the stakeholders. A summary completes the presentation. The summary is a full account of the total study and is distinctly different from the abstract. While the abstract contains only the main message, the summary recapitulates the full content with objectives, procedures, results, and conclusions in a concise and concentrated form not exceeding a few pages. The abstract is well suited as reference material for literature databases. The summary serves as an overview for the swift reader who does not want to spend much time reading the full account of the work but still wants to know how the authors proceeded. Literature references are listed at the end of the report. In order to fulfill the requirements of reproducibility, it may be appropriate to attach an annex or compact disk containing data sheets with databases and calculations.

The summary also serves as the foundation for the executive summary, even though there are important differences between the two. The executive summary is written for an audience that is not familiar with MFA. Thus, no technical jargon can be used. For example, it cannot be assumed that the reader of an executive summary knows and appreciates the differences between goods and substances. Also, the summary is not the place to explain this difference. Hence, the art of writing an executive summary is to choose the words and language the audience understands while conveying the MFA message in a clear and transparent way to the customer.

To maximize the impact of the MFA findings, figures that visualize the results and conclusions are of fundamental importance. Several standard graphs — flowcharts, partitioning diagrams, etc. — are proven means of illustrating the results. A flowchart is a drawing that comprises all processes, stocks, material flows, and imports and exports entering and leaving the system. All processes, stocks, and flows must be named and quantified. To help the reader rapidly assess the importance of a flow, the width of all flows should be drawn proportionally to the numerical value of the flow. This type of illustration, known as a Sankey diagram, is used to display material, energy, and cash flows. The figure should identify the system boundaries and the units of flows and stocks. The figure also provides an overview of the system's behavior by showing the sum of all imports at the top left-hand corner (above the system boundary), the sum of all exports at the top right-hand corner (above the system boundary), and the total stock and the changes in stock between these two numbers (see Figure 2.19). This method of presentation allows the reader to check at a glance whether materials are accumulated or depleted in the system and to identify which sources, pathways, and sinks are most important. In the summary, n+1 such flowcharts can be drawn, 1 for the flow of goods and n for the various substance flows.

Transfer coefficients of single processes can be visualized by partitioning diagrams (see Figure 2.12). Such diagrams supply easily understandable information about the different behaviors of substances in a process or system, especially when many substances are investigated.

2.2.7 MATERIALS ACCOUNTING

As defined in Section 2.1.11, materials accounting stands for the routine, usually long-term analysis of material flows and stocks in a defined system by measuring as few key flows and stocks as possible. Such materials accounting allows early recognition of harmful loadings and depletions of substances in stocks. Examples include regional accounting of heavy metals in order to assess long-term soil pollution, accounting for plastic flows to detect future recycling potentials, accounting for carbon to support decisions regarding climate change, accounting for metals (iron, copper, lead) in order to identify future resource potentials, and others.

2.2.7.1 Initial MFA

In order to identify the key processes, stocks, and flows in a system, an initial MFA has to be carried out. If forthcoming materials accounting is the objective of the

Methodology of MFA

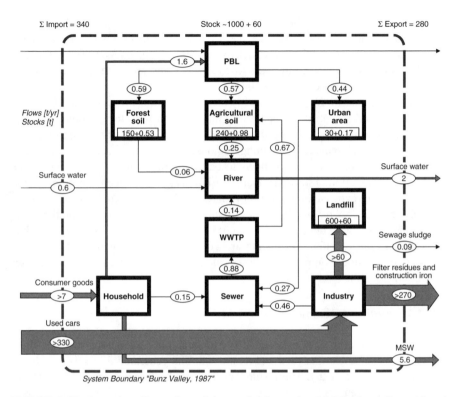

FIGURE 2.19 Exemplary illustration of the result of a regional MFA. Lead flows (t/year) and stocks (t) in the Bunz valley (see Case Study 1 in Chapter 3.1.1). The system boundaries are given in space and time. The sums of imports and exports are displayed as well as the regional stock and its changes. The size of the arrows is proportional to the mass flows. (From Brunner, P.H., Daxbeck, H., and Baccini, P., Industrial metabolism at the regional and local level: a case study on a Swiss region, in *Industrial Metabolism — Restructuring for Sustainable Development*, Ayres, R.U. and Simonis, U.E., Eds., United Nations University Press, Tokyo, 1994. With permission.)

MFA, it is important to consider that some procedures will be repeated in the future. Transparency and efficient reproducibility are especially important and should be carefully considered when planning and implementing the initial MFA. Whenever possible, labor-intensive tasks such as data acquisition should be automated, and routines should be established. This includes data mining and linkage with existing databases for financial accounting, environmental protection, waste management, water resources management, and the like.

2.2.7.2 Determination of Key Processes, Flows, and Stocks

Based on the results of the initial MFA, key processes, flows, and stocks are identified. First, the goals for future materials management must be defined, such as reduction or increase of single flows, the improvement of pollution control (transfer coefficients) by new or optimized technology, a change in the management of stocks, or combinations of these. The next step is to select those processes and flows that are best suited to determine whether the goals have been achieved. "Best suited" means those processes or flows that deliver the highest accuracy at the lowest cost. "To determine" can mean to measure or to assess using any other method such as indirect calculations based on some intrinsic system property.

In general, measurements are a costly means of monitoring a system. Therefore, it is better to find key processes and flows that are measured for reasons other than MFA, such as gross domestic product (GDP) calculations, national environmental statistics, corporate financial accounting, regional waste management, etc. An important criterion for the selection of flows to be monitored is their homogeneity. The accurate determination of flows with large variations and heterogeneous composition requires intense, expensive monitoring programs. Key flows should be as stable as possible in order to be monitored accurately and cost-effectively. For the same reason, processes with constant transfer coefficients are superior candidates for key processes. If the transfer coefficients are constant for varying throughputs, then knowledge of a single output flow allows calculation of the total input (see Chapter 3, Section 3.3.1.2).

2.2.7.3 Routine Assessment

Once the key processes, stocks, and flows for materials accounting are identified, the system is routinely analyzed and monitored. The generated data sets establish a time series $(t_0, t_1, t_2,...,t_n)$. The results are discussed, measures are evaluated with respect to the given goal (eventually given as a function), and if necessary, adaptations must be made (see Figure 2.20). After several routines, it might be advantageous to carry out another full MFA to check assumptions and provisions made (Section 2.2.7.2) and to readjust the materials accounting procedures.

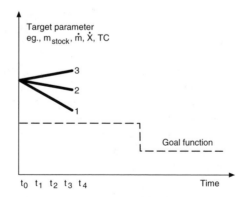

FIGURE 2.20 Time series $(t_0, t_1, ..., t_n)$ established through materials accounting. Initial MFA carried out at t_0. The broken line gives the goal function prescribed by the decision makers. Trend 1 is on track. Trend 2 needs adaptation. Trend 3 calls for reconsideration of taken measures.

TABLE 2.8
Mean Substance Concentrations in Sewage Sludge, mg/kg dry matter

N	Cl	F	S	PCB	Cd	Hg	As	Co
28,000	360	100	14,000	0.2	2	3	20	15

Ni	Sb	Pb	Cr	Cu	Mn	V	Sn	Zn
800	10	150	100	300	500	30	30	1500

2.2.8 Problems — Section 2.2

Problem 2.9: Design a concept for the management of sewage sludge. First, select the substances you want to include in the study. The average sewage sludge composition is known from a previous study and is given in Table 2.8. (a) Which substances do you include in the study? Give reasons for your choice. (b) Present the concept.

Problem 2.10: Which time span do you select as a temporal system boundary when your objective is to determine your average daily turnover of materials? Which spatial system boundary do you choose?

Problem 2.11: Bottom ash is the main solid residue of MSW incineration. In some countries, mechanically processed bottom ash (e.g., crushing, sieving, separation of bulky materials and iron scrap) is used as a substitute for gravel. (a) Assess the contribution of this practice to resource conservation (i.e., conservation of gravel) by an MFA system qualitatively. Assume that all MSW is incinerated. (b) Evaluate the contribution of bottom ash to the conservation of gravel quantitatively. First, determine the data you need. Second, find these data by an Internet search within 15 to 20 min. Third, compare your results with the data given in the solution to this problem.

Problem 2.12: Concentrations of selected elements in MSW incineration bottom ash and in genuine gravel are given in Table 2.9. Discuss gravel substitution by bottom ash from a substance-based point of view (see Problem 2.11).

TABLE 2.9
Mean Concentrations of Selected Elements in MSW Incineration Bottom Ash (BA) and Gravel (G)

	Si, %	Al, %	Fe, %	Mg, %	Ca, %	K, %	C_{org}, %	Cu, g/kg	Zn, g/kg	Pb, g/kg	Cd, mg/kg	Hg, mg/kg
BA	20	4	3	1	10	1	2	2	10	6	10	0.5
G	20	4	2	0.16	20	0.17	<<0.1	0.01	0.06	0.015	0.3	0.05

Problem 2.13: Quantify the stock of wastes (mass and volume) that is built up during an average lifetime. Consider all wastes such as sewage sludge, construction and demolition waste, end-of-life vehicles, and the like. (Use data from waste-management literature or from the World Wide Web.) Do not consider loss of mass by incineration, biological digestion, and so on. Make three assumptions: (a) waste generation remains constant, (b) waste generation increases linearly, and (c) waste generation increases by 2% per year.

Problem 2.14: Let us assume that legislation on the application of sewage sludge on land limits the increase of heavy metals to less than 5% (relative) within 100 years. Further, assume for simplicity that erosion of soil, eluviation, leaching, uptake by plants, etc. can be neglected. How much sewage sludge can be applied to land? Discuss the example of zinc: Zn concentration in sewage sludge is 1500 mg per kg dry matter; Zn concentration in soil is 30 mg/kg. Is there enough agricultural soil to spread all sewage sludge?

Problem 2.15: Consider the system given in Figure 2.21. It was developed to investigate the resource efficiency of metals management. Discuss the system boundary in space: is it chosen appropriately?

Problem 2.16: Take an environmental report of a regional company of your choice (e.g., pulp and paper mill, electricity supplier, cement manufacturer, car manufacturer, food producer; many reports can be downloaded from the World Wide Web) and analyze it from the point of view of materials balances. What are the most important goods and substances (quantitatively and qualitatively)? Is it possible to establish material balances based on the

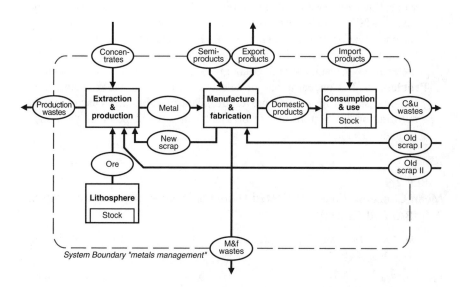

FIGURE 2.21 Flowchart of the stocks and flows of a metal.

Methodology of MFA

information provided? If not, which data are missing? Find issues for improvement by MFA.

The solutions to the problems are given on the Web site, www.iwa.tuwien.ac.at/MFA-handbook.

2.3 DATA UNCERTAINTIES

OLIVER CENCIC

Statistics is a tool that can extract the best possible knowledge of an entity based on data from several inaccurate samples. When several data samples for the same objective entity are available, the mean value and standard deviation can be calculated. It is also possible to get an idea of the distribution of the samples. If there are occasional strongly deviating data or a skewed distribution, the median value and interpercentiles can be used instead of the mean value and the standard deviation.

All measured values represent variables with at least random deviations. It is often the case that the only available data are single values from measurements, interviews, or historical sources. In these cases, the uncertainty of a value can be estimated roughly by evaluating the source or the context from which the data were acquired.[20] If nothing is known about the quantity before obtaining the observations, the classical statistical standard methods might be used. If prior information about the quantity is available, based on whatever knowledge or reasoning, it is possible to combine these prior opinions with the observed evidence and use Bayesian updating methods to obtain posterior distributions.[21]

2.3.1 PROPAGATION OF UNCERTAINTY

If random variables are used within a function, the result is another random variable. Its probability distribution can be determined exactly only if the probability distributions of the input values are known. But even if they are given, it is difficult to compute the results. In the following sections, two common and straightforward methods are presented to incorporate the propagation of uncertainties into the calculated results of an MFA.

2.3.1.1 Gauss's Law

One approach to evaluate the propagation of uncertainties through a system is to apply Gauss's law of error propagation: a function is expanded into a Taylor series that is cut off after the linear part (first-order term). In this way, a linear approximation of the function for a certain development point (result of the function when the mean values μ_i of the random input variables X_i are used) is created. The smaller the distance from the development point, the better are the results obtained by the approximation. With the help of the Taylor series, it is possible to determine approximately the expected value and the deviation of the function's result. The formula for calculating the deviation is called Gauss's law of error propagation.[21]

$$Y \approx f(X_1, X_2, \ldots, X_n)$$

$$\mathrm{Var}(Y) \approx \sum_{i=1}^{n} \left(\mathrm{Var}[X_i] \cdot \left[\frac{\partial Y}{\partial X_i} \right]_{X=\mu}^{2} \right) + \qquad (2.6)$$

$$2 \cdot \sum_{j=1}^{n} \sum_{j=i+1}^{n} \left(\mathrm{Cov}[X_i, X_j] \cdot \left[\frac{\partial Y}{\partial X_i} \right]_{X=\mu} \cdot \left[\frac{\partial Y}{\partial X_j} \right]_{X=\mu} \right)$$

where
μ_i, $E(X_i)$ = mean value of X_i
σ_i = standard deviation of X_i
σ_i^2, $\mathrm{Var}(X_i)$ = variance of X_i
σ_{ij}, $\mathrm{Cov}(X_i, X_j)$ = covariance of X_i and X_j

There are certain restrictions for the application of Gauss's law: the random variables have to be normally distributed, and the uncertainties have to be small. Reasonable results can be expected for these conditions only.[22]

EXAMPLE 1

The output flow Y of a process is the sum of the two input flows X_1 and X_2 (Figure 2.22).

$$Y = X_1 + X_2$$

where

$$E(X_1) = 100, \; \sigma_{X1} = 10 \; (= 10\% \text{ of } X_1)$$

$$\mathrm{Var}(X_1) = \sigma_{X1}^2 = 100$$

and

$$E(X_2) = 200, \; \sigma_{X2} = 20 \; (= 10\% \text{ of } X_2)$$

$$\mathrm{Var}(X_2) = \sigma_{X2}^2 = 400$$

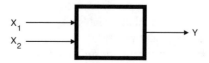

FIGURE 2.22 Process with two input flows and one output flow.

FIGURE 2.23 Process with one input flow and two output flows.

The application of Gauss's law (Equation 2.6) leads to

$$E(Y) \approx E(X_1 + X_2) = E(X_1) + E(X_2)$$

$$\text{Var}(Y) \approx \text{Var}(X_1 + X_2) = \text{Var}(X_1) + \text{Var}(X_2) + [2 \times \text{Cov}(X_1, X_2)]$$

If the variables X_1 and X_2 are independent, their covariance is zero. This leads to

$$\text{Var}(X_1 + X_2) = \text{Var}(X_1) + \text{Var}(X_2)$$

Result:

$$E(Y) \approx 100 + 200 = 300$$

$$\text{Var}(Y) \approx 100 + 400 = 500 \rightarrow \sigma_Y = \sqrt{500} = 22.4 \ (= 7.5\% \text{ of } Y)$$

EXAMPLE 2

The output flow Y_1 is a certain percentage (given by the transfer coefficient TC_1) of the input flow X (Figure 2.23).

$$Y_1 = TC_1 \times X$$

where

$$E(X) = 100, \ \sigma_X = 5 \ (= 5\% \text{ of } X)$$

$$\text{Var}(X) = \sigma_X^2 = 25$$

and

$$E(TC_1) = 0.40, \ \sigma_{TC1} = 0.04 \ (= 10\% \text{ of } TC_1)$$

$$\text{Var}(TC_1) = \sigma_{TC1}^2 = 0.0016$$

The application of Gauss's law (Equation 2.6) leads to

$$E(Y_1) \approx E(TC_1 \times X) = E(TC_1) \times E(X)$$

$$\text{Var}(Y_1) \approx \text{Var}(TC_1 \times X) = [E(X)^2 \times \text{Var}(TC_1)] + [E(TC_1)^2 \times \text{Var}(X)] + [2 \times E(TC_1) \times E(X) \times \text{Cov}(TC_1, X)]$$

If the variables TC_1 and X are independent, their covariance is zero. This leads to

$$E(Y_1) \approx E(TC_1 \times X) = E(TC_1) \times E(X)$$

$$\text{Var}(Y1) \approx \text{Var}(TC_1 \times X) = [E(X)^2 \times \text{Var}(TC_1)] + [E(TC_1)^2 \times \text{Var}(X)]$$

Results:

$$E(Y_1) \approx 0.40 \times 100 = 40$$

$$\text{Var}(Y_1) \approx 100^2 \times 0.0016 + 0.40^2 \times 25 = 20 \rightarrow \sigma_{Y1} = \sqrt{20} \approx 4.5\ (=11.3\%\ \text{of}\ Y_1)$$

2.3.1.2 Monte Carlo Simulation

If the data are not normally distributed or if deviations are too large, a different approach might be more useful: the Monte Carlo simulation. In this method, the statistical distributions of the input variables are assumed to be known. For each input parameter, a computer algorithm creates a random number according to its distribution. These numbers are used to calculate the result of a function. If this procedure is repeated, e.g., 1000 times, then 1000 possible results will be obtained. These results are then evaluated to get information about their statistical distribution (e.g., mean, standard deviation).[23]

For many quantities (lengths, concentrations, flows, etc.), the normal distribution is used even though it is theoretically inappropriate because negative values are allowed. As long as the coefficient of variation σ/μ is less than about 0.2, this problem can be safely ignored in most applications, as the probability of obtaining observations more than five standard deviations away from the mean is quite small.[21] However, using a normal distribution is also problematic when dealing with quantities that have an upper limit (e.g., $0 \leq TC \leq 1$). Because the normal distribution stretches to infinity on both sides of the mean, it is possible that the random creation of a number yields a value that is outside of the allowed range. In these cases, the probability of obtaining an observation beyond the limit can be estimated by calculating

$$c = \left| \frac{z - \mu_Z}{\sigma_Z} \right| \tag{2.7}$$

and looking up the corresponding probability in Table 2.10 (z is the limit, and μ_Z and σ_Z are the parameters of a random variable Z). The variable c represents the distance of the limit from the mean in terms of the standard deviation. If the quantity is limited on both sides, repeat the calculation for the lower and the upper limit and add the corresponding probabilities.

Methodology of MFA

TABLE 2.10
Probability of Obtaining an Observation beyond $\mu + (c \times \sigma)$

c	0.5	1	1.5	2	2.5	3	3.5	4
Probability, %	30.854	15.866	6.681	2.275	0.621	0.135	0.023	0.003

If a random quantity is, for example, limited on one side and c is greater than four, the possibility of a randomly created value lying beyond the limit is less than 0.003%, i.e., it is very unlikely to get such a value.

EXAMPLE 1

The output flow Y of a process is the sum of the two input flows X_1 and X_2 (Figure 2.24).

$$Y = X_1 + X_2$$

where

$$E(X_1) = 100, \sigma_{X1} = 10, \text{ normally distributed}$$

$$E(X_2) = 200, \sigma_{X2} = 20, \text{ normally distributed}$$

The normal distribution is chosen here in order to compare the results of the Monte Carlo simulation to the results gained by using Gauss's law in Section 2.3.1.1, Example 1. Using Equation 2.7 leads to $c_{X1} = c_{X2} = 10$. Because of this, it is very unlikely that negative values are obtained, and the normal distribution can be considered to be a good approximation of the real ones.

Using, e.g., the Microsoft® Excel® function "norminv(rand();mean;standard deviation)," a random number is obtained that follows a normal distribution defined by its mean and its standard deviation. In this way, random values for X_1 and X_2 are created. The value of Y is calculated by adding the values of X_1 and X_2. This procedure is repeated 1000 times. In the end, the mean and the standard deviation of X_1, X_2, and Y can be computed out of the 1000 values of each variable. Table 2.11 shows one possible result of the Monte Carlo simulation with 1000 repetitions. The higher the number of repetitions, the better will the results fit the given parameters of X_1 and X_2 and the calculated parameters of Y obtained by Gauss's law (see Table 2.12).

FIGURE 2.24 Process with two input flows and one output flow.

TABLE 2.11
Monte Carlo Simulation of Y (= $X_1 + X_2$) with 1000 Repetitions

No.	X_1	X_2	Y
1	84.57	205.31	289.88
2	114.84	194.26	309.10
3	98.58	219.39	317.96
4	97.57	187.04	284.61
5	94.47	193.82	288.28
6	90.99	185.28	276.27
7	106.70	227.59	334.29
8	104.00	205.54	309.54
9	94.42	177.08	271.50
10	96.69	226.57	323.26
...	...	...	...
999	94.94	201.30	296.24
1000	99.71	204.69	304.41
Mean	99.56	200.23	299.79
Standard deviation	10.07	20.14	22.66

TABLE 2.12
Given Parameters (Mean and Standard Deviation) of X_1 and X_2 and Calculated Parameters of Y (= $X_1 + X_2$) according to Gauss's Law

	X_1	X_2	Y
Mean	100	200	300
Standard deviation	10	20	22.4

Note: Taken from Chapter 2.3.1.1, Example 1.

EXAMPLE 2

The output flow Y_1 is a certain percentage (given by the transfer coefficient TC_1) of the input flow X (Figure 2.25).

$$Y_1 = TC_1 * X$$

where

$E(X) = 100$, $\sigma_X = 5$, normally distributed

$E(TC_1) = 0.40$, $\sigma_{TC1} = 0.04$, normally distributed

Methodology of MFA

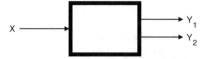

FIGURE 2.25 Process with one input flow and two output flows.

TABLE 2.13
Monte Carlo Simulation of $Y_1 = (TC_1 \times X)$

No.	X	TC_1	Y_1
1	93.24	0.44	40.97
2	102.77	0.43	44.34
3	94.46	0.34	31.94
4	110.59	0.42	45.94
5	106.42	0.44	47.11
6	101.09	0.30	30.57
7	95.83	0.40	38.68
8	99.85	0.42	42.19
9	104.38	0.39	40.22
...	...	...	...
999	99.06	0.42	41.27
1000	102.24	0.38	39.32
Mean	99.98	0.40	40.08
Standard deviation	5.00	0.04	4.36

The normal distribution is chosen here in order to compare the results of the Monte Carlo simulation with the results gained by using Gauss' law in Section 2.3.1.1, Example 2. Using Equation 2.7 leads to $c_X = 20$ and $c_{TC1} = 10$. Because of this, it is very unlikely that negative values are obtained, and the normal distribution can be considered to be a good approximation of the real distributions.

Using, e.g., the Excel function "norminv(rand();mean;standard deviation)," a random number is obtained that follows a normal distribution defined by its mean and its standard deviation. In this way, random values for TC_1 and X are created. The value of Y is calculated by multiplying the values of TC_1 and X. This procedure is repeated 1000 times. In the end, the mean and the standard deviation of TC_1, X, and Y can be computed out of the 1000 values of each variable. Table 2.13 shows one possible result of the Monte Carlo simulation with 1000 repetitions. The higher the number of repetitions, the better will the results fit the given parameters of TC_1 and X and the calculated parameters of Y obtained by Gauss's law (see Table 2.14).

2.3.2 LEAST-SQUARE DATA FITTING

If linear equations describe a material flow system in an overdetermined way (number of equations > number of variables), it is possible to calculate the best-fitting values by using the method of least squares of Gauss.

TABLE 2.14
Given Parameters (Mean and Standard Deviation) of X and TC_1 and Calculated Parameters of Y_1 (= $TC_1 \times X$) according to Gauss's Law

	X	TC_1	Y_1
Mean	100	0.40	40
Standard deviation	5	0.04	4.5

Note: Taken from Chapter 2.3.1.1, Example 2.

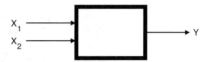

FIGURE 2.26 Process with two input flows and one output flow.

EXAMPLE

The output flow of a process is the sum of the two input flows: $Y = X_1 + X_2$ (Figure 2.26). All flows have been measured.

1. $X_1 = 30$
2. $X_2 = 60$
3. $Y = X_1 + X_2 = 105$

There is a discrepancy in the three values because $X_1 + X_2 \neq Y$. The values can be adjusted with the method of least squares. Using this method, the best values of $\hat{X}_i$ are reached when the sum of squares of the residues r_i between best values of $\hat{X}_i$ and observations X_i is a minimum. The residues r between best values of $\hat{X}_i$ and observations X_i can be written as $\hat{X}_i - X_i = r$;

1. $\hat{X}_1 - 30 = r_1$
2. $\hat{X}_2 - 60 = r_2$
3. $\hat{X}_1 + \hat{X}_2 - 105 = r_3$

There are two common approaches to find a solution:

2.3.2.1 Geometrical Approach[24]

Using the vectors $\bar{a} = (1/0/1)$, $\bar{b} = (0/1/1)$, $\bar{c} = (30/60/105)$, and $\bar{r} = (r_1/r_2/r_3)$, the linear equation system can be written as

Methodology of MFA

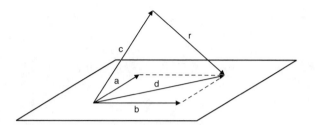

FIGURE 2.27 Geometrical approach of least-square data fitting.

$$\begin{pmatrix}1\\0\\1\end{pmatrix}\cdot\hat{X}_1 + \begin{pmatrix}0\\1\\1\end{pmatrix}\cdot\hat{X}_2 - \begin{pmatrix}30\\60\\105\end{pmatrix} = \begin{pmatrix}r1\\r2\\r3\end{pmatrix}$$

$$\bar{a}\cdot\hat{X}_1 + \bar{b}\cdot\hat{X}_2 + \bar{c} = \bar{r}$$

The vector of residues $\bar{r}$ can be seen as the difference of vector $\bar{d}$ and the vector of constant $\bar{c}$. Vector $\bar{d}$ can be written as the linear combination of the vectors $\bar{a}\cdot x$ and $\bar{b}\cdot y$ and lies in the plane going through $\bar{a}$ and $\bar{b}$ (Figure 2.27).

The best solution from a geometric point of view is where the vector of residues $\bar{r}$ has its minimum. The Euclidian length of $\bar{r}$ is a minimum when $\bar{r}$ is perpendicular to the plane going through $\bar{a}$ and $\bar{b}$. This is also the optimum in the sense of the method of least squares, because the Euclidian length of the vector $\bar{r}$ ($\sqrt{r_1^2 + r_2^2 + r_3^2}$) and thus the value of $r_1^2 + r_2^2 + r_3^2$ is at a minimum.
If

- $\bar{r}$ is perpendicular to $\bar{a}$, then $\bar{r}\cdot\bar{a} = 0$
- $\bar{r}$ is perpendicular to $\bar{b}$, then $\bar{r}\cdot\bar{b} = 0$
- $\bar{r}$ is perpendicular to the plane going through $\bar{a}$ and $\bar{b}$ then,

1. $\bar{r}\cdot\bar{a} = \left[\begin{pmatrix}1\\0\\1\end{pmatrix}\cdot\hat{X}_1 + \begin{pmatrix}0\\1\\1\end{pmatrix}\cdot\hat{X}_2 - \begin{pmatrix}30\\60\\105\end{pmatrix}\right]\cdot\begin{pmatrix}1\\0\\1\end{pmatrix} = 0$

2. $\bar{r}\cdot\bar{b} = \left[\begin{pmatrix}1\\0\\1\end{pmatrix}\cdot\hat{X}_1 + \begin{pmatrix}0\\1\\1\end{pmatrix}\cdot\hat{X}_2 - \begin{pmatrix}30\\60\\105\end{pmatrix}\right]\cdot\begin{pmatrix}0\\1\\1\end{pmatrix} = 0$

This leads to a set of linear equations with two unknown variables:

1. $2\cdot\hat{X}_1 + \hat{X}_2 = 135$
2. $\hat{X}_1 + 2\cdot\hat{X}_2 = 165$

Solving this system leads to:

$$\hat{X}_1 = 35$$

$$\hat{X}_2 = 65$$

$$\hat{Y} = \hat{X}_1 + \hat{X}_2 = 100$$

2.3.2.2 Analytical Approach[25]

The relationship between observations $\bar{x}$ and best-values $\bar{x}_0$ can be written in vector format as

$$\bar{x} = \bar{A} \cdot \bar{x}_0 + \bar{r}$$

Without proof, the solution for this problem in the sense of the method of least squares is

$$\bar{x}_0 = (\bar{A}^T \bar{A})^{-1} \bar{A}^T \bar{x}$$

EXAMPLE

$$\begin{pmatrix} X_1 \\ X_2 \\ Y \end{pmatrix} = \begin{pmatrix} 1 & 0 \\ 0 & 1 \\ 1 & 1 \end{pmatrix} \cdot \begin{pmatrix} \hat{X}_1 \\ \hat{X}_2 \end{pmatrix} + \begin{pmatrix} r_1 \\ r_2 \\ r_3 \end{pmatrix}$$

Solution:

$$\begin{pmatrix} \hat{X}_1 \\ \hat{X}_2 \end{pmatrix} = \left(\begin{pmatrix} 1 & 0 & 1 \\ 0 & 1 & 1 \end{pmatrix} \cdot \begin{pmatrix} 1 & 0 \\ 0 & 1 \\ 1 & 1 \end{pmatrix} \right)^{-1} \cdot \begin{pmatrix} 1 & 0 & 1 \\ 0 & 1 & 1 \end{pmatrix} \cdot \begin{pmatrix} 30 \\ 60 \\ 105 \end{pmatrix}$$

$$\begin{pmatrix} \hat{X}_1 \\ \hat{X}_2 \end{pmatrix} = \frac{1}{3} \cdot \begin{pmatrix} 2 & -1 \\ -1 & 2 \end{pmatrix} \cdot \begin{pmatrix} 135 \\ 165 \end{pmatrix} = \begin{pmatrix} 35 \\ 65 \end{pmatrix}$$

There are also approaches that take uncertainties into account, where flows are corrected according to their uncertainty. Larger deviations allow bigger corrections and vice versa. In every case, the uncertainty of the corrected values is smaller than the original uncertainty.[22]

2.3.3 SENSITIVITY ANALYSIS

In a sensitivity analysis, the influence of individual parameters on system variables is examined. There are different forms of sensitivity to distinguish:[26]

Absolute sensitivity:

$$\frac{\Delta Y_i(X_j)}{\Delta X_j} \tag{2.8}$$

Relative sensitivity:

$$\frac{\Delta Y_i(X_j)}{\Delta X_j} \cdot \frac{1}{Y_i} \tag{2.9}$$

Absolute sensitivity per 100%:

$$\frac{\Delta Y_i(X_j)}{\Delta X_j} \cdot X_j \tag{2.10}$$

Relative sensitivity per 100%:

$$\frac{\Delta Y_i(X_j)}{\Delta X_j} \cdot \frac{X_j}{Y_i} \tag{2.11}$$

The absolute sensitivity (Equation 2.8) shows the absolute change, ΔY_i, of the variable Y_i in relation to a change, ΔX_j, of the parameter X_j. The relative sensitivity (Equation 2.9) shows the relative change, $\Delta Y_i/Y_i$, of the variable Y_i in relation to a change ΔX_j of the parameter X_j.

Equation 2.10 and Equation 2.11 are used to compare sensitivities for parameters with different units or magnitudes, respectively. They show the absolute and the relative change of the variable Y_i in reaction to a fictitious change of 100% of parameter X_j.

Within a sensitivity analysis, all input parameters are altered systematically. The model can be visualized as a machine with wheels to adjust the parameters. By turning one wheel, while all others remain fixed, the sensitivity of the result concerning the parameter of interest can be identified.

EXAMPLE

The output flow Y_1 is a certain percentage of the input flow X_1 (given by the transfer coefficient TC_1) and a certain percentage of the input flow X_2 (given by the transfer coefficient TC_2). The transfer coefficients are considered to be constant (Figure 2.28). What is the absolute sensitivity of Y_1 with reference to X_1 and X_2?

FIGURE 2.28 Process with two input flows and two output flows.

$$Y_1 = (TC_1 \times X_1) + (TC_2 \times X_2) = f(TC_1, X1, TC_2, X_2)$$

where
  $TC_1 = 0.5$
  $TC_2 = 0.3$
  $X_1 = 50$
  $X_2 = 200$

First, Y_1 is calculated by holding TC_1, TC_2, and X_2 constant while altering X_1 by dX. In a second step, Y_1 is calculated by holding TC_1, TC_2, and X_1 constant while altering X_2 by dX. The results can be seen in the second and third column of Table 2.15. The graph with the steepest gradient points out the parameter to which the result is most sensitive. Figure 2.29 shows that a change in X_1 has a greater effect on the result of Y_1 than a change in X_2.

In many cases it is sufficient to calculate the partial derivative $\partial Y/\partial X$ instead of the exact value of $\Delta Y/\Delta X$.[22]

$$\frac{\Delta Y_1}{\Delta X_1} \approx \frac{\partial Y_1}{\partial X_1} = TC_1 = 0.5$$

$$\frac{\Delta Y_1}{\Delta X_2} \approx \frac{\partial Y_1}{\partial X_2} = TC_2 = 0.3$$

The partial derivative with the largest gradient points out the parameter to which the result is most sensitive.

2.4 SOFTWARE FOR MFA

OLIVER CENCIC

2.4.1 GENERAL SOFTWARE REQUIREMENTS

Software applications are used mainly to support and facilitate procedures and calculations that otherwise would be time consuming. There are certain requirements that such software products should fulfill:[27,28]

- **Documentation:** Documentation is comprehensive, detailed, and easy to understand. It includes an installation guide, a user manual, and on-line help.

TABLE 2.15
Sensitivity Analysis of $Y_1 = (TC_1 \times X_1)$
$+ (TC_2 \times X_2) = f(TC_1, X1, TC_2, X_2)$

dX	$f(X_1 + dX)$	$f(X_2 + dX)$
−10	80.0	82.0
−9	80.5	82.3
−8	81.0	82.6
−7	81.5	82.9
−6	82.0	83.2
−5	82.5	83.5
−4	83.0	83.8
−3	83.5	84.1
−2	84.0	84.4
−1	84.5	84.7
0	85.0	85.0
1	85.5	85.3
2	86.0	85.6
3	86.5	85.9
4	87.0	86.2
5	87.5	86.5
6	88.0	86.8
7	88.5	87.1
8	89.0	87.4
9	89.5	87.7
10	90.0	88.0

Note: $X_1 = 50$, $X_2 = 200$, $TC_1 = 0.5$, $TC_2 = 0.3$, and dX varies between −10 and +10.

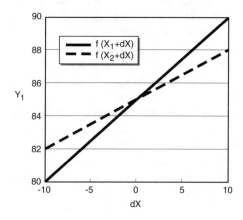

FIGURE 2.29 Sensitivity analysis of $Y_1 = (TC_1 \times X_1) + (TC_2 \times X_2)$.

- **User friendliness:** The application menus are similar to those used in popular software products. The application is based on a widely used operating system such as Microsoft Windows®. The software is, as far as possible, self-explanatory and easy to use. The program is available in different languages.
- **Support and maintenance:** The producer of the software offers comprehensive support (via e-mail, telephone, in person). Product maintenance is guaranteed; upgrades and add-ins are available via the World Wide Web.
- **Stability:** The software is stable and reliable. Known bugs have been cleared and there are no conflicts with other applications.
- **Cost benefit:** The price is reasonable in relation to the benefits of using the software. A free test version is available.
- **Calculation speed and accuracy:** The application generates accurate calculations within an acceptable time span.
- **Compatibility with other software applications:** It is easy to import and export data and figures from and to other applications in various formats. Creation of specialized interfaces is possible.
- **Flexibility and automatization:** The program can be adapted to individual requirements without knowing a programming language. Advanced users can automate special or repeatedly used routines with the help of a programming language (e.g., Visual Basic®, Pascal®, JavaScript®).

2.4.2 Special Requirements for Software for MFA

In addition to the general requirements listed in Section 2.4.1, MFA software must meet the following special demands:[27,28]

- **Terminology:** The terminology is compatible with MFA, i.e., it is possible to deal with materials on different levels (goods and substances) and relations (concentrations).
- **Methodology:** The methodology is compatible with MFA, i.e., it can create material flow systems such as networks by assembling elements in an interactive way. The standard symbols are easily adjustable. It should be possible to define subsystems to enhance understanding and clarity. The software can deal with feedback loops (goods are sent back to a point in a system where they have already been) and storages with delayed output.
- **Data:** The software has the following data-handling capabilities:
  - *Input:* Data in the form of values, functions, or graphs can be imported from databases (Microsoft Access®, DBase®, Oracle®) or accessed with the help of scripting languages. All parameters can be easily changed. Data from remote databases can be requested and incorporated.
  - *Data storage:* The entire input information and all results can be stored in databases. This facilitates the comparison of different scenarios and the exchange of data with other databases.

Methodology of MFA

- *Output:* The user can present the results in various ways (tables, figures, graphs, flowcharts, Sankey diagrams). The user can freely choose layout and content of the display.
- *Uncertainties:* The software allows consideration of data uncertainties and application of different kinds of statistical distributions. The propagation of these uncertainties is taken into account by using Gauss's law-of-error propagation and/or Monte Carlo simulation methods. Sensitivity analysis and optimization of measured data for an overdetermined system are featured.
- **Calculations:** The software can provide the following information:
  - Description of the actual state of a given system "model."
  - Simulation of scenarios with new processes, new goods, or new input data (e.g., mass flow, chemical composition, transfer coefficients).
  - Comparison of scenarios with the original system (deviation).
  - Simulation of the *dynamic* behavior of a system as well as its *static* behavior. Single processes as well as whole systems can be balanced.

2.4.3 SOFTWARE CONSIDERED

The software marketplace has many applications designed to simulate physical processes. At the moment, Simbox® is the only software product that has been especially developed for MFA and adheres strictly to MFA methodology.[22] Simbox has been developed in German and is now available in English, too. It is not possible to get a test version. Thus, it is not discussed here further. Instead, two other software products — Umberto® and GaBi® — are presented here to illustrate the use of software that can be used in MFA. These products were chosen because they are:

- Well established
- Available in English
- Available as test or evaluation versions (GaBi for free; Umberto for 300)
- Well supported and maintained
- Based on different calculation models (GaBi: linear equation system; Umberto: Petri nets)

To round out the comparison, the following case study is also modeled with Microsoft Excel. The potential uses and limitations of these three software tools — Umberto, GaBi, and Excel — are compared in the following sections.

2.4.4 CASE STUDY

The case study presented here involves a simplified model of PVC flows and stocks in Austria (Figure 2.30). The objective is to simulate the dynamic behavior of the system from 1950 to 1994. In the following discussion, flows and processes are written in italic letters. Three processes are to be considered:

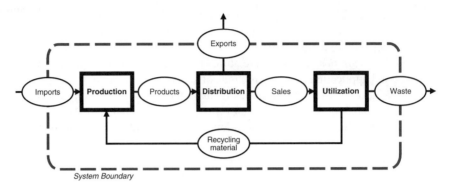

FIGURE 2.30 "PVC Austria": a simplified model of PVC flows and stocks in Austria.

1. *Production: Imports* (raw material, semiproducts) and *recycling materials* are used to create *products*.
2. *Distribution:* The *products* are distributed to *exports* and *sales*.
3. *Utilization:* The process *utilization* can be split into three subprocesses: *consumption*, *stock*, and *collection*. The *products* are consumed. Parts of the *products* (articles with a utilization time shorter than a year) are directly collected afterwards as *waste I* (determined by the waste rate), while the rest is stored as *inventory* of a *stock*. After a mean utilization time of 22 years, the *inventory* is released as *waste II*. *Waste I* and *waste II* are collected. A fraction of it is used as *recycling material* (determined by the recycling rate), while the rest is discarded as *waste*.

The following data about flows and transfer coefficients are given:

Sales: the amount of PVC sold is given by a time series (1950–1994) taken from Fehringer and Brunner.[29] It is approximated by a third-order polynomial function (Figure 2.31):

$$\text{sales of PVC} = (-0.8633 \times \text{years}^3) + (94.271 \times \text{years}^2) + (484.57 \times \text{years})$$

where years = period – 1950 (= number of years since 1950)

Exports: the amount of PVC exported from the system is estimated to be 47% of the amount of *sales*.

Waste rate (WR): the WR is the percentage of PVC goods that are not incorporated into the *stock* but disposed of as *waste I*. The assumption is that the WR changes linear from 1950 to 1994, starting with WR = 0.50 in 1950 and reaching WR = 0.20 in 1994. The linear interpolation between 1950 and 1994 leads to:

$$WR = 0.50 - [(0.50 - 0.20)/(1994 - 1950)] \times (\text{period} - 1950)$$

Methodology of MFA

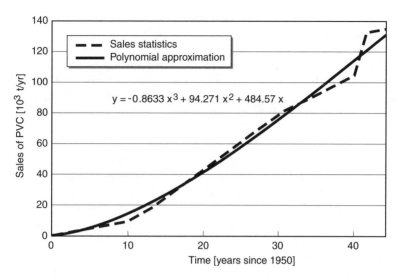

FIGURE 2.31 Polynomial approximation of the time series of PVC sales in Austria from 1950 to 1994.

Recycling rate (RR): the RR is the percentage of collected PVC goods going back to *production* as *recycling material*. From 1950 to 1989, RR is zero. It is assumed that starting with RR = 0.01 in 1990, RR increases by 0.002 annually:

$$RR = 0.01 + (period - 1990) \times 0.002$$

Mean utilization time: the mean utilization time of PVC goods is estimated to be 22 years.

2.4.5 Calculation Methods

Each material flow system consists of processes and flows. A process contains detailed instructions of what happens to its input and output flows. These instructions are normally based on a set of linear equations containing unknown quantities. The material flow system is described completely when the linear equations of all processes are united to form a single set of linear equations. If the number of unknown quantities is equal to the number of linear equations, the system can be solved by using, for example, Gauss's elimination as a standard method. If the number of equations is smaller than the number of unknown quantities, it is not possible to find a solution. A complete solution is only possible when all necessary information has been gathered. This is the approach used by the GaBi software.

Usually, the acquisition of data and the construction of the material flow system are much more time-consuming than the calculation itself, especially when dealing with complex systems. In this case, it could be better to use easily available data first to calculate only parts of the system in order to obtain new information. This is the approach used by the Umberto software, which uses terminology and methodology that are close to those used in Petri nets. The network contains processes called "transitions" and flows called "arrows." The difference with respect to MFA methodology is that transitions cannot be connected directly. Instead, the transitions are separated by objects called "places." Thus, flows can only go from places to transitions or vice versa. On one hand, places are used to separate transitions (storages), and on the other hand, they are used to link them (flow points). The calculation is based on the possible derivation of new information from given or calculated information. Each transition is repeatedly checked to see if its flows can be computed entirely from given or recently calculated data in its direct vicinity. If yes, the flows are calculated and the adjacent places are checked. If there is no contradiction to flows on the other side of the respective places, the calculated transition will not be considered anymore during the computation procedure. When all transitions are calculated, or when there is not enough data left to calculate the remaining transitions, the computation stops. The calculated and given values can be displayed. A log file contains warnings that point out where problems occurred. In this way, it is possible to gain new information even when not all of the necessary information to calculate the entire system is available.[30]

2.4.6 MICROSOFT EXCEL

Microsoft Excel® is a spreadsheet and analysis program that is available as part of Microsoft Office®. It is possible to perform MFA using Excel, even though it has not been specially designed for this purpose. If a user wants to model a system that contains only a few processes (<20), it might be appropriate to use Excel. The main advantage of this approach is that many users are already familiar with this software. Moreover, it is often installed on personal computers, and there is no need to purchase additional software for MFA.

Excel's draw utilities are used to create a clearly structured flowchart (Figure 2.32). The user manually defines which cells display processes, flows, or other parameters (e.g., transfer coefficients). The user then has to state formulas that show how to calculate unknown quantities to compute a flow quantity and balance an adjacent process. For example, to balance the process *production* (assuming no change of stock), one must calculate the value of *products* (= *imports + recycling material*), and the formula in cell E12 must be "= A12 + G25." If there is a change in the stock of a process, it can be calculated from the difference between inputs and outputs of the process. For example, to calculate the value of *change of stock* (= *inventory – waste II*), the formula in cell K19 must be "= K15 – K22."

The creation of formulas can be simplified by giving names to the cells of importance (e.g., name of K15 = "inventory"):

1. Select a cell.
2. Choose "Insert → Names → Define" from the main menu of Excel.
3. Enter a name and click "OK."
4. Defined names can be used afterward to state formulas.

Methodology of MFA

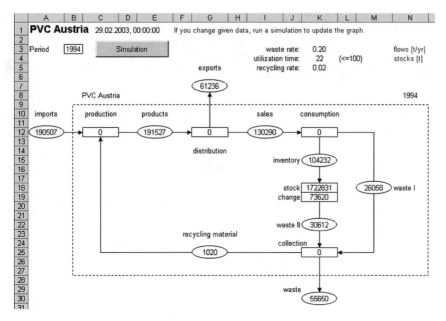

FIGURE 2.32 MFA system of PVC Austria created with Excel. The "simulation" button is used to automatically calculate all system quantities from the given time series.

If the user is familiar with VBA (Visual Basic for applications), the dynamic behavior of a system can be simulated easily. VBA code (Figure 2.33) can be used to read data from a database (e.g., a spreadsheet containing a time series as in Figure 2.34), perform calculations, and store the calculated values. The calculated data can then be used to produce the graphs and charts needed (Figure 2.35). This approach is not as user friendly as working with a specially designed program, but it offers a lot of flexibility.

```
(Allgemein)                              DataImport

Option Explicit

Function DataImport(Head As Range, Periode As Integer)

Dim os As Integer
os = Periode - 1950 + 1

If Periode >= 1950 Then
    DataImport = Worksheets("Data").Cells(Head.Row, Head.Column) _
    .Offset(os, 0).Value
Else
    DataImport = 0
End If

End Function
```

FIGURE 2.33 VBA code to import data from a data spreadsheet.

	A	B	C	D	E	F	G	H	I	J	K	L	M
1	period	years	sales	exports	RR	WR	stock	invent.	waste I	waste II	waste	import	recyc. m.
2	1950	0	0	0	0,000	0,50	0	0	0	0	0	0	0
3	1951	1	578	272	0,000	0,49	293	293	285	0	285	850	0
4	1952	2	1339	629	0,000	0,49	981	688	651	0	651	1969	0
5	1953	3	2279	1071	0,000	0,48	2167	1186	1093	0	1093	3350	0
6	1954	4	3391	1594	0,000	0,47	3955	1788	1603	0	1603	4985	0
42	...	...	...	...	...	...	...	...	...	...	...	...	...
43	1990	40	114965	54034	0,010	0,23	1329142	88837	26128	21317	46971	168524	474
44	1991	41	118837	55854	0,012	0,22	1421781	92639	26198	23493	49095	174095	596
45	1992	42	122686	57662	0,014	0,21	1518257	96476	26210	25769	51251	179620	728
46	1993	43	126505	59457	0,016	0,21	1618599	100342	26164	28142	53437	185094	869
47	1994	44	130290	61236	0,018	0,20	1722831	104232	26058	30612	55650	190507	1020
48													

FIGURE 2.34 Spreadsheet containing given and calculated data.

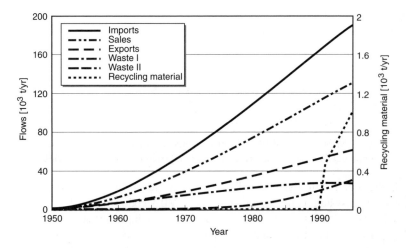

FIGURE 2.35 Graph created with data from "PVC Austria."

A commonly encountered problem is that of circular references, which occur when a formula in a cell refers to the same cell. For example, if the formula in A1 is "= A1 + 1," Excel does not know how to calculate the result, since it could keep adding "1" forever. It is also possible to have a circular reference created by more than one cell. For example, A1 could contain "= B1 + 3," and B1 could contain "= A1 + 4." Again, Excel does not know how to calculate the values. If a system contains a feedback loop, circular references of the second type may be created, and Excel displays a warning. In this case, Excel's iteration feature can be set to solve the problem:

1. Choose "Tools → Options" in the main menu.
2. Click the "Calculation" tab.
3. Select the "Iteration" check box.
4. Click "OK."

If a circular reference occurs, Excel will now try to solve the problem iteratively.

Methodology of MFA

2.4.7 UMBERTO

2.4.7.1 Program Description

Umberto was developed by the Institute for Energy and Environmental Research Heidelberg Ltd. (IFEU) in cooperation with the Institute for Environmental Informatics Hamburg Ltd. (IFU). It was first presented at CEBIT Hannover in 1994. Since then, over 350 licenses have been sold. Umberto 4 is available in different versions (consult/business/educ/test). For more detailed information or a free demo version (with restricted functionality), contact the Umberto Web site (www.umberto.de/english/).

Umberto serves to visualize material- and energy-flow systems. Data are taken from external information systems or are calculated within a model. With its comfortable graphical user interface (GUI), even complex structures can be handled: production facilities of a company, process chains, or life-cycle assessments (LCAs). With Umberto, flows and stocks can be evaluated using performance indicators. Scaling per unit of products or per period is possible. In addition, the environmental costs of the system can be displayed and analyzed. Cost accounting and cost allocation are some of the outstanding features of Umberto. The software cannot only be used to show relevant flows and environmental effects, but it is also helpful in identifying possibilities for enhancing the system to achieve economic and ecological objectives.

Umberto is useful for companies that want to establish an environmental management system. It is also a flexible and versatile tool for research institutions and consultancies, e.g., for material flow analysis or for life-cycle assessments of products.

2.4.7.2 Quickstart with Umberto

This introduction is not intended to be a complete manual. It points out only a small range of the potential of Umberto, and there might be different ways to reach the same results. For further information, consult the Umberto manual.[31] The case study was implemented with a test version of Umberto 4.0.

◆ Wherever this sign appears, it indicates detailed instructions on how to implement the case study from Section 2.4.4.

2.4.7.2.1 Starting the Program

To start the program, double-click the program icon on the desktop or choose Umberto from the start menu. The "Login" window opens (Figure 2.36).

Umberto (i.e., the InterBase database) allows the administration of users with different access rights. By default "SYSDBA" (System Database Administrator) is set as login name and "masterkey" (in lowercase characters) as password.

1. Enter the user name and password or leave the default settings unchanged.
2. Click "OK."
3. The "Umberto 4" window (Figure 2.37) opens.

Start Umberto.

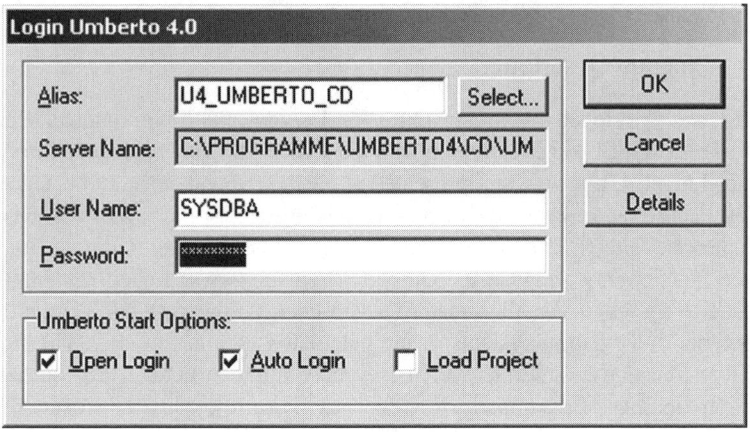

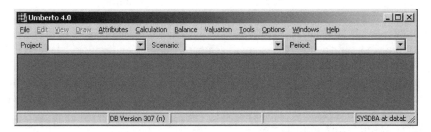

FIGURE 2.36 "Login" window.

FIGURE 2.37 The "Umberto 4" window contains the main menu and the speed bar. The speed bar can be used to select a project, scenario, and period.

2.4.7.2.2 Creating a Project

The user can create individual projects. Each project is characterized by a freely definable and expandable list of products, raw materials, pollutants, forms of energy, etc., all referred to as materials. They are managed in a hierarchically structured material list.

1. Select "File → New → Project…" from the main menu. The "New Project" window (Figure 2.38) opens.
2. Enter a name into the project name field and add a description.
3. Confirm by clicking "OK."
4. The "Materials" window (Figure 2.39) opens.

◆ Create a new project. Enter "PVC Austria" as project name and "Simplified PVC-household of Austria between 1950 and 1994" as description.

2.4.7.2.3 Creating a Scenario

Within a project, several scenarios can be created. A scenario is characterized by a graphical network structure.

Methodology of MFA

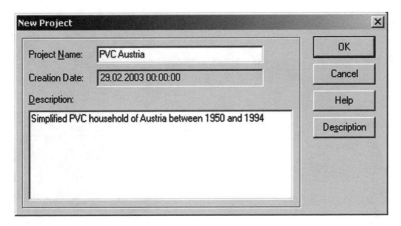

FIGURE 2.38 The "New Project" window contains the project name and a description.

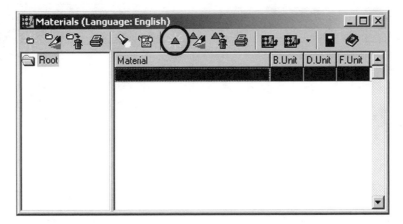

FIGURE 2.39 The material list of the project is managed in the "Materials" window, which contains all materials occurring in a specific project. When opened for the first time, it is empty. The "New Material" button on the toolbar is marked with a circle.

1. Select "File → New → Scenarios…" from the main menu. The "New Scenario" window (Figure 2.40) opens.
2. Enter a title for the scenario in the "Scenario Name" field. Press the "Enter" key or click the "OK" button.

◆ Create a new scenario. Enter "Scenario1" as scenario name.

2.4.7.2.4 Building a Material Flow Network

After you create a scenario, the "Network" window (Figure 2.41) appears on the screen. In its working area, a material flow network can be modeled by inserting the required elements. The toolbar contains buttons to support this task.

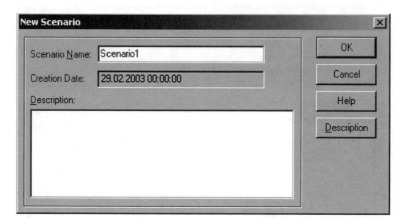

FIGURE 2.40 The "New Scenario" window contains the scenario name and a description.

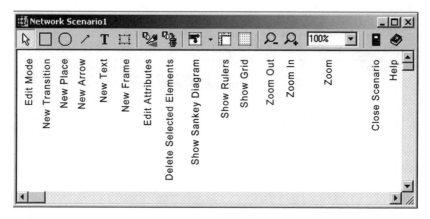

FIGURE 2.41 The "Network" window containing a toolbar and a working area. This figure shows a description of the toolbar buttons.

2.4.7.2.5 Inserting a Transition

"Transitions" represent processes where manipulations of materials occur. In Umberto, transitions are represented by a square. According to the methodology of Petri nets, they can only be connected via "places."

1. Click the "New Transition" button on the tool bar.
2. Position the transition by clicking on the working area of the "Network" window (see also Section 2.4.7.2.9, "Relocating and Adjusting of Places and Transitions"). It is automatically labeled by an identifier consisting of a "T" followed by a number.
3. Each additional click creates a new transition.
4. Return to the editing mode by clicking the right mouse button (called right-click).

Methodology of MFA

2.4.7.2.6 Naming of a Transition
To name a place:

1. Click the label of a transition. It will be marked. To edit it directly, click again.
2. Alternatively, a double-click opens the "Label Attributes" window in which text can be entered as well as formatted.
3. Enter the name of the transition and press "Enter."

◆ Create three transitions and name them (*T1: production*; *T2: distribution*; *T3: utilization*).

2.4.7.2.7 Inserting a Place
"Places" are elements in a network where no transformation of materials occurs, but which are used for other reasons. They can be stocks of materials, input or output places connecting the material flow network with its environment ("import" and "export" flows in MFA terminology), or places where the output stream of one process becomes the input of another ("internal" flows in MFA terminology). In Umberto places are represented by a circle.

1. Click the "New Place" button in the tool bar.
2. Position the place by clicking in the "Network" window (see also "Relocating and Adjusting of Places and Transitions"). It is automatically labeled by an identifier consisting of a "P" followed by a number.
3. Each additional click creates a new place.
4. Return to the editing mode by right-clicking.

2.4.7.2.8 Naming of a Place
To name a place:

1. Click the label of a place. It will be marked. To edit it directly click again. Alternatively a double-click opens the "Label Attributes" window in which text can be entered as well as formatted.
2. Enter the name of the place and press "Enter."

◆ Create six places and name them (*P1: imports; P2: products; P3: sales; P4: waste; P5: exports; P6: recycling material*).

2.4.7.2.9 Relocating and Adjusting of Places and Transitions
The position of symbols can be rearranged.

1. Click the "Edit Mode" button or right-click in the "Network" window.
2. Position the mouse pointer over the object that should be moved. The pointer symbol changes into a hand.
3. Drag (includes holding down the left mouse button while moving the mouse and then releasing the button) the object to the desired position.

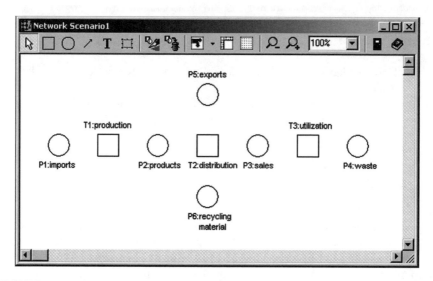

FIGURE 2.42 Arrangement of processes and places needed in the PVC case study.

2.4.7.2.10 Shifting the Labels

Labels can be relocated in the same way as network elements.

◆ Arrange the inserted transitions, places, and labels as shown in Figure 2.42.

2.4.7.2.11 Inserting Arrows between Places and Transitions

So far, the network consists of places and transitions. However, they are not linked yet. This link is to be established with arrows.

1. On the toolbar click the "New Arrow" button.
2. Click the source element (place/transition).
3. Click the target element (transition/place). The connection will be established.
4. The appearance of an arrow can be changed by clicking it. Black marks will appear that can be used to modify the course of the selected arrow. Drag the marks to the desired positions.

◆ Repeat the proceedings from above until the material flow network looks as shown in Figure 2.43.

2.4.7.2.12 Assigning Place Types

According to the different functions that places can have in a material flow network, Umberto distinguishes among four different types of places:

1. Input places ("imports" in MFA terminology)
2. Output places ("exports" in MFA terminology)

Methodology of MFA

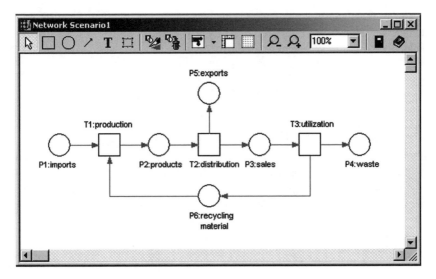

FIGURE 2.43 Material flow network of the PVC case study.

3. Storage places
4. Connection places ("internal flows" in MFA terminology)

One important function of places is that they delimit the material flow network from its environment; they are points of contact to the world that lies outside the system boundaries.

Whenever materials enter the system ("imports" in MFA terminology), they do so through places of the "input" type. They can supply the material flow network with (theoretically infinite) quantities of materials. The symbol for input places is a circle with an additional mark on the left:

Whenever materials leave the system, i.e., whenever they are emitted to the outside world ("exports" in MFA terminology), they do so through places of the "output" type. They are sinks that take up infinite quantities of materials. The symbol for output places is a circle with an additional mark on the right:

A third type of places is labeled "storage." It serves to model stocks: incoming materials increase the stock; materials that are withdrawn decrease it. If there is not enough material in stock to supply a subsequent process, this process cannot run.

In terms of the material flow network, this means that the calculation routine cannot calculate the entire network. Consequently a warning message will be displayed. The symbol for storage places is a circle:

The fourth type of places is called "connection." It links one process to another without using any intermediate storage (e.g., two machines running synchronously in a production line). The place forms the link from the output of one process to the input of the following process ("flows" in MFA terminology). The symbol for places of the "connection" type has two concentric circles:

The characteristic of a "connection" place is that the amount of materials entering the place must equal the amount leaving it. Thus, the stock inside a "connection" place cannot change. In material flow networks, this rule is used to determine unspecified material flows. If all flows of a material to or from a "connection" place except one are given, the unknown flow can be determined from the mass balance between input and output.

1. To assign the "Input" or "Output" type, right-click the place.
2. Select "System Boundaries" from the context menu. Choose between None, Input, Output, and Input/Output (Figure 2.44).

◆ For *P1: imports* choose "Input" as system boundary. For *P4: waste* and *P5: exports*, select "Output" as system boundary.

1. To assign the "Connection" type, right-click the place.
2. Choose "Calculation Flags → Connection" from the context menu.

◆ For *P2: products*, *P3: sales*, and *P6: recycling material*, choose "Connection" as calculation flag.

The graphical structure of the model has been designed now. In the next section a material list is created, and the transitions are specified.

2.4.7.2.13 Defining New Basic Units

In Umberto, the term *material* is used in a very broad sense. Both substances and forms of energy are referred to as materials. The term *basic unit* refers to the unit into which all values are converted when a calculation is conducted. The unit "kilogram" (kg) is the basic unit for all substances, and "kilojoule" (kJ) is the basic unit for all forms of energy. If needed, additional basic units can be added.

Methodology of MFA

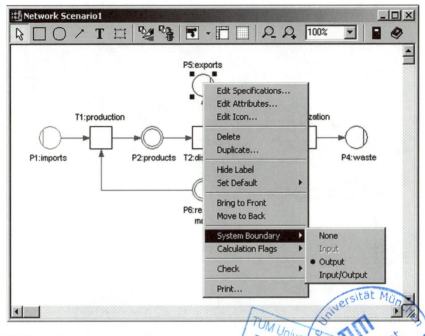

FIGURE 2.44 Assignment of place types.

1. Select "Options → Materials…" from the main menu. The "Material Options" window will appear.
2. Click the "Basic Unit" tab.
3. Click "New."
4. Enter the new basic unit. It is possible to add a description.
5. Select the "Show Unit," "Show Sum," and "Check Balance" check boxes.
6. Click "OK."

◆ Define "metric tons per year" (t/year) as an additional basic unit. Enter "Material flow unit" as description (Figure 2.45).

2.4.7.2.14 Creating a Material List

In the "Materials" window, the list of materials is managed. The material list is the base for the specification of the material flow network.

1. Click the "New Material" button inside the "Materials" window. If it is not clear which button is meant, let the pointer rest over it and a "hint" will appear.
2. The "New Material" window opens (Figure 2.46).
3. Write the name of the new material into the "Material Name" field.
4. Choose a basic unit. All internal calculations will use this unit.
5. The selection of a "Material type" (good, neutral, or bad) does not have any influence on the calculation.

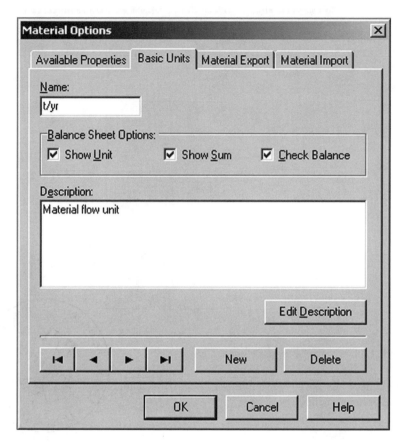

FIGURE 2.45 "Material Options" window.

6. To add another material, click "Insert."
7. To finish the inputs, click "OK."
8. The new material will appear inside the "Materials" window (Figure 2.47).

▶ Define nine materials (*imports, exports, products, sales, inventory, recycling material, waste I, waste II,* and *waste*). Choose "t/year" as a basic unit.

2.4.7.2.15 Editing a Period

When a new scenario is created, the present year is automatically set as the default period. Of course, it is possible to change this setting:

1. Choose "Attribute → Period..." from the main menu.
2. Select a period by using the "Navigation" keys.
3. Enter the first and the last day of the period.
4. Assign a short name to the period (e.g., "1950").
5. Click "OK."

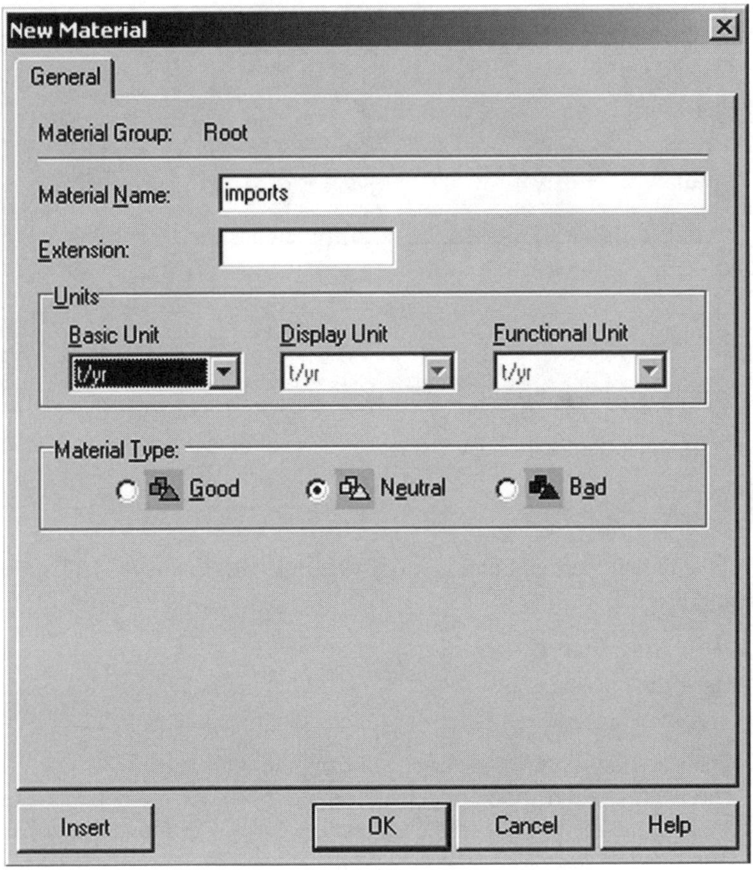

Methodology of MFA 99

FIGURE 2.46 "New Material" window.

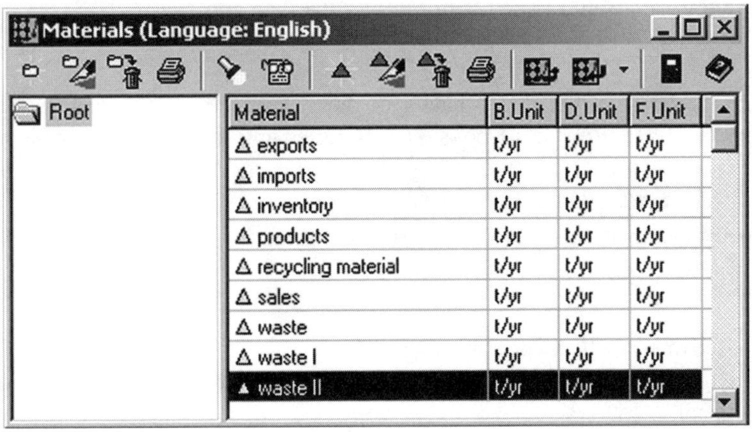

FIGURE 2.47 "Materials" window containing a list of materials needed within the PVC case study.

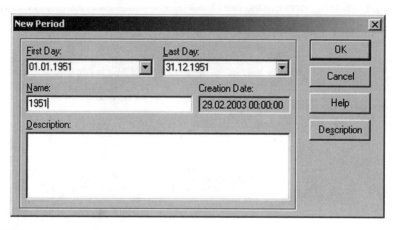

FIGURE 2.48 "New Period" window.

◆ Edit the default period. Enter first day (01.01.1950), last day (31.12.1950), and the name (1950).

2.4.7.2.16 New Time Period
To set a new time period:

1. Choose "File → New → Period…" from the main menu.
2. The "New Period" window opens (Figure 2.48). As the default setting, the first and the last day of the following year can be found. Enter name, first day, and last day of the new period.
3. Close the window.
4. Alternatively, choose "Attribute → Period…" from the main menu and click the "New Period" button. Enter the data and click "OK."

◆ To simulate the dynamic behavior of the system, create additional periods of time (1951–1994). Repeat the procedure 44 times until all years are covered.

2.4.7.2.17 Specifying a Transition
To specify a transition means to describe the relations between input and output flows of a process.

1. Activate the "Network" window.
2. Right-click a transition. Choose "Edit Specifications…" from the context menu.
3. Alternatively, double-click the transition.
4. The "Transition Specification" window opens (Figure 2.49).

Methodology of MFA

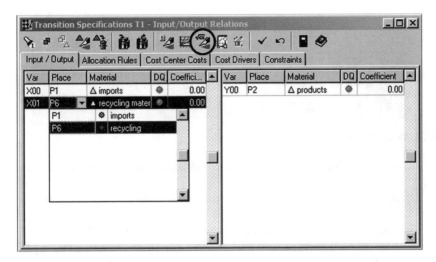

FIGURE 2.49 "Transition Specification" window. The "Function" button on the toolbar is marked with a circle.

2.4.7.2.18 Insert Materials

To specify a transition, insert materials on the input and/or output side of the "Transition Specification" window.

1. Drag the material to be inserted from the "Materials" window to the "Transition Specification" window.
2. Each material is represented by a variable (e.g., X00). The name of the variable that is assigned automatically can be edited by double-clicking it.

2.4.7.2.19 Changing the Place Identifier

Check that the identifiers of the places are displayed correctly. If needed, correct it as follows:

1. Activate the material of concern by clicking the line it is located in.
2. Click the arrow that appears next to the "Places" field. A list of all places that are connected to the transition will be displayed. Choose the correct one.

There are different ways to specify a transition in Umberto:

- To import a transition specification from the library
- To enter the linear mass ratio between inputs and outputs
- To enter user-defined mathematical functions
- To define a subnet
- To import a given subnet from the library

2.4.7.2.20 Entering User-Defined Mathematical Functions

To enter a user-defined mathematical function:

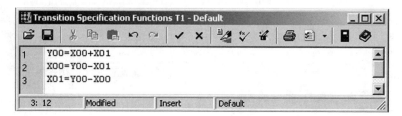

FIGURE 2.50 "Transition Specification" window.

1. Click the "Function" button on the tool bar of the "Transition Specifications" window.
2. A warning will appear that all specifications made by coefficients will be overwritten. Click "OK."
3. The "Transition Specification Functions" window opens (Figure 2.50).
4. Express each variable with a function in a separate line (e.g., Y00 = X00 + X01).
5. Close the "Transition Specification Functions" window.
6. Close the "Transition Specification" windows. If the transition is correctly specified, its symbol inside the network diagram will be displayed with a blue frame.

◆ Define the transitions *T1: production* and *T2: distribution*.

- T1: *production*
  Inputs: *imports* (X00), *recycling material* (X01)
  Outputs: *products* (Y00)
  Functions:
   Y00 = X00 + X01
   X00 = Y00 − X01
   X01 = Y00 − X00

- T2: *distribution*
  Inputs: *products* (X00)
  Outputs: *sales* (Y00), *exports* (Y01)
  Functions:
   X00 = Y00 + Y01
   Y00 = X00 − Y01
   Y01 = X00 − Y00

2.4.7.2.21 Creating a Subnet
To create a subnet:

1. Right-click the transition that shall be defined by a subnet. Choose "Create Subnet" from the context menu.

Methodology of MFA

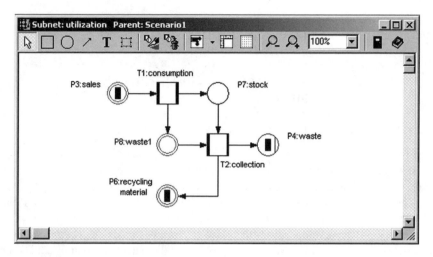

FIGURE 2.51 "Subnet" window.

2. If the transition has already been specified before, a warning will appear that the existing specification will be deleted. Click "OK."
3. A "Subnet" window opens (Figure 2.51). Places that are connected to the transition on parent level will also be displayed in the subnet. They work as interfaces. The symbols of these places contain a brown box.
4. Design the subnet using transitions, places, and arrows.
5. Specify the transitions of the subnet.

◆ Define the transition *T3: utilization* with a subnet that looks like the following graph. *P7: stock* serves as a storage that releases material after its utilization time of 22 years.

◆ Specify the transitions of the subnet. For *T1: consumption* and *T2: collection*, time-dependent functions are used to describe the time series of the waste rate and the recycling rate.

- *T1: consumption* (subnet)
  Inputs: *sales* (X00)
  Outputs: *inventory* (Y00), *waste 1* (Y01)
  Functions:
  $WR = 0.50 - [(0.50 - 0.20)/(1994 - 1950)] \times (fdy - 1950)$
  $Y00 = X00 \times (1 - WR)$
  $Y01 = X00 \times WR$

The waste rate (WR) is modeled with a linear function (0.50 in 1950 and 0.20 in 1994). The parameter "fdy" represents the year of the active period (1950 to 1994).

- *T2: collection* (subnet)
  Inputs: *inventory* (X00), *waste I* (X01)
  Outputs: *recycling material* (Y00), *waste* (Y01)
  Functions:
     RR = if(<(fdy,1990),0,0.01 + (fdy − 1990) × 0.002)
     XG = X00 + X01
     Y00 = XG × RR
     Y01 = XG × (1 − RR)

It is necessary that materials entering and leaving a place be of the same type. Thus, *inventory* instead of *waste II* is used as input of *T2: collection*.

The recycling rate (RR) is also modeled with a linear function (0 until 1990, 0.01 in 1991 with a constant rise of 0.002 per year).

Finally, a delayed output from the place *P7: stock* is modeled. The *inventory* is released after its utilization time of 22 years, i.e., the calculated output equals the input into the *stock* 22 years ago. Insert the following lines at the beginning of the transition specification of *T2: collection*.

Functions:
   WR0 = 0.50 − [(0.50 − 0.20)/(1994 − 1950)] × (fdy − 1972)
   X00 = if(<(fdy,1973),0,(1 − WR0) × {−0.8633 × EXP[3 × LN(fdy − 1972)] + [94.271 × SQR(fdy − 1972)] + [484.57 × (fdy − 1972)]} × gwfy)

WR0 is the waste rate, and X00 is the calculated input of the place *P7: stock* 22 years ago. The expression EXP[3 × LN(x)] is used because Umberto cannot cope with x^3. EXP(x) stands for e^x, LN(x) for ln(x), and SQR(x) for $\sqrt{x}$. Umberto scales input values in case of a leap year. To avoid this, the parameter "gwfy" is necessary.

At last, select "Edit Attributes" from the context menu of the *T2: collection* transition. Click the "Calculation" tab and select the "Check Transition for Calculation" check box inside the "Calculation Process" group box. This is necessary to calculate a transition without knowing adjacent flows.

2.4.7.2.22 Definition of Start Values

To define the start values:

1. Double-click a connection or right-click and choose "Edit Specifications…" from the context menu.
2. The "Arrow Specification" window opens.
3. Drag a material from the "Materials" window to the "Arrow Specification" window.
4. Click the "Set Quantity" button inside the "Arrow Specification" window and enter the quantity. Press "Enter" or click "OK."
5. Alternatively enter the quantity directly in the "Arrow Specification" window.
6. Click "Close" or close the window.

Methodology of MFA

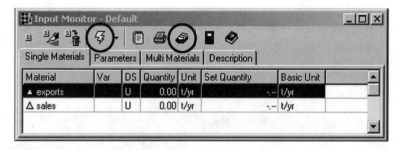

FIGURE 2.52 "Input Monitor" window. The "Set Input Vector" button (left) and the "Edit/Select Input Vector" button (right) on the toolbar are marked with a circle.

To simulate the dynamic behavior of a system, enter the values of given flows for each period manually or use the "Input Monitor." The "Input Monitor" helps to simplify the assignment of values by creating an input vector that serves as container for all input data.

1. Choose "Tool → Input Monitor…" from the main menu. The "Input Monitor" window opens (Figure 2.52).
2. Click the "Edit/Select Input Vector" button. Click "New…" Enter a name for the new input vector. Click "OK" twice. The new input vector will automatically be selected.
3. Drag the materials to set from the "Materials" window to the "Input Monitor."
4. To enter the values for a material, double-click it. The "Edit Coefficient" window opens (Figure 2.53).
5. Enter data for the chosen material as user input, function, or SQL statement.
6. To define data with the help of a function, click in the "Data Source" field on "Function." Enter the function that describes the value of the material in the "Function" field.
7. Click the "Referenced Arrows" tab. Click "Insert." The "Insert Reference" window opens (Figure 2.54).
8. Choose the scenario and the network to which the arrow belongs.
9. Choose the period to which the calculated value belongs or select the "Insert References for All Periods" check box.
10. Choose the correct connection from the "Arrow" collection.
11. Click "OK" twice.
12. Click the "Set Input Vector" button.
13. Close the "Input Monitor."

◆ Create a new input vector and call it "Data for scenario 1." Drag the materials *sales* and *exports* from the "Materials" window to the "Input Monitor." Enter the following functions and referenced arrows:

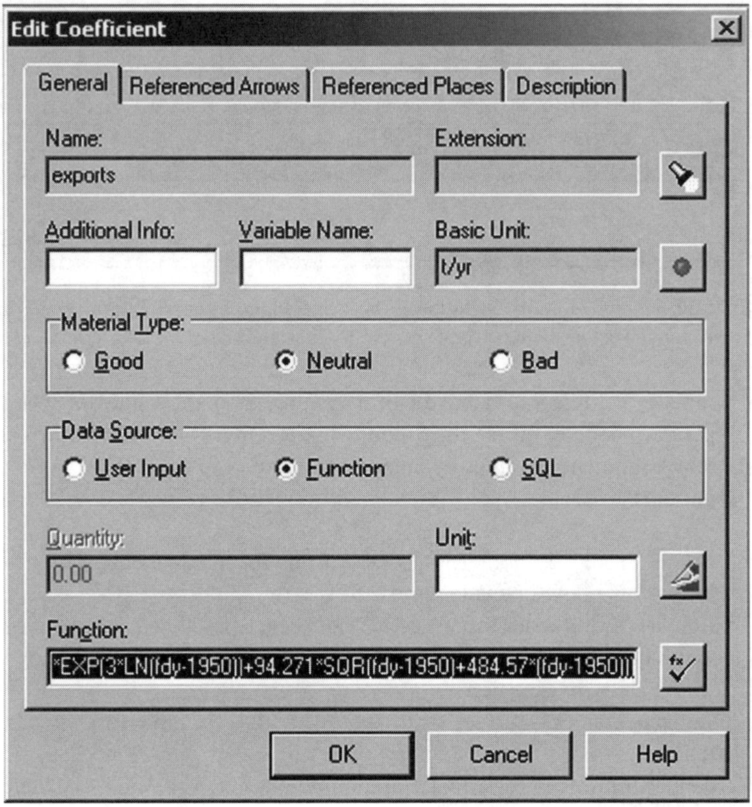

FIGURE 2.53 "Edit Coefficient" window.

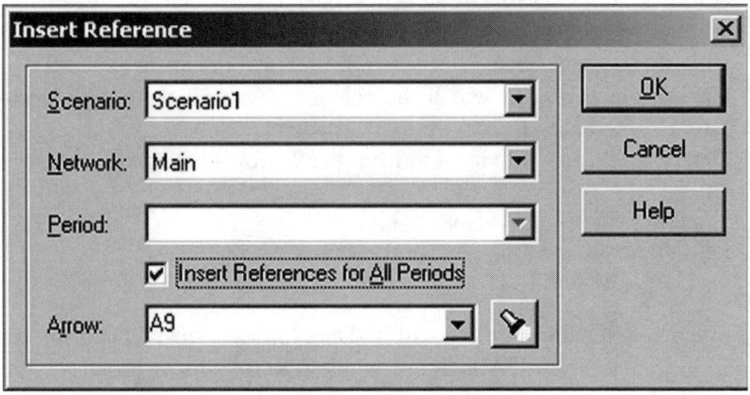

FIGURE 2.54 "Insert Reference" window.

- *Sales*
  Function: if(<(fdy,1951),0,–0.8633 × EXP[3 × LN(fdy – 1950)] + [94.271 × SQR(fdy – 1950)] + [484.57 × (fdy – 1950)])
  Referenced arrows: Scenario2/Main/Insert Reference for all Periods/T2>P3
- *Exports*
  Function: if(<(fdy,1951),0,0.47 × {–0.8633 × EXP[3 × LN(fdy – 1950)] + [94.271 × SQR(fdy – 1950)] + [484.57 × (fdy – 1950)]})
  Referenced arrows: Scenario2/Main/Insert Reference for all Periods/T2>P5

The parameter "fdy" stands for the year of the active period (1950 to 1994). The expression EXP[3 × LN(x)] is used because Umberto cannot cope with x^3.

2.4.7.2.23 Performing a Calculation
To perform a calculation:

1. Choose "Calculation → Network → All Periods" from the main menu.
2. If errors occur during the calculation, the "Log-File" indicates where the origin of errors can be found. Check the system and repeat the calculation.

2.4.7.2.24 Viewing the Calculated Results
To view the calculated results:

1. Select the period of interest in the speed bar.
2. Double-click a connection.
3. Alternatively right-click and choose "Edit Specifications…" from the context menu.
4. The "Arrow Specification" window opens and the materials that are transported along this connection together with their quantities will be displayed.

Calculated materials are marked with a "C," while materials that have been entered manually are labeled with an "M."

In Umberto, it is possible that more than one material is flowing along a connection, so it is not necessary to draw a connection for each material flow. This contributes to the clarity of a material flow network.

1. By double-clicking a place symbol, the "Place Specification" window opens. There the stock inside a place at the beginning and at the end of the time period considered is displayed and can be altered.

2.4.7.2.25 Defining Sankey Diagrams
Umberto can display the width of the arrows according to the quantity of the material flow. This is called a Sankey diagram. Several materials in one arrow will be displayed in different colors. Materials with no color assigned will be displayed in grey. This feature is available in the business/consulting version only.

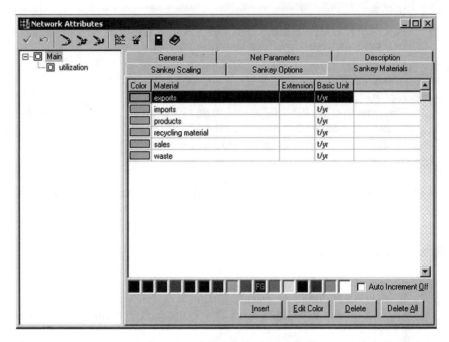

FIGURE 2.55 "Network Attributes" window.

1. Choose "Attributes → Network…" from the main menu.
2. The "Network Attributes" window opens (Figure 2.55).
3. The Sankey materials have to be defined separately for the main net and all of its subnets. Select the desired net or subnet from the tree structure on the left side.
4. Click the "Sankey Materials" tab.
5. To choose the materials that are to be displayed in a colored Sankey diagram, click "Insert." Select the desired material from the "Search Material" window and click "OK."
6. The materials chosen will be inserted into the list of Sankey materials. On this occasion, a color is automatically assigned to each material. The color can be changed by double-clicking the colored area next to the material name.
7. Close the "Network Attributes" window.

2.4.7.2.26 Displaying Sankey Diagrams

To display the Sankey diagram:

1. Click the "Show Sankey Diagram" button in the tool box of the "Network" window.
2. Display Sankey diagrams for different periods by selecting the desired period from the "Period" pull-down menu on the speed bar.

Methodology of MFA

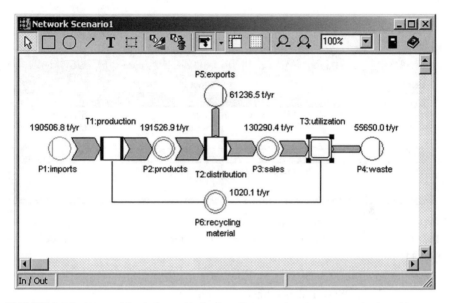

FIGURE 2.56 "Network" window with Sankey diagram showing the results for 1994.

3. To display the value of a flow, let the pointer hover over the respective arrow. A note will appear that shows the amount of the material flow.
4. The calculated values can be displayed continuously by right-clicking the arrow of interest and selecting "Show Sankey Label" from the context menu (Figure 2.56).
5. To return to the regular display mode, click the "Show Network" button.

2.4.7.2.27 Creating a Balance

In the "Balance Sheet" window of Umberto, input/output balances can be drawn for the system under examination.

1. Select a period in the speed bar.
2. Choose "Balance → Preview → Default Boundaries" from the main menu to create a balance sheet. During this action, the "Network" window has to be active.
3. The "Balance Sheet Preview" window opens (Figure 2.57).

To display the internal flows of the system:

1. Select "Options → Balance Sheet…" from the main menu.
2. Click the "General" tab and select the "Show Internal Flows" check box.
3. Click "Apply" and close the window.
4. An "Internal Flows" tab is added to the "Balance Sheet Preview" window.

In order to compare balances, they must first be saved:

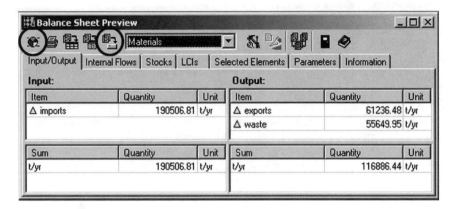

FIGURE 2.57 "Balance Sheet Preview" window showing the input/output balance for 1994. The "Chart" button (left) and the "Save Balance" button (right) on the toolbar are marked with a circle.

1. Click the "Save Balance Sheet" button.
2. Enter the name of the balance in the "Comment" field.
3. Save the balance by clicking "OK." Close the window.
4. Repeat this procedure for all calculated periods of interest.

2.4.7.2.28 Comparing Balances

To compare balances:

1. Choose "Balance → Select" from the main menu.
2. The "Select Balance Sheets" window opens (Figure 2.58).
3. Select the check boxes of the balances that are to be compared.
4. Click "OK."

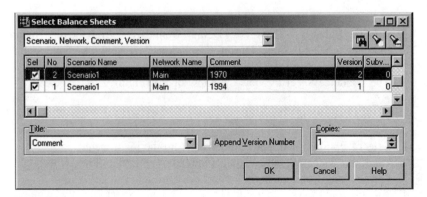

FIGURE 2.58 "Select Balance Sheets" window. Two balances are available. Both are selected to be compared.

Methodology of MFA

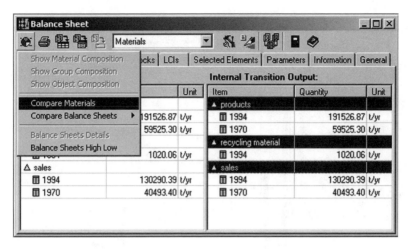

FIGURE 2.59 "Balance Sheet" window showing the internal flows of 1970 and 1994.

2.4.7.2.29 Creating a Chart

To create a chart:

1. In the "Balance Sheet" window mark the materials to be displayed in a chart (Figure 2.59). Multiple selections are possible by clicking while holding the "CTRL" key. Multiple materials can be selected only if they belong to the same side of the balance sheet.
2. Click the "Chart" button. Choose "Compare Materials."
3. The "Umberto Chart" window opens (Figure 2.60).
4. Format the chart by using the various options on the toolbar.

◆ Create balances for 1970 and 1994. Compare these balances with regard to the internal flows of *products, recycling material,* and *sales.*

2.4.7.2.30 Finishing the Session

To finish the session, close all open windows:

1. To close a scenario, close the "Network" window.
2. To close a project, close the "Materials" window.
3. To terminate the application, close the Umberto main window.

2.4.7.3 Potential Problems

When working with Umberto, the following problem might appear. A simple system with a feedback loop (Figure 2.61) is chosen to illustrate the problem:

In T1, the flows A1 and A6 are added to form A2 (A2 = A1 + A6). Via a "connection," place A2 is linked to A3 (i.e., A2 = A3), as is A6 to A5 (A5 = A6). In T2, the flow A3 is divided into A4 and A5 according to a given ratio (A4 = A3

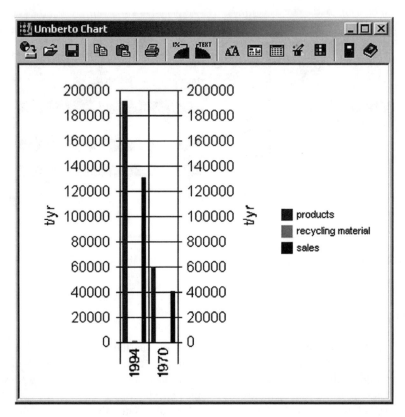

FIGURE 2.60 "Umberto Chart" window showing one possible chart displaying the internal flows of 1970 and 1994.

× TC34 and A5 = A3 × TC35). During the calculation procedure, each transition is repeatedly checked to see if its flows can be computed entirely from given or recently calculated data in its direct vicinity. If yes, the flows are calculated, and the transition is not considered anymore during the computation procedure. When all transitions are determined, or when there are not enough data left to calculate the remaining transitions, the computation stops.

For example, if A3 is given, then A4 and A5 can be calculated. Thus, T2 is completely specified and will not be considered anymore in the calculation procedure. Because P2 and P4 are places of the "connection" type, A2 = A3 and A6 = A5. With this information, A1 (= A6 − A2) can be calculated. Thus, T1 is specified completely, too. The calculation procedure works flawlessly. But what happens if A1 is given instead of A3? Umberto tries to calculate the flows of T1. This is not possible because there is only one equation given (A2 = A1 + A6) that contains two unknown quantities. The calculation of T1 is cancelled. T2 also cannot be computed because no additional information has been gained. The calculation stops.

How to cope with this situation? Check what material is supposed to come from P4, enter a special amount of this material as stock in P4 (e.g., zero), and allow the

Methodology of MFA

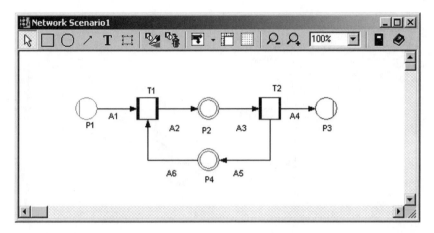

FIGURE 2.61 Simple system with a feedback loop.

place to create a flow (context menu of P4: "Calculation Flags → Create Flows") in the beginning of the calculation procedure. With A1 and the created flow A6, A2 can be calculated. Thus, T1 is completely specified. With A3 = A2, T2 could be calculated completely, too. But there is a problem with A5. There is a discrepancy between the value of A5 calculated from A5 = A6 and the one calculated from A3 × TC35. A warning concerning this error will appear. The calculation of T2 cannot be completed.

This problem can be solved by introducing a new material (e.g., "help") as an additional input of the transitions. It is important that these additional inputs be defined so that they cannot be calculated from the transitions' specifications. If this condition is met, then it is not possible to finish the complete calculation of T1 and T2 with the normal calculation procedure. If the user chooses the option of allowing "Incomplete Calculations," it is not necessary to determine a transition completely in order to use its calculated flows for further calculation. The respective transition will be considered again later. Because the additional material cannot be calculated by the specification of the transition, the computer will remain in an endless cycle, stopping only when the values of the flows that can be calculated do not change anymore. A warning will appear that the "Help" flows cannot be calculated. But all other flows will be displayed correctly (Figure 2.62).

"Incomplete Transition Calculation" can be set by choosing "Edit Attributes" from the context menu of a transition. Click the "Calculation" tab and select the "Incomplete Transition Calculation" checkbox.

2.4.8 GABI

2.4.8.1 Program Description

The GaBi 4 software system for life-cycle engineering was developed by the Institute for Polymer Testing and Polymer Science (IKP) at the University of Stuttgart in

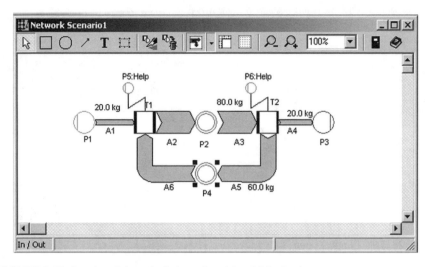

FIGURE 2.62 Results of the calculation when A1 = 20 kg is given.

cooperation with PE Europe GmbH (PE) in Leinfelden-Echterdingen (Germany). GaBi 4 reflects experience in applied modeling work since 1989; the first version was completed in 1992. Through 2002, over 300 licenses had been sold.

GaBi 4 (stands for "ganzheitliche bilanzierung" or, in English, "life-cycle engineering") is available in different versions (professional/lean/academy/educational). For additional information or a free 90-day trial version (with full functionality), contact the GaBi Web site (www.gabi-software.com).

GaBi is an LCA tool that is used in industry, academia, and consultancy. In contrast to conventional LCA software tools and as an extension of the first versions, GaBi 4 offers the additional opportunity to carry out a consistent and detailed cost evaluation of the system investigated (life-cycle costing [LCC]) and to use the software in support of on-site environmental management and audit schemes (EMAS). The implemented "cost assistants" allow the user to develop precise models of costs for material/energy, personnel, and machinery. The software also enables the user to address socioeconomic aspects of a project with the life-cycle working time (LCWT) methodology on working environment that was developed and implemented by IKP. The methodology includes such issues as working time, qualification, accidents, child work, equal opportunity, etc.

The multifunctional features of GaBi 4 also make it useful as a tool for simply and quickly modeling and analyzing complex and data-intensive problems for MFA projects. A user-friendly graphical interface ensures that GaBi 4 can be used intuitively and that it does not take long to learn how to use it. The case study in Section 2.4.8.2 was implemented with a full version of GaBi 4.

2.4.8.2 Quickstart with GaBi

This introduction is not intended to be a complete manual. It highlights only a small range of GaBi's potential. There might be different ways to obtain the same

Methodology of MFA

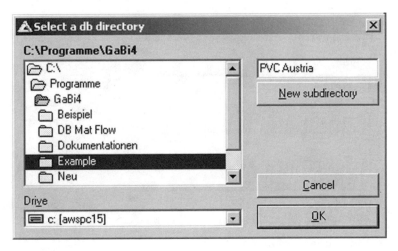

FIGURE 2.63 Select a database directory.

results. For further information, consult the GaBi manual or contact the distributor for support.[32]

◆ Wherever this sign appears, it indicates detailed instructions on how to implement the case study from Section 2.4.4.

2.4.8.2.1 Starting the Program

To start the program, double-click the program icon on the desktop or choose GaBi from the start menu. The GaBi database (DB) manager opens. The "object hierarchy" displays all databases that are connected to GaBi. It is possible to connect GaBi to other databases or create new ones.

2.4.8.2.2 Creating a New Database

To create a new database:

1. Select "Database → Create new database" from the main menu.
2. Select a directory on the hard drive or create a new directory (by entering a name and clicking "New subdirectory") where the new database should be stored (Figure 2.63).
3. Click "OK."

The new database will be displayed within the "object hierarchy." In the beginning, the database contains folders with predefined units, quantities, and flows (such as ores, goods, and emissions with some properties) but no processes or plans.

◆ Start GaBi. Create a new database in a subdirectory of the GaBi 4 directory. Call it "PVC Austria."

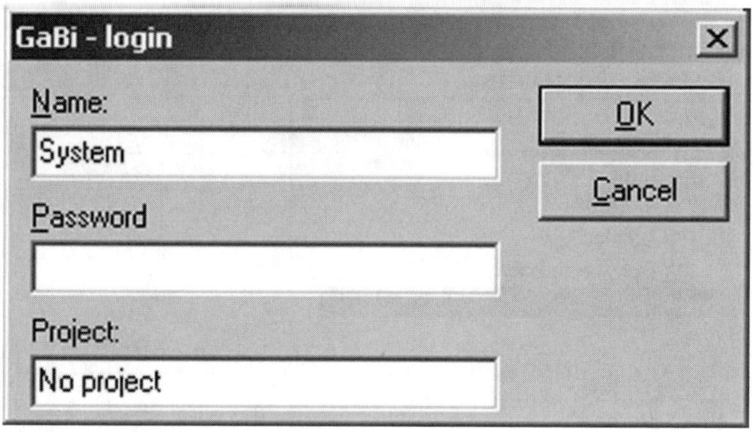

FIGURE 2.64 "Login" window of GaBi.

2.4.8.2.3 Activating a Database

To work with one of the connected databases, it must first be activated.

1. Right-click the database within the "object hierarchy" that is to be opened.
2. Choose "Activate" from the context menu.
3. Enter name and password, or leave the default settings unchanged (Figure 2.64).
4. Click "OK."

GaBi works on an object-oriented basis. Objects of different types are linked hierarchically to form layers of rising aggregation of information (units → quantities → flows → processes → plans). "Plans" are the top-level object type in a GaBi model. They are a kind of working table where material flow chains and nets can be modeled. In order to clarify the interrelations of the different object types, the description starts at the lowest end of the hierarchy, named "units." In practical work, one typically works with objects of higher hierarchy, such as "plans" and "processes." New "units" are rarely created; new "quantities" are seldom created; and new "flows" are created only occasionally.

The "object hierarchy" within the GaBi DB manager (Figure 2.65) contains all databases that have been connected to GaBi. Only one of these databases can be active at a time. Within the activated database are folders for each object type. Folders are designated by a folder symbol within the object-type icon. The structure is analogous to that used in Windows Explorer®: Folders marked by a "+" sign contain subfolders. These subfolders are displayed by clicking the "+" sign or double-clicking the folder. Objects are displayed by the object-type icon without a folder symbol. The objects and subfolders contained within a folder are found on the right-hand side of the GaBi DB manager.

Methodology of MFA

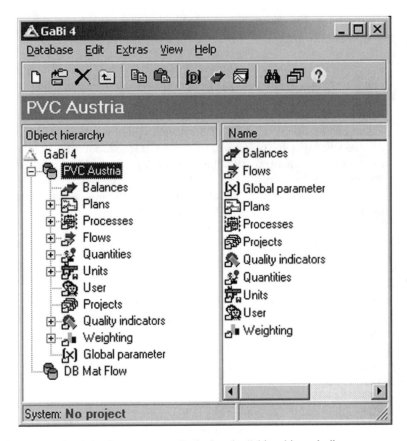

FIGURE 2.65 GaBi database manager displaying the "object hierarchy."

2.4.8.2.4 Creating Additional Folders

To create additional folders, proceed as follows:

1. Right-click the folder or subfolder where you want to add a new folder.
2. Choose "New folder" from the context menu.
3. Enter the name of the new folder and press "Enter."

2.4.8.2.5 Searching an Object

If a previously defined object is to be used, it can be searched manually within the folders of the "object hierarchy," or the "Search for" option of GaBi can be used.

1. Select the object type or one of its subfolders in the object tree.
2. Open the "Search for" window by selecting "Search" from the context menu of the folder symbol or by clicking the spyglass in the main menu.
3. Type a string of letters of the object's name (the first letters or some letters from the middle — the result is the same in either case) and click "Start."

The "Search for" option can also be used within text fields where the name of an object is required:

1. Type a string of letters of the object's name into a text field of an object of higher hierarchy (first letters or some letters from the middle) and press "Enter."
2. Search results are displayed in the lower part of the "Search for" window.
3. Select one by double-clicking. If the search option has been used in the database manager, the object will be opened. If the search option has been used directly from a text field, the name of the object will be inserted in the respective location.
4. It is also possible to drag one of the listed objects into objects that are higher in the hierarchy (e.g., a flow into the inputs of a process).

There are further search options that can be explored by clicking the respective tabs of the search window.

2.4.8.2.6 Creating a Project

If the results of a project are to be credible, the project's documentation must adhere to the standards specified by ISO 14040. The goal of the project, the functional unit, the system boundaries, data quality requirements, comparisons between systems, and critical review considerations can be defined in the project window. A project must also record all objects that have been created or used since the project was activated. To create and activate a project:

1. Right-click "Projects" in the "object hierarchy."
2. Choose "New" from the context menu.
3. Enter a name, describe the project, and click "Activate project."
4. Click the "Object list" tab to display a list of objects used within the active project.
5. Click the "Save" button (disc symbol) and close the window.

◆ Create a new project with the name "PVC Austria" and activate it.

2.4.8.2.7 Creating Units

Units are the actual units of measurement, e.g., kg or kg/year.

1. Select "Units" or one of its subfolders from the "object hierarchy" where the new unit is to be located.
2. Click "New" on the toolbar.
3. Enter the name and the standard unit of the new unit (Figure 2.66).
4. Click the "Save" button and close the window.

◆ Create a new unit. Use the name "Unit of mass per time" and designate "t/year" as a standard unit.

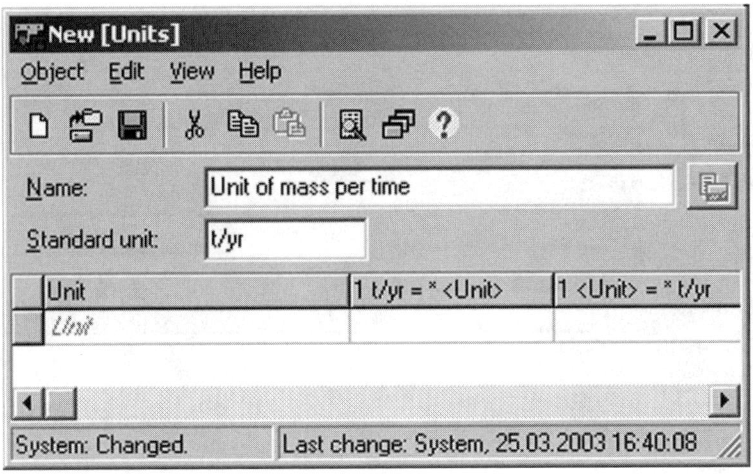

Methodology of MFA

FIGURE 2.66 Creating new units.

2.4.8.2.8 Creating Quantities

"Quantities" are used to measure flows in the above-described units, e.g., mass or mass/time measured in units of kg or kg/year, respectively. Quantities are also used to define additional properties of a flow, e.g., its specific volume, element content, calorific value, etc.

1. Select "Quantities" or one of its subfolders from the "object hierarchy" where the new quantity is to be located.
2. Click "New" on the toolbar.
3. Enter the name and the unit of the new quantity (Figure 2.67).
4. The "Search for" option can be used within the unit field.
5. Click the "Save" button and close the window.

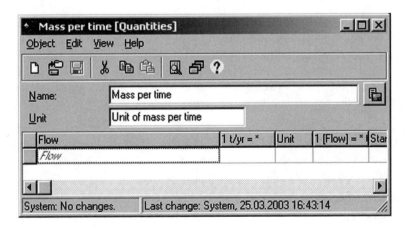

FIGURE 2.67 Creating new quantities.

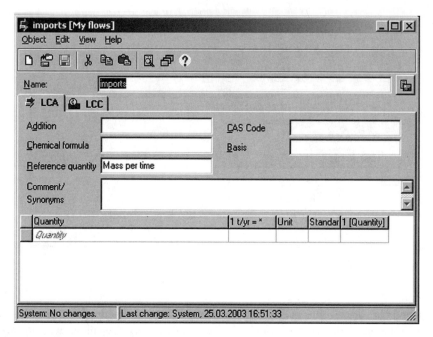

FIGURE 2.68 Creating new flows.

◆Create a new quantity with the name "Mass per time" and use "Unit of mass per time" as unit.

2.4.8.2.9 Creating Flows

"Flows" are typically real flows such as materials (e.g., iron), mixed materials (e.g., household waste), energy carriers (e.g., diesel fuel), etc. They are measured in the above-described quantities and corresponding units.

1. Select "Flows" or one of its subfolders from the "object hierarchy" where the new flow is to be located.
2. Click "New" on the toolbar.
3. Enter the name and the reference quantity of the new flow (Figure 2.68).
4. The "Search for" option can be used within the reference quantity field.
5. Click the "Save" button and close the window.

◆Create a subfolder with the name "My flows." Create flows with the names *imports, exports, products, sales, inventory, recycling material, waste I, waste II,* and *waste.* As a reference quantity for all flows, enter "Mass per time."

2.4.8.2.10 Creating Processes

"Processes" can be conversion processes (e.g., fuel is converted to electricity and emissions) or transport processes (e.g., a good and some diesel fuel enter a lorry and the good is transported some distance). Flows enter a process or leave it.

TABLE 2.16
"Internal" Processes of PVC Austria

Type of Flow	Processes				
	Production	Distribution	Consumption	Stock	Collection
Input flows	imports	products	sales	inventory	waste I
	recycling material	—	—	—	waste II
Output flows	products	exports	waste I	waste II	recycling material
	—	sales	inventory	recycling material	waste

1. Select "Processes" or one of its subfolders from the "object hierarchy" where the new process is to be located.
2. Click "New" on the toolbar.
3. Enter the name of the process in the "Name" field.
4. Click "Flow" in the "Inputs" or "Outputs" section. Enter the name of the flow or at least some of its letters to use the "Search for" option within the "Flow" field.
5. Repeat the last step until all flows needed to specify the process are inserted.
6. In order to use these flows later to connect processes, check the "Tracked flows" field in the line of the respective flow name. Click once inside this field. An "X" will appear.
7. Click the "Save" button and close the window.

◆ Create a subfolder with the name "My processes." Create five processes using the names and flows in Table 2.16. Additionally, create three processes (with the listed flows) representing the import and export locations of the system (Table 2.17).

2.4.8.2.11 Specifying a Process

There are two ways to specify a process. The relation between flows can be defined directly by using coefficients or indirectly by using parameters.

TABLE 2.17
"External" Processes of PVC Austria

Type of Flow	Processes		
	Import	Export	Disposal
Input flows	—	exports	waste
Output flows	imports	—	—

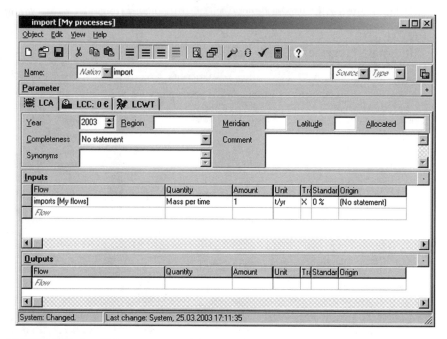

FIGURE 2.69 Specifying a process without parameters.

2.4.8.2.11.1   Direct Specification
To specify a process by direct specification:

1. Open the process by double-clicking it in the "object hierarchy."
2. For each flow, enter a value in the "Amount" field.
3. Continue until the relations of all flows are specified.
4. Click the "Save" button and close the window.

◆ The processes *import*, *export*, and *disposal* contain only one flow. Enter "1" in the "Amount" field (Figure 2.69).

2.4.8.2.11.2   Specification via Parameters
To specify a process by specification via parameters:

1. Open the process by double-clicking it in the "object hierarchy."
2. Double-click "Parameter" inside the process window.
3. Enter the name of a parameter in the first free "Parameter" field.
4. If the formula field is left empty and the parameter is directly defined by a value, the (free) parameter can later be varied locally on any plan where this "basis version" of the process is used or in the GaBi-Analyst.
5. Click in the "Alias" field to the left of the input or output flows where the parameter should be used.
6. Click the appearing drop-down arrow and select the desired parameter.

Methodology of MFA

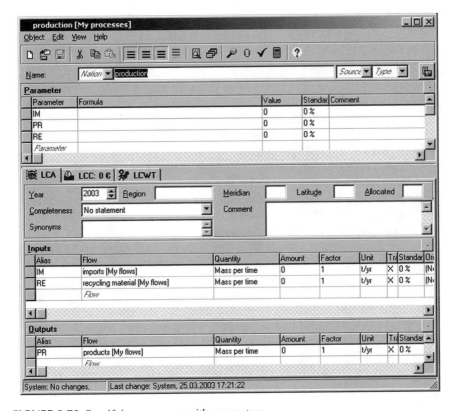

FIGURE 2.70 Specifying a process with parameters.

7. Enter a factor inside the "Factor" field. The value of the parameter stated under "Alias" multiplied by this factor yields the coefficient for the considered flow. If only the parameter to define the quantity of the flow is wanted, set the "Factor" to 1; otherwise use it as an additional multiplier. The value of flows without parameters is entered directly in the "Amount" field.
8. Continue until all flows are specified.
9. Click the "Save" button and close the window.

◆ Create parameters for the processes *production, distribution, consumption, collection,* and *stock.* For each flow of a process, define a parameter consisting of the first two letters of the flow's name (e.g., IM [imports], EX [exports], PR [products], SA [sales], IN [inventory], RE [recycling], W1 [waste I], W2 [waste II], and WA [waste]). Leave the formula field empty (the value of the parameter will be assigned later). Use the parameter in the "Alias" field of the respective flow. Enter "1" inside the "Factor" field (Figure 2.70).

The idea of preparing "0" parameters or "0" processes makes sense only if the "Plan parameter" function is used later. For "normal" processes, the desired values for the parameters would be set directly in the process.

2.4.8.2.12 Creation of Plans

"Plans" are the working places where processes are connected via their input and output flows. In this way, a model of the material flow network is created.

1. Select "Plans" or one of its subfolders in the "object hierarchy" to locate the new plan.
2. Click "New" on the toolbar.
3. Edit the name of the appearing plan by clicking on the name in the upper left corner of the "Plan" window.
4. Resize the GaBi DB manager window and the "Plan" window in a way that both windows are located side by side.
5. In the GaBi DB manager, select the process to be used within the plan. Drag it to the "Plan" window. An instance of the chosen process will appear on the plan window displayed by a rectangle.
6. Repeat this procedure until all processes needed are found in the plan.
7. Rearrange the process icons until they are in the desired positions.
8. To connect processes, click the source process. Two bars will appear inside the selected process icon representing the input (left, red) and the output (right, brown) interfaces.
9. Drag the right bar of the source process to the target process.

If the source process contains an output flow with the same name as an input flow of the target process, a connection will be established. Otherwise, the user is asked to select manually which flows should be connected. The connection is displayed by an arrow. Remember: only those flows can be connected where the "Tracked flow" field is marked by an "X" in the source and sink process! To change a process that is used in a plan, select "Details" from its context menu. This opens the basic process (the database object). By double-clicking it instead, only the local settings of the process can be edited.

◆ Create a new plan with the name "PVC Austria." Drag the processes *import*, *production*, *distribution*, *consumption*, *collection*, *stock*, *export* and *disposal* from the GaBi DB manager to the "PVC Austria" plan. Rearrange the process icons and connect them until the plan looks like Figure 2.71).

2.4.8.2.13 Visualization Options

Plans in GaBi are automatically displayed in Sankey diagrams. The widths of flows are automatically adjusted to match their quantities. Flows with the value "0" are shown as dotted lines. The displayed quantity can be changed by double-clicking the plan surface. Select the desired quantity from the list of available ones (i.e., those that have been defined before and that have been assigned to flows). The color of any flow can be changed by double-clicking the respective flow. Default colors can be assigned to any flow folder in the DB manager by right-clicking; they will be used in all plans established thereafter. The amount of the flow and also the flow

Methodology of MFA

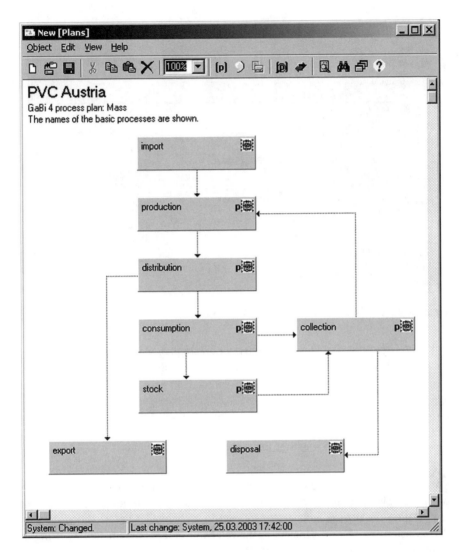

FIGURE 2.71 Creating a plan.

name and further visualization options can be selected from the plan's pull-down menu "View."

2.4.8.2.14 Defining Plan Parameters

Plan parameters are an additional modeling function. The user can create plan parameters to assign them to process parameters.

1. Click the "Parameter" button (p) in the tool bar of the "Plan" window.
2. The "Plan parameter" window opens, displaying a list of process parameters belonging to processes used in the actual plan.

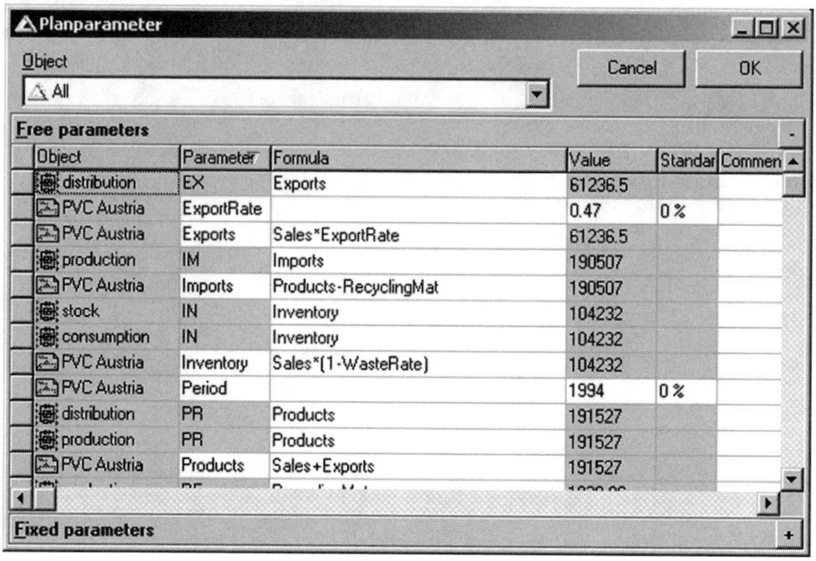
Planparameter
Object
All
Cancel
OK
Free parameters
Object
Parameter
Formula
Value
Standar
Commen
distribution EX Exports 61236.5
PVC Austria ExportRate 0.47 0 %
PVC Austria Exports Sales*ExportRate 61236.5
production IM Imports 190507
PVC Austria Imports Products-RecyclingMat 190507
stock IN Inventory 104232
consumption IN Inventory 104232
PVC Austria Inventory Sales*(1-WasteRate) 104232
PVC Austria Period 1994 0 %
distribution PR Products 191527
production PR Products 191527
PVC Austria Products Sales+Exports 191527
Fixed parameters

FIGURE 2.72 Defining and assigning plan parameters.

3. Assign values to these parameters or create plan parameters at the end of the list and use their names to support the assignment.
4. Click "OK."

Tip: define the free parameters first (those without formulas). It is then easier to insert the formulas of the depending parameters by choosing the available parameters from the context menu of the respective formula fields.

GaBi was not designed to perform dynamic simulations. Hence, time has to be one of the plan parameters that are used to calculate other time-dependent parameters. Define the plan parameters (displayed in alphabetical order) shown in Table 2.18. Assign the plan parameters to the referring process parameters by using the same name in the formula field (Figure 2.72).

2.4.8.2.15 Calculation

Calculations are automatically performed in GaBi. When the system is completely specified, the result is displayed. A warning appears if any mistake or incorrect definition has been made. In this case, check and correct the specification of processes and parameters.

On each plan, typically one process has to be fixed to get a system of linear equations that is explicitly solvable. This is done by assigning a "Scaling factor" to the process and fixing this factor.

1. Inside the "Plan" window, double-click the process that is to be fixed.

TABLE 2.18
Plan Parameters of PVC Austria

Plan Parameter	Formula	Value	Comment
Exports	ExportRate*Sales	—	Exported goods in relation to sale
ExportRate	—	0.47	Export rate
Imports	Products–RecyclingMat	—	Imported goods
Inventory	(1–WasteRate)*Sales	—	Part of sale that goes to stock
Years	Period–ZeroPeriod	—	Years after period zero
Products	Sales+Exports	—	Produced goods
RecyclingMat	(Waste1+Waste2)*RecyclingRate	—	Recycled goods
RecyclingRate	If(Period> = 1990;0,01+ (Period–1990)* 0,002;0)	—	Rate of recycling
Sales	X3*Years^3+X2*Years^2+X1*Years	—	Sold goods
UtilTime	—	22	Utilization time of products
Waste	(Waste1+Waste2)*(1–Recycling Rate)	—	Total amount of waste
WasteRate	0,5-(0,5-0,2)/(1994–ZeroPeriod)*Years	—	Waste rate
WasteRate1	0,5-(0,5-0,2)/(1994–ZeroPeriod)* (Years–UtilTime)	—	Waste rate when material entered the stock
Waste1	WasteRate*Sales	—	Part of sale that goes directly to collection after use
Waste2	if(Years>Utiltime;(1–WasteRate1)* (X3*(Years–Utiltime)^3+X2* (Years–Utiltime)^2+X1* (Years–Utiltime));0)	—	Inventory that is released out of stock after utilization time
X1	—	484.57	Coefficient 1 of purchase function
X2	—	94.271	Coefficient 2 of purchase function
X3	—	–0.8633	Coefficient 3 of purchase function
Period	—	1994	Actual period
ZeroPeriod	—	1950	Period before start of time series

2. In the appearing process window, enter "1" as the scaling factor and select the "Fixed" check box.
3. Click "OK."

◆ Fix the process *distribution* and set the "Scaling factor" to "1" (Figure 2.73).

2.4.8.2.15 Displaying the Results
To display the results:

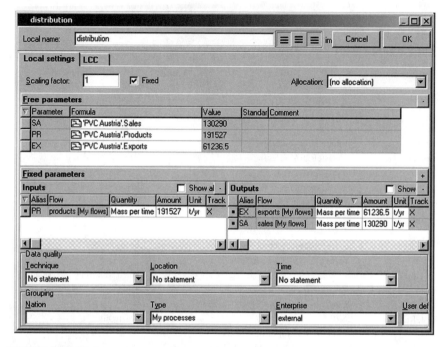

FIGURE 2.73 Fixing a process.

1. Select "Edit – Properties…" from the menu of the "Plan" window.
2. Select "Flow" in the "Quantity" drop-down list.
3. Click "OK."
4. Select "View → Show flow quantities" from the menu of the "Plan" window.

2.4.8.2.16 Changing Parameters

To change parameters:

1. Click the "Parameter" button (p) in the tool bar of the "Plan" window.
2. The "Plan parameter" window opens (Figure 2.72).
3. Change the values of the desired parameters.
4. Click "OK."
5. The Sankey diagram in the "Plan" window will be automatically updated.

◆ To show the results of the system for the year 1994, set the plan parameter "Year" (1950 ≤ Year ≤ 1994) to "1994." Display the results (Figure 2.74).

2.4.8.2.17 Creating a Balance

To create a balance:

1. Click "Balance calculation…" on the toolbar of the "Plan" window.
2. The "Balances" window opens.

Methodology of MFA

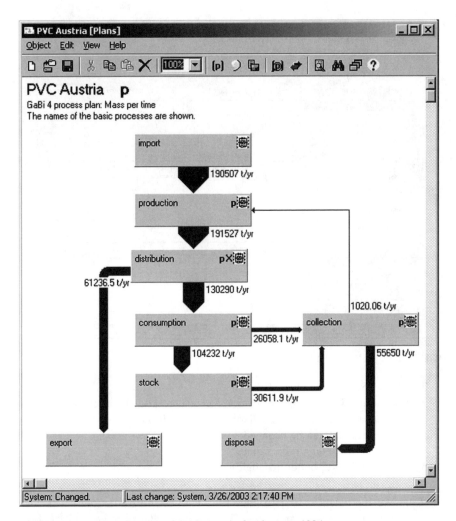

FIGURE 2.74 Sankey diagram of PVC Austria for the year 1994.

3. Clear the "In/out aggregation" check box.
4. Select "Flow" in the "Quantity" pull-down menu.
5. Select "All" in the "Rows" and in the "Columns" pull-down menu.
6. Choose between different forms of data display: absolute values, relative contributions, columns relative, or rows relative.

◆ Create a balance for the displayed Sankey diagram (Figure 2.75).

2.4.8.2.18 Comparing Balances

To compare balances of different scenarios:

1. Click the "GaBi Analyst" button (showing two graphs) in the tool bar of the "Balances" window.

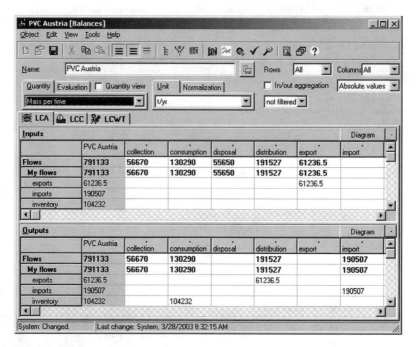

FIGURE 2.75 Balance of PVC Austria for the year 1994.

2. Select "Scenario analysis."
3. Right-click the "Parameter" section and select "Add all parameters" from the appearing context menu. Two scenarios will be added displaying all parameters that can be changed.
4. Edit the parameters to be varied.
5. Click the "Start" button (triangle).

◆ Compare the balances of 1970 and 1994 with regard to the internal flows of *products*, *recycling material*, and *sales*. The only parameter to be varied is "Period." Enter "1970" for Scenario 1 and "1994" for Scenario 2 (Figure 2.76).

- Every line of the "Result values" (Figure 2.77) represents a flow that will be displayed in the diagram below.
- To edit a line, click on the field that is to be changed (balance table, balance column, balance row, quantity, or unit) and select the desired parameter from the pull-down menu. Proceed from left to right.
- To delete a line that is not needed, click the square (grey) on the left side of the line. Press the "Del" key.
- The diagram can be edited in various ways. Choose the desired options.

◆ Set the parameters shown in Figure 2.77 (under "Result values") to display the flows *products*, *recycling material*, and *sales* within the chart. Delete the lines that are not needed.

Methodology of MFA

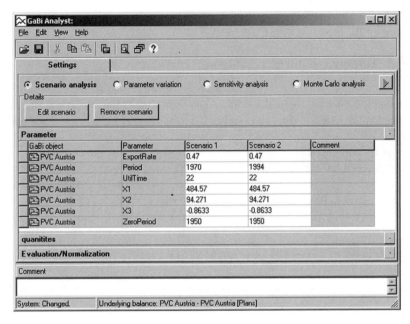

FIGURE 2.76 "GaBi Analyst" window showing the parameter settings for different scenarios.

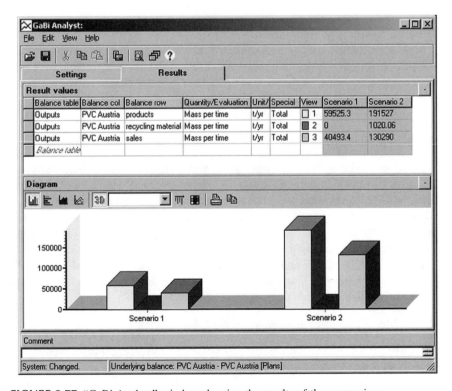

FIGURE 2.77 "GaBi Analyst" window showing the results of the comparison.

2.4.8.2.19 Finishing the Session
To close GaBi, close the GaBi DB manager.

2.4.8.3 Potential Problems

In GaBi, it is not possible to calculate stocks automatically by balancing a process. Thus, the amount of material stored in *stock* can only be estimated roughly by the integration of {function (*inventory*) – function (*waste II*)}. A better solution is to manually add the values of (*inventory – waste II*) beginning from 1950 to the period of interest. This is achieved by copying and pasting the appropriate values from the GaBi Analyst to a spreadsheet program (e.g., Excel) and performing the necessary calculations in that application. The analyst settings and results can be saved and reopened later. Any analyst setting is related to the underlying balance table, which has to be saved, too.

When creating plan parameters, circular references must be avoided.

2.4.9 COMPARISON

All software products have been tested on a personal computer operating under Microsoft Windows 2000. The software products available by the end of 2002 were Microsoft Excel 2002®, Umberto 4.0, and GaBi 4.

2.4.9.1 Trial Versions

GaBi is available as a 90-day trial version with full functionality, free of charge, and can be downloaded from the GaBi Web site (www.gabi-software.com).

An Umberto demo version with restricted functionality can be downloaded from the Umberto Web site (www.umberto.de/english/). A 30-day trial version with full functionality is available for the price of €300. Half the price (€150) will be refunded if the program is sent back completely and in time.*

Microsoft does not offer a trial version of Excel.

2.4.9.2 Manuals and Support

It must be stressed that Excel is a spreadsheet and analysis program, while GaBi and Umberto are LCA/LCE tools (life-cycle assessment/life-cycle engineering). Because of that, only GaBi and Umberto provide instructions on how to deal with material flows in their manuals and on-line help. The GaBi and Umberto manuals provide detailed descriptions of how to construct and calculate a material flow system. In Excel, the user is challenged to employ the software's capacity to perform an MFA on his own.

The support for GaBi and Umberto is excellent and was tested via email, telephone, and in person. Excel has its advantages in the huge amount of printed information available as hard copy or on the World Wide Web.

2.4.9.3 Modeling and Performance

While testing the programs with the case study (fewer than 10 processes), no stability problems were encountered.

---

* If a full version is ordered later, the €300 will be deducted from the full price.

Methodology of MFA

In contrast to GaBi, the calculations in Umberto must be started manually and require more time due to the different calculation algorithms (Petri nets). The calculation speed of Excel depends on the capability of its user.

Umberto offers the possibility of implementing nonlinear processes. In Umberto, it is also possible to use script languages (Python®, VBScript®, JavaScript, PerlScript®), for example, to specify transitions. The objects of Umberto (transitions, places, flows, etc.) can be addressed and altered via program code from other applications. Data can be imported from databases by using SQL (structured query language) statements. This programmability is missing in GaBi 4, where at the moment it is only possible to transfer data via copy and paste. Excel is fully programmable by using its internal VBA programming language.

While it is easy to change or enlarge existing systems in GaBi and Umberto, this can be a laborious task in Excel because the system is not based on objects that can be inserted and moved.

Using these applications, it is possible to perform static simulations. While only Umberto was designed to perform dynamic simulations, it is also possible but difficult to do this in GaBi. In Excel, this feature can be realized by using VBA.

In GaBi it is possible to consider data uncertainties and their propagation within the system. Sensitivity analysis can be performed as well as Monte Carlo simulations. Scenarios can be compared and parameters varied to show the course of effect. In Umberto, uncertainty is only considered qualitatively in the form of data quality. Still, it is possible to implement error propagation by (1) introducing the uncertainty of flows as additional materials and (2) adding the manually calculated rules of error propagation (using Gauss's law) into the specification of transitions. In Excel, these features can be realized by using VBA.

GaBi and Umberto offer huge LCI databases (life cycle inventory) with pre-defined processes or transitions. If there is no need to perform real LCAs, the versions with fewer LCI datasets (GaBi lean/edu/academy and Umberto business/educ) are sufficient and hence recommended.

While GaBi is closer to MFA terminology/methodology than Umberto, only Excel can be trimmed by users (programming experience is of advantage) to fit the requirements of MFA.

Because none of the tested software products was developed specifically for MFA, none of them is the perfect choice. GaBi and Umberto can be used to perform MFA, but they are much better suited for LCA. Excel is the most flexible tool of all and it is a good choice if students want to get a first impression of MFA.

Table 2.19 is a summary of the comparison of the three software products. For more detailed information regarding functions and prices, refer to the home pages of the various software products.

2.5 EVALUATION METHODS FOR MFA RESULTS

The results of an MFA are quantities of flows and stocks of materials for the system of study. Aside from analytical and numerical uncertainties, these are *objective* quantities derived from analyses, measurements, and the principle of mass conservation. On the assumption that the study has been carried out in detail, carefully,

TABLE 2.19
Comparison of the Suitability of Excel, GaBi, and Umberto for MFA

	Excel	GaBi 4	Umberto 4
Installation guide	+	+	+
User manual	+	+	+
On-line help	+	+	+
Operating system	Microsoft Windows, Mac OS	Microsoft Windows	Microsoft Windows
Languages	several	German/English/Japanese	German/English/Japanese
User friendliness	+	+	+
Support	+	+	+
Stability	+	+	+
Trial version	—	+ (free/90 days)	+ (€300/30 days)
Speed	+/–	+	+/–
Programmability	+	—	+
Data import/export	+	+/–	+
Static simulation	+	+	+
Dynamic simulation	+	+/–	+
Uncertainties	+/–	+	+/–
Sensitivity analysis	+/–	+	—
Monte Carlo error simulation	+/–	+	—
MFA terminology/methodology	+	+/–	+/–

Note: + = good, +/– = average, — = not available.

and comprehensibly, there is usually little or no discussions about the numerical results. On the other hand, and in contrast to the measurement of mass flows and concentrations, the interpretation and evaluation of MFA results is a *subjective* process, too: it is based on social, moral, and political values. For instance, a depletion time for a reservoir of a nonrenewable resource of, say, 50 years may be considered as sufficiently long or alarmingly short, depending on the person one may consult. Another example: the "eco-indicator95" evaluation method attributes the same weight to the death of one human being per million and to the damage of 5% of an ecosystem.[33] It is obvious that such valuations and weightings cannot be based exclusively on scientific/technical principles. Social and ethical aspects play an important role, too. Assessment is a matter of values, and values can change over time and may vary among societies and cultures. Hence, assessment is and will remain a dynamic process that must be considered as a result of the according era.

Another problem when dealing with MFA results arises when alternative scenarios for a single system or from different systems are to be compared. A common situation is the following: consider a system with 5 processes and 20 flows. As the result of an optimization step, 10 flows and 3 processes are changed. Some flows may have become "better;" some have become "worse." How much has the system improved, relative and/or absolute? A measure and a scale are needed.

Methodology of MFA

To bring objectivity and comparability into the evaluation process, so-called indicators need to be applied. The Organization for Economic Cooperation and Development (OECD) provides the following definition for an (environmental) indicator: a parameter or a value derived from parameters that provides information about a phenomenon. The indicator has significance that extends beyond the properties directly associated with the parameter value. Indicators possess a synthetic meaning and are developed for a specific purpose.[34] According to the World Resource Institute, indicators have two defining characteristics: (1) They quantify information so that its significance is more readily apparent. (2) They simplify information about complex phenomena to improve communication.[35] In a wide sense, an indicator can be considered as a metric that provides condensed information about the state of a system. When applied to time series, the information is about the development of a system. Indicators should convey information that is meaningful to decision makers, and it should be in a form that they and the public find readily understandable. This implies policy relevance and, in cases of complex systems, a certain degree of aggregation. The more an indicator is based on appropriate scientific laws and principles, the more it can be considered as objective. However, as will be shown in Section 2.5.1, none of the available evaluation methods fully satisfies these demands at present.

Generally, to be comprehensive, assessment should consider *resource* as well as *human* and *ecotoxicological* aspects. From this requirement, it can already be concluded that "the one and only" indicator may not exist. As a result of continuously increasing knowledge, the rating of nonrenewable resources and the toxicity of substances are constantly being revised. The same is true for the weighting between the importance of resources and toxicity. It is certain that 1 kg of zinc and 1 kg of dioxin will be rated differently in 50 years compared with today. Hence, it is clear that indicators cannot relieve decision making in environmental, resource, and waste management from all types of subjectivity.

2.5.1 EVALUATION METHODS

Selected evaluation methods are briefly described and discussed in this section. The selection is based on the potential for application to MFA results. These methods are based on different ideas, philosophies, or concepts, and therefore each has certain advantages and shortcomings. In most cases, none of them can be considered complete and sufficient for a comprehensive assessment. On the other hand, most of the introduced methods are constantly undergoing further development regarding standardization, reliability, and completeness. The choice to apply a certain method is usually determined by the kind of specific problem to be investigated. Another motive may be the preference of the client and the performer of the study. In cases where the results of the evaluation process give reason for doubts, the application of another, complementary method is advisable. Generally, the MFA results themselves are the best starting point to analyze and evaluate a system. This requires some practice and experience and will be demonstrated in the case studies presented in Chapter 3.

2.5.1.1 Material-Intensity per Service-Unit

The concept of material intensity per service unit (MIPS) was developed by Schmidt-Bleek[36–39] and colleagues from the Wuppertal Institute, Germany. MIPS measures the total mass flow of materials caused by production, consumption (e.g., maintenance), and waste deposal/recycling of a defined service unit or product. Examples for a service unit are: a haircut, the washing cycle of a dishwasher, a person-kilometer, the fabrication of a kitchen, a power pole.[38,40,41] The total mass flow for a service unit can consist of overburden, minerals, ores, fossil fuels, water, air, biomass; i.e., MIPS employs a life-cycle perspective and also considers the "hidden" flows of a service unit. This "ecological rucksack" comprises that part of the material input that is not incorporated within the products or materials directly associated with the service unit. The material intensity of 1 t of copper from primary production is 350 t of abiotic materials, 365 t of water, and 1.6 t of air. Other examples are the 3000 t of soil that have to be moved to produce 1 kg gold in the U.S.[36] or the 14 t of processed materials that are necessary to produce a personal computer.[39] Table 2.20 gives examples of the material intensity for various materials and products.

TABLE 2.20
Material Intensity for Materials and Products Compiled Using the MIPS Concept

	Abiotic Materials, t/t	Biotic Materials, t/t	Water, t/t	Air, t/t	Soil, t/t	Electricity, kWh/t
Aluminum	85	0	1380	9.8	0	16,300
Pig iron	5.6	0	22	1	0	190
Steel (mix)	6.4	0	47	1.2	0	480
Copper	500	0	260	2	0	3000
Diamonds [a]	5,300,000	0	0	0	0	n.d.[b]
Brown coal	9.7	0	9.3	0.02	0	39
Hard coal	2.4	0	9.1	0.05	0	80
Concrete	1.3	0	3.4	0.04	0.02	24
Cement (Portland)	3.22	0	17	0.33	0	170
Plateglass	2.9	0	12	0.74	0.13	86
Wood (spruce)	0.68	4.7	9.4	0.16	0	109
Paper clip	0.008	0	0.06	0.002	n.d.[b]	n.d.[b]
Shirt	1.6	0.6	400	0.06	n.d.[b]	n.d.[b]
Jeans	5.1	1.6	1200	0.15	n.d.[b]	n.d.[b]
Toilet paper	0.3	0	3	0.13	n.d.[b]	n.d.[b]
Tooth brush	0.12	0	1.5	0.028	n.d.[b]	n.d.[b]

Note: Updated data may soon be available at www.mips-online.info.

[a] Overburden and mining.
[b] n.d. = not determined.

Source: From Schmidt-Bleek, F., *Das MIPS-Konzept*, Droemer Knaur, Munich, 1998.

MIPS only considers input flows to avoid double counting, since input equals output. Also, there are fewer inputs than outputs for the industrial economy, which facilitates accounting. In order to get more structure into the approach, Schmidt-Bleek and colleagues suggest grouping inputs into five categories, namely, biotic and abiotic materials, Earth movements, water, and air.[37,39] Energy demands for the supply of the service unit are also accounted for on a mass basis. In later works, electricity and fuels are listed as a sixth category to provide further information (see Table 2.20). Another characteristic is that MIPS does not discriminate among different materials. The indicator assigns the same relevance to 1 kg of gravel and 1 kg of plutonium. A rationale for this assumption is given in Schmidt-Bleek[42] and Hinterberger et al.[37] It is mainly based on the insight that it is difficult to determine the ecotoxicity of a given substance due to unknown long-term impacts and unknown synergistic and antagonistic effects of substances. Hence, it is impossible to determine the ecotoxicity of 100,000 or more chemicals. A detailed discussion of this rationale is provided by Cleveland and Ruth.[43]

MIPS plays an important role in the discussion about dematerialization. Since one-fifth of the world's population consumes some 80% of the resources, the developed economies have to cut down their turnover of materials and energy if equality for all societies and countries is a goal and if less-developed countries are to have similar chances to prosper. Discussions about dematerialization are also known as ecoefficiency and the factor-X debate. Results suggest that reduction factors from 4 to 50 are needed.[44] Applied to the material flows of large economies as well as to single services and products, the MIPS concept is regarded as a useful tool for monitoring progress in dematerialization. MIPS in MFA can be applied to mass balances at the level of goods. The input into a system can be aggregated according the above-mentioned rules. In most cases, it will be possible to derive a reasonable service unit from the system investigated.

2.5.1.2 Sustainable Process Index

The sustainable process index (SPI) was developed by Narodoslawsky and Krotscheck[45,46] at the Graz University of Technology, Austria. The basic concept of the SPI is to calculate the area that is necessary to embed a process or service into the biosphere under the constraint of sustainability. The idea is that all mass and energy flows that the process extracts or emits can be translated into area quantities by a precisely defined procedure and, ultimately, aggregated to a final value (A_{tot}). The lower the A_{tot} for a given process, the lesser is the impact on the environment. The rationale for using area as the normative value is that, in a sustainable economy, the only real input that can be utilized over an indefinite period of time is solar energy. The utilization of solar energy is bound to the surface of the Earth. Furthermore, area can be considered as the limiting resource for supply and disposal in a world of growing population.

The SPI concept considers the consumption of raw materials (A_R); energy (A_E); the requirements for infrastructure (again mass and energy) and the area to set up the infrastructure for the process (A_I); as well as the necessary area to assimilate the products, wastes, and emissions of the process (A_P). In cases where labor-intensive processes are investigated, an area for the staff can also be allocated (A_{ST}).

$$A_{tot} = A_R + A_E + A_I + A_P + (A_{ST}) \qquad (2.12)$$

Three different types of raw materials are distinguished: renewable raw materials, fossil raw materials, and mineral raw materials. A_{RR} is the area for renewable raw materials and is given as

$$A_{RR} = \frac{F_R \cdot (1+f_R)}{Y_R} \qquad (2.13)$$

F_R is the consumed flow (mass per unit of time, normally one year) of the considered raw material. The factor f_R takes into account how much "grey" area has been exerted downstream to provide F_R. The factor f_R is sometimes designated as cumulative expenditure or "rucksack" (see Section 2.5.1.1). Y_R is the yield for the renewable material given as mass per area and time.

The area for fossil raw materials (A_{FR}) is derived from a formally identical equation as A_{RR}. F_F (instead of F_R) is the flow of fossil raw materials into the process, and f_F considers the area of the "rucksack" (e.g., energy expenditure for refining and transporting fossil fuels). Y_F stands for the "yield" of sedimentation of carbon in the oceans (ca. 0.002 kg/m²/year). The rationale for the "sedimentation yield" is that as long as no more carbon is emitted as can be fixed by oceans, the global carbon cycle is not changed relevantly and sustainability is guaranteed.

A_{MR}, the area for minerals, is defined by the following equation:

$$A_{MR} = \frac{F_M \cdot e_D}{Y_E} \qquad (2.14)$$

F_M is the flow of mineral raw material consumed by the process. The energy demand to provide one mass unit of the considered mineral is e_D. Y_E, the yield for industrial energy, is dependent on the mix of energy-transformation technologies in a country (e.g., hydropower or fossil or nuclear sources). For a sustainable energy system, Y_E is approximately 0.16 kWh/(m²·year). A_E, the area for electricity consumption, is given by

$$A_E = \frac{F_E}{Y_E} \qquad (2.15)$$

with F_E representing the electricity demand in kWh/year.

The infrastructure area A_I often contributes only a small part to A_{tot}. Hence a rough assessment usually is sufficient. In contrast, the process-dissipation area A_P is usually decisive. The SPI concept assumes a renewal rate for the assimilation capacity of any environmental compartment. Sustainable assimilation occurs when

Methodology of MFA

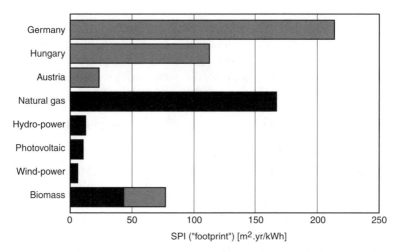

FIGURE 2.78 Sustainable process index (SPI) of the energy supply of selected national economies and various energy systems.[48] Differences between Germany, Hungary, and Austria are due to the mix of hydropower, fossil fuel, and nuclear energy that make up the national energy supply. The range for biomass stems from different technologies such as pyrolysis, gasification, and combustion. (From Krotscheck et al., *Biomass Bioenergy*, 18, 341, 2000. With permission.)

the emissions of the process are outweighed by the renewal rate and the elemental composition of the compartment is not changed. Note the similarity to the A/G approach (Section 2.5.1.7) in this point. The following equation calculates A_P:

$$A_{Pci} = \frac{F_{Pi}}{R_c \cdot c_{ci}} \tag{2.16}$$

where F_{Pi} denotes the mass per year of substance i in product/emission flow P (e.g., kg Cd/year). R_c stands for the renewal rate of compartment c in mass per area and year (e.g., kg soil/m²/year), and c_{ci} is the natural (geogenic) concentration of substance i in compartment c (e.g., kg Cd/kg soil).

The final step is to relate A_{tot} to the product or service provided by the process. The SPI has been applied to various processes such as transport, to aluminum and steel production,[47] to pulp and paper production, to energy from biomass,[48] as well as to entire regional economies.[49,50] Figure 2.78 gives a qualitative example of SPIs for energy-supply systems.

The SPI can by applied to any MFA result. The consumed area ("footprint") of the investigated system can be compiled if data about the various yield factors and other nonspecific MFA data such as energy demand are available. The advantage of the SPI is that resource consumption is considered in a more differentiated way than is done by MIPS. Emissions and wastes are also included in the assessment. Determination of the SPI can be demanding and labor-intensive, but the indicator can be regarded as one of the most universal, holistic, and comprehensive metrics.

2.5.1.3 Life-Cycle Assessment

Life-cycle assessment (LCA) is a tool that was developed during the 1980s and 1990s in Europe and the U.S. The Society of Environmental Toxicology and Chemistry (SETAC) soon served as an umbrella organization with the aim of further developing LCA and standardizing and harmonizing procedures.[51,52] These activities finally led to the development of a series of ISO LCA standards (the 14040 series of the International Organization for Standardization). Accordingly, LCA consists of four steps:[53]

1. "Goal and scope definition," where the goal of the study is formulated; the scope is defined in terms of temporal, spatial, and technological coverage; and the level of sophistication in relation to the goals is fixed. Additionally, the product(s) of study are described and the functional unit is determined.
2. The "inventory analysis," which results in a table that lists inputs from and outputs to the environment ("environmental interventions") associated with the functional unit. This requires the setting of system boundaries, selection of processes, collection of data, and performing allocation steps for multifunctional processes (e.g., a power plant producing energy not only for a single product).
3. The "impact assessment," during which the inventory table is further processed and interpreted in terms of environmental impacts and societal preferences. This means that impact categories such as depletion of resources, climate change, human- and ecotoxicity, noise, etc. have to be selected. "Classification" designates the step where the entries of the inventory table are qualitatively assigned to the preselected impact categories. In the "characterization" step, the environmental interventions are quantified in terms of a common unit for that category (e.g., kg CO_2 equivalents for climate change), allowing aggregation into a single score for that category: the category indicator result. Additional and optional steps are "normalization" and "weighting" of impact categories that lead to a single final score.
4. The "interpretation" of the results, which comprises an evaluation in terms of soundness, robustness, consistency, completeness, etc., as well as the formulation of conclusions and recommendations.

LCAs have been carried out for a multitude of goods ranging from batteries,[54] PET bottles,[55] paper,[56] tomato ketchup,[57] catalytic converters for passenger cars,[58] fuel products,[59] different floor coverings,[60] a rock crusher,[61] to steel bridges.[62] One of the first LCAs was for packaging materials.[63] Fewer studies have been undertaken on the process[64–66] and system level.[67–72] Despite well-defined rules and recipes on how to execute an LCA, studies are sometimes disputed. Most objections concern data consistency and the reliability of the impact assessment.[73] Ayres provides a concise discussion about potential problems concerning LCA. He concludes that often studies are too focused on the impact assessment, and analysis and control of

Methodology of MFA 141

basis data are neglected.[74] Since LCAs can be labor-intensive and therefore costly, Graedel suggests an approach on how to streamline LCA in order to make it more attractive to companies.[75]

MFA can be regarded as a method to establish the inventory for an LCA. This is especially true when LCA is applied to systems rather than to single goods. Hence, the impact assessment of LCA can be applied to MFA results. A certain discrepancy will be that LCA strives for assessing as many as possible substances and compounds to guarantee completeness while MFA is directed towards reducing the number of substances of study as much as possible to maintain transparency and manageability.

2.5.1.4 Swiss Ecopoints

The Swiss ecopoints (SEP) approach belongs to the family of impact-assessment methodologies in LCA such as the SETAC method,[76] the CML method,[77] or the eco-indicators by Goedkoop.[33] SEP is based on the idea of critical pollution loads, an idea first published by Müller-Wenk in 1978.[78] Later it was further developed and concretized by members of the service sector, the industry, the administration, and the academia in Switzerland.[79–81] The SEP score for an environmental stressor (emissions to air, water, and soil) is calculated using the following formula:

$$SEP_i = F_i \cdot \frac{1}{F_{crit}} \cdot \frac{F_{Sys}}{F_{crit}} \cdot 10^{12} \qquad (2.17)$$

F_{crit} stands for the critical (in the meaning of maximal acceptable) flow of the stressor in a defined region (e.g., Switzerland). F_{sys} is the actual flow of the stressor within the same region. The first ratio in Equation 2.17 normalizes the stressor flow F_i of the source of study and determines its importance. The second ratio weights the stressor with regard to its importance for the region. The prominent introduction of F_{crit} — it appears two times as a reference factor — also brings in social and political aspects. This is the main difference of the SEP approach compared with other toxicity- and effects-based impact assessment methodologies. F_{crit} can be fixed differently from region to region with regard to time, condition of the environment, technological standards, economic development, etc. (e.g., see the different reduction targets for countries in the Kyoto protocol) and stands for an environmental-quality goal. Scores of different stressors F_i can be added up to a final score. The higher the score, the higher the environmental burden of the investigated product or process for the system.

Besides emissions to the environmental compartments, the SEP approach also considers the quantity of waste produced on a mass basis and the consumption, energetically and as feedstock, of scarce energy resources (mainly fossil fuels, uranium, potential energy). Other resource consumption is not considered. The rationale is that minerals are not scarce, since matter does not vanish. However, the availability of minerals can diminish (e.g., through declining ore grades), which results in increasing environmental impacts when such resources are mined and processed. Those impacts (emissions) are considered in the impact analysis.

TABLE 2.21
Impact Assessment of Selected Air Pollutants by SEP, by HTP as an Impact Category in LCA, and by Exergy

	Swiss Ecopoints (SEP),[134] SEP/g	Human Toxicity Potential (HTP)[135] kg 1,4-DCB eq./kg [a]	Exergy,[89,136] kJ/g
Particulates	60.5	0.82	7.9
C_6H_6	32	1,900	42.3
NH_3	63	0.1	19.8
HCl	47	0.5	2.3
HF	85	2,900	4.0
H_2S	50	0.22	23.8
SO_2	53	0.096	4.9
NO_2	67	1.2	1.2
Pb	2,900	470	—
Cd	120,000	150,000	—
Hg	120,000	6,000	—
Zn	520	100	—

Note: Different results may be obtained, depending on which assessment method is applied.

[a] 1,4-Dichlorobenzene equivalent.

Hertwich[82] and colleagues give an illustrative example for a discrepancy between the SEP approach and the established rating method for greenhouse gases: The ecopoints calculated for the U.S. for 1 kg of CO_2, CH_4, and N_2O are 1.14, 59,700, and 3890, respectively. The global warming potentials (GWP) for a time span of 20 years are, according to the IPCC, 1, 63, and 270, respectively. The authors impute the difference to the equal and linear valuation of the different stressors. Ahbe et al.[80] discuss the pros and cons of other nonlinear valuation functions, such as logistic and parabolic functions for ecopoints.

The SEP method has been applied to a multitude of problems ranging from packaging[83] to MSW incineration.[84] Table 2.21 gives SEP scores for selected air pollutants and compares the results with the human toxicity potential (as used in LCA) and with the exergy concept, which is described in Section 2.5.1.5.

2.5.1.5 Exergy

Exergy is a measure of the maximum amount of work that can theoretically be obtained by bringing a resource (energy or material) into equilibrium with its surroundings through a reversible process (i.e., a process working without losses such as friction, waste heat, etc.). The surroundings, the reference environment, or simply the reference state must be specified, i.e., temperature and pressure. In cases in which materials are considered, the chemical composition must be known, too. For material flow studies, the environment usually consists of the atmosphere, the ocean, and the

Methodology of MFA

Earth's crust, as suggested by Szargut et al.[85] The term *exergy* for "technical working capacity" was coined by Rant in 1956.[86] Other terms have been used synonymously, such as *available work*, *availability*, and *essergy* (essence of energy). Exergy is an extensive property* and has the same unit as energy (e.g., J/g). Unlike energy, there is no conservation law for exergy. Rather, exergy is consumed or destroyed, due to irreversibility in any real process.

Consider the following example: approximately 3.1 kWh of electricity (P) is required at minimum to heat water from 10°C to 37°C for a 100-l bathtub. This is calculated using the first law of thermodynamics, i.e., $P = Q_W + L$, with loss (L) assumed to be ≈ 0. The energy content (Q_W) of the water (m = 100 kg) in the tub is $Q_W = m \cdot c_p \cdot (T_W - T_0) = 11,300$ kJ ≈ 3.1 kWh (specific heat capacity of water: c_p = 4.18 kJ/kg/K; T_W = 310.15 K; reference temperature: T_0 = 283.15 K). While the exergy content of electricity is 100% ($P = E_1$), which means that electricity can be transformed into all other kinds of energy (heat, mechanical work, etc.), the exergy content of the water in the tub is $E_2 = Q_W[1 - (T_0/T_W)] = 0.27$ kWh. The difference between E_1 and E_2 ($\approx 91\%$ of E_1) is the exergy loss of the process "water heating." E_2 is the maximum amount of energy that could be transformed back into work (e.g., again electricity) from the water. E_1 and E_2 can be regarded as measures that quantify the "usefulness" or "availability" of 3.1 kWh of electricity or warm water. First-law efficiency of the process is $\eta_I = Q_W/P = 100\%$, second-law efficiency is $\eta_{II} = E_2/E_1 = 8.7\%$.

The exergy content of a solid material can be compiled from standard chemical exergy values $e^0_{ch,j}$ as introduced by Szargut et al.[85] The $e^0_{ch,j}$ values are substance-specific values (j) that are calculated for the standard state (T_0, p_0) and related to the mean concentration of the reference species of substance j in the environment. The assumption is that there is only one reference species for each element. Consider the example Fe_3O_4. The reference species of Fe is assumed to be Fe_2O_3 in the Earth's crust. The standard chemical exergy of reference species in the Earth's crust is calculated using Equation 2.18,

$$e^0_{ch,j} = -R \cdot T_0 \cdot \ln x_j \qquad (2.18)$$

with x_j being the average mole fraction of the reference species in the Earth's crust ($x_{Fe_2O_3} = 1.3 \cdot 10^{-3}$); $e^0_{ch,Fe_2O_3} = -8.31 \cdot 298.15 \cdot \ln 1.3 \cdot 10^{-3} = 16.5$ kJ/mol = 0.1033 kJ/g. The reference species for O_2 is O_2 in the atmosphere and is calculated using Equation 2.19,

$$e^0_{ch,j} = R \cdot T \cdot \ln \frac{p_0}{p_{j,0}} \qquad (2.19)$$

with p_0 = 101.325 kPa (mean atmospheric pressure) and $p_{j,0}$ = 20.4 kPa (partial pressure of O_2 in the reference state); e^0_{ch,O_2} = 3.97 kJ/mol.

---

* An extensive property is dependent on the size (mass, volume) of the system. Intensive properties are, e.g., temperature, pressure, and chemical potentials.

The standard chemical exergy of Fe can now be calculated from

$$2Fe + 3/2\ O_2 \rightarrow Fe_2O_3$$

and

$$e_{ch}^0 = \Delta G_f^0 + \sum_{el} n_{el} \cdot e_{ch,el}^0 \qquad (2.20)$$

where e_{ch}^0 is the standard chemical exergy of the target compound (e.g., Fe_2O_3), ΔG_f^0 is Gibbs's energy of formation (e.g., tabulated in Barin,[87] $\Delta G_{f,Fe_2O_3}^0 = -742.294$ kJ/mol), n_{el} is the number of moles of the elements in the target compound, and $e_{ch,el}^0$ are the standard chemical exergy values of the elements. Equation 2.20 yields

$$16.5 = -742.294 + 2 \cdot e_{ch,Fe}^0 + 3/2 \cdot 3.97$$

and

$$e_{ch,Fe}^0 = 376.4\ kJ/mol$$

The standard chemical exergy of Fe_3O_4 is now (using again Equation 2.20, $\Delta G_{f,Fe_3O_4}^0 = -1015.227$ kJ/mol)

$$3Fe + 2O_2 \rightarrow Fe_3O_4$$

$$e_{ch,Fe_3O_4}^0 = -1015.227 + (3 \times 376.4) + (2 \times 3.97) = 121.9\ kJ/mol = 0.5265\ kJ/g$$

If a hypothetical type of iron ore consists of 60% Fe_2O_3, 30% Fe_3O_4, and 10% other minerals (having a standard chemical exergy of ≈ 2 kJ/g), then the standard chemical exergy of the iron ore is

$$e_{ch,ironore}^0 = (0.6 \times 0.1033) + (0.3 \times 0.5265) + (0.1 \times 2) = 0.42\ kJ/g$$

Applying the standard chemical exergy values, which are tabulated for many common substances,[85,88] and having information about the chemical composition of materials, the exergy of materials can be calculated, and exergy balances for combined materials/energy systems can be established.

Initially, the exergy concept was applied to energetic systems such as heat and turbo engines in order to understand which processes cause major losses (e.g., cooling, throttling) and to learn how to improve energy efficiency. Since exergy can be calculated theoretically for all materials and energy flows, it can be applied to any materials balance. Hence, it is a useful tool for resource accounting because it aggregates materials and energy to one final exergy quantity. For instance, the production and fabrication industry can be described as a system that uses exergy in the form of fossil fuels and raw materials to produce consumer goods and wastes of lower exergy. Moreover, the technical efficiency of any system can be expressed as exergy efficiency. Studies for single branches of industry, as well as for entire national economies, have been carried out.[89–94]

In addition, exergy is considered to be a useful indicator for environmental impacts of emissions and wastes.[95,96] The rationale for this assertion is as follows: the higher the exergy of a material or energy flow, the more the flow deviates from the thermodynamical and chemical state of the environment, and the higher the potential to cause environmental harm. On the other hand, the correlation between exergy and environmental impact is not very strong. For example, the exergy values of substances emitted to the atmosphere are not proportional to their toxicity.[97] The exergy value for PCDD/F (dioxins and furans) is ca. 13.0 kJ/g, and for carbon monoxide it is 9.8 kJ/g.[89] Hence, exergy of both substances ranges in the same order of magnitude. Yet the emission limits for MSW incinerators within the European Union are 0.1 ng/m^3 for PCDD/F and 50 mg/m^3 for CO,[98] a difference of eight orders of magnitude.

Another characteristic of the exergy concept is that exergy balances are often dominated by energy flows, and materials (e.g., wastes, emissions) seem to play a minor role. Consider the following example. The emission of 1 kg PCDD/F corresponds to an exergy value of 13 MJ. (For comparison, the estimated total dioxin and furan emissions for Germany in 1990 were between 70 and 950 g TEQ*/year.[99]) This is equivalent to the release of some 500 l of warm water (Q_W = [4.18 × 500 × (55 − 10)]/1000 = 94 MJ; $E_W = Q_W[1 − (T_0/T_W)]$ = 13 MJ). Such examples show that the exergy concept must be carefully considered when applied to materials as well as combined materials and energy systems. Detailed information about the theory of exergy and application in resource accounting can be found in Wall,[100] Baehr,[101] Ayres and Ayres,[88] and Szargut et al.[85]

2.5.1.6 Cost–Benefit Analysis

The concept of cost–benefit analysis (CBA)** dates back more than 150 years to the work of J. Dupuit, who was concerned with the benefits and costs of constructing a bridge.[102] Since then, the concept of CBA has been constantly refined and focused. In the late 1950s, an extensive literature on the foundations of CBA emerged. Most of the published works focused on how to assess the net economic value of public works projects. Of special interest were water-resource developments that withdrew productive factor inputs such as land, labor, capital, and materials from the economy to produce tangible outputs such as water, hydroelectric power, and transportation (Johansson,[102] Hanley and Spash[103]).

---

* TEQ: Toxic Equivalent: PCDD/F occur as a mixture of different individual compounds (cogeners) which have different degrees of toxicity. The emission of each cogener is multiplied by a weighting factor (referred to as a Toxic Equivalent Factor (TEF)). The weighted values are then added together to give the TEQ of the mixture.
** Sometimes referred to as "benefit–cost analysis."

CBA has its roots in welfare economics, in the theory of public goods, and in microeconomic investment appraisal.[104] Generally, CBA is a tool to determine quantitatively the total advantages (benefits) and disadvantages (costs) of alternative projects or measures. The goal is to determine whether and how much a public project can contribute to national economic welfare, which of several options should be selected for action, and when the investment is to be executed. Benefits and costs are quantified in monetary units (e.g., $, €) and can therefore be balanced against each other. This is the crucial advantage of the method, since decision makers are already familiar with the measure (in contrast to metrics such as exergy efficiency [%], ecopoints [—], eutrophication [kg PO_4 equivalents], etc.). CBA also has some obvious shortcomings. Many effects, be they costs or benefits, cannot be exactly quantified in monetary terms (the beauty of a landscape or the life of a human being). On the other hand, several methods such as the contingent valuation method, the hedonic price method, and the travel cost method have been developed to convert problematic effects and environmental impacts into costs.[102,103]

Another approach to overcome such deficiencies — developed by Döberl and colleagues[106] — combines cost-effectiveness analysis and multicriteria analysis in a method known as modified cost-effectiveness analysis (MCEA). MCEA subdivides general goals into concrete subgoals. For example, the general goal "protection of human health and the environment" can be subdivided in a first step into (1.1) protection of air, (1.2) protection of water, and (1.3) protection of soil quality. In a second step, goal 1.1 can be subdivided into the subgoals (1.1.1) reduction of impact by regionally important pollutants, (1.1.2) reduction of the anthropogenic greenhouse effect, and (1.1.3) reduction of damage to the ozone layer. In contrast to the abstract goal "protection of human health and the environment," each of the latter subgoals can be described by single indicators (e.g., global-warming potential for 1.1.2 and ozone-depletion potential for 1.1.3), and targets for reduction can be quantified. This procedure may result in a multitude of subgoals and indicators of different importance and public preference. One way to make them comparable and amenable to aggregation is to assign a specific weight to each indicator. The weights can be obtained from a ranking process carried out by a group of experts or by stakeholders from a variety of interests. Finally, MCEA compares costs with the efficiency of reaching the defined targets.

According to Hanley and Spash,[103] a CBA comprises the following eight steps:

1. Definition of the project, which includes identifying the boundaries of the analysis and determining the population over which costs and benefits are to be aggregated.
2. Identification of all impacts resulting from the implementation of the project (required resources [materials, labor], effects on local unemployment levels, effects on local property prices, emissions to the environment, change to the landscape, etc.).
3. Determination of which impacts are to be counted based on certain rules and conventions.
4. Determination of the physical amounts of cost and benefit flows for a project and identification of when they will occur in time.

Methodology of MFA

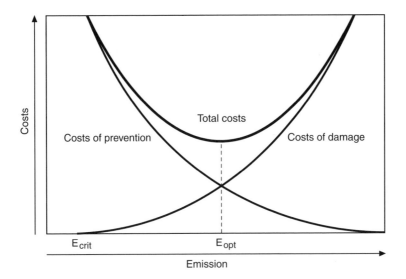

FIGURE 2.79 CBA of an emission problem: E_{crit} stands for the minimum emission load where environmental damage is expected to occur. E_{opt} is the emission where the total of costs of prevention (e.g., for filter technology) and costs of damage (e.g., treatment of respiratory diseases) are minimal. Note that the economically optimized emission (E_{opt}) accepts a certain degree of environmental burden.

5. Valuation of the physical measures of impact flows in monetary units. (This includes predicting prices for value flows extending into the future, correcting market prices where necessary, and calculating prices where none exist.)
6. Conversion of the monetary amounts of all relevant costs and benefits into present money values. (This is achieved by discounting, a method that makes costs and benefits comparable regardless of when they occur.)
7. Comparison of total costs (C) and total benefits (B). (If B > C, the project is qualified for acceptance or at least improves social welfare in the theory of neoclassical welfare economics.)
8. Performance of a sensitivity analysis to assess the relevance of uncertainties.

This is a good place to mention that the latter step is a requirement for all assessment methods. Currently, the combination of CBA and MFA is being further developed and applied to problems mainly in waste management by Schönbäck and colleagues from the Vienna University of Technology and the GUA (consultants in Vienna).[104–106] Figure 2.79 gives an example of a typical CBA solution.

2.5.1.7 Anthropogenic vs. Geogenic Flows

The anthropogenic vs. geogenic flow (A/G) approach is derived from the *precautionary principle* (P2) and a possible definition for the concept of sustainability. The Wingspread Conference in 1998 defined the P2 as follows (Hileman[107]): "When an activity raises threats of harm to human health or the environment, precautionary measures should be taken, even if some cause and effect relationships are not fully established scientifically."

Aspects of the P2 can be traced back centuries, even millennia. For example, precaution as a management guideline can be found in the historical oral traditions of indigenous people of Eurasia, Africa, the Americas, Oceania, and Australia.[108] Haigh mentions the British Alkali Act of 1874, which required that emissions of noxious gases from certain plants should be prevented without any need to demonstrate that the gases were actually causing harm in any particular case.[109] In the 20th century, the principle emerges in Scandinavian legislation in the early 1970s[110] and a few years later in Germany, when large-scale environmental problems such as acid rain, pollution of the North Sea (seals dying, carpets of algae, bans on swimming, etc.), and global climate change became evident. Since then, the P2 has been used in other international agreements and legislation,[110] notably the Rio Declaration on Environment and Development[111] of 1992. In that document, principle 15 says that "in order to protect the environment, the precautionary approach shall be widely applied by States according to their capabilities. Where there are threats of serious or irreversible damage, lack of full scientific certainty shall not be used as a reason for postponing cost-effective measures to prevent environmental degradation." The P2 is more widely accepted in Europe, where it has been regularly used in cases with less-than-certain scientific information for decision makers, especially in E.U. legislation (e.g., the Maastricht Treaty). In the U.S., the P2 has often been criticized on the basis of implementation costs[112] and the lack of comprehensive and authoritative definition.[113] Nevertheless, elements of the principle can be found in U.S. environmental laws.[114]

The approach "anthropogenic vs. geogenic flows" (A/G), first mentioned in Güttinger and Stumm,[115] is also part of the definition of ecological sustainability as given in SUSTAIN[116] and cited in Narodoslawsky and Krotscheck[46] (see also Daly[117]):

1. Anthropogenic material flows must not exceed the local assimilation capacity and should be smaller than natural fluctuations of geogenic flows.
2. Anthropogenic material flows must not alter the quality and the quantity of global material cycles.
3. The natural variety of species and landscapes must be sustained or improved.

Applying the A/G approach to an MFA system means to determine both the materials balances of the anthropogenic system and the corresponding materials balances of the environment into which the anthropogenic system is embedded. Of particular relevance is the extent to which anthropogenic flows alter geogenic flows and stocks. Examples are given in Figure 2.80. Since many cause/effect relationships between external material flows and environmental compartments are not entirely known and understood, the P2 is applied. Exclusively conservative (i.e., small) alterations of geogenic flows and stocks are considered to be acceptable. A trivialized summary of the A/G approach might go as follows: "As long as a human activity does not affect the environmental compartments significantly, there is no harm to the environment. The activity can be regarded as ecologically sustainable." Since the cause/effect relationship is not known, a "nonsignificant" change cannot be determined on scientific terms. Significance has to be defined, rather, on political and

Methodology of MFA

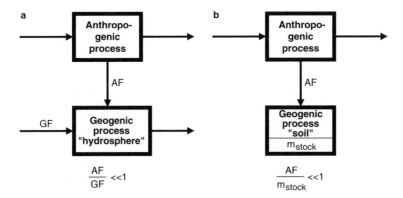

FIGURE 2.80 "Anthropogenic vs. geogenic flow" approach: (a) alteration of a geogenic flow by an anthropogenic process; (b) alteration of a geogenic stock. The smaller the ratios of AF/GF and AF/m_{stock}, the greater is the sustainability of the anthropogenic process.

ethical grounds. This may be regarded as the weak point of the A/G approach. On the other hand, this reflects the inherent problem of evaluation as a subjective process.

In order to determine the ratios AF/GF and AF/m_{stock} as illustrated in Figure 2.80, the natural concentrations of geogenic flows and stocks have to be known. Sometimes it is difficult to determine real geogenic concentrations, since virtually all ecosystems show traces of anthropogenic activities. Moreover, geogenic concentrations can vary considerably between regions. On the other hand, it is not essential to know the exact concentration to apply the approach, and abundant data are available on trace element concentrations in unpolluted air, surface- and groundwaters, soils, etc. Generally, the A/G approach is a reliable and handy tool for MFA evaluation, since its application requires only a little additional work. Figure 2.81 shows how the A/G approach can be employed to evaluate an MFA system.

2.5.1.8 Statistical Entropy Analysis

2.5.1.8.1 Background

The idea of applying statistical entropy to MFA results was developed at the Vienna University of Technology.[118] So far, it is the only evaluation method that has been tailor-made for MFA. While other methods do not consider all aspects of an MFA* or are capable of processing more information about the system than is provided by the MFA,** statistical entropy analysis (SEA) uses all information*** and requires little additional computing. For this reason, SEA is discussed here in some detail.

SEA is a method that quantifies the power of a system to concentrate or dilute substances. As will be shown in Chapter 3, Section 3.2.2, this is an essential feature of any material flow system. Today, especially in waste management, the importance

---

* For example, MIPS evaluates only inputs of goods. An LCA evaluates emissions but not the composition of solid goods.
** Examples include the sustainable process index (SPI) or CBA.
*** There is only one exception: the magnitude of a stock is not considered. This is discussed in Chapter 3, Section 3.2.2.

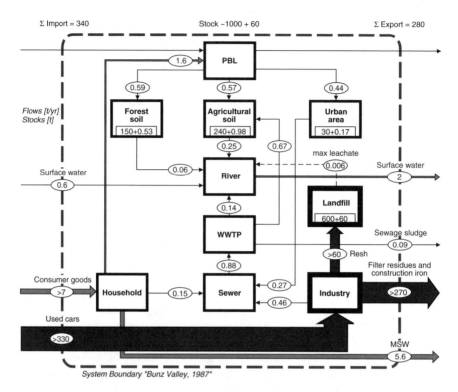

FIGURE 2.81 Lead balance in t/year for a 66-km² region in Switzerland with 28,000 inhabitants.[125] The main flow of lead is induced by the import of used cars that are treated in a large shredder. Some 60 t of lead are contained in the waste stream of the shredder (resh), which is landfilled. The product (scrap) from the shredder is processed in a regional steel mill, where lead is concentrated in the filter dust and exported. The stock of lead in the landfill is assessed at some 600 t and represents the largest and fastest growing reservoir of lead within the region. Assuming that the river entering the system can be regarded as unpolluted (geogenic concentration), the A/G approach limits the leachate from the landfill to a nonrelevant impact of, say, 1%, which equals 0.006 t/year. Such an emission means that a maximum of 0.001% of the lead in the landfill may be emitted per year. It is evident that the lead flows from WWTP (wastewater treatment plant) and soils exceed this limit (see discussion in Chapter 3.1.1 and Chapter 3.4.1).

of concentrating* resources as well as pollutants is not yet fully understood. For example, one argument against state-of-the-art incineration is that it produces a concentrate of hazardous substances designated as fly ash. On the other hand, another concentrating process is commonly and rightly considered as positive, namely the concentrating of paper, plastics, metals, glass, etc. Indeed this process is designated as *collection* of valuable resources for recycling purposes, though it is a typical *concentrating process*. An incinerator also collects resources in the fly ash and can

---

* *Concentrating* is used instead of the more common term *concentration* to stress that concentrating designates a transformation process or action.

Methodology of MFA

be designated as a valuable *collection process* (see Chapter 3, Section 3.3.2.3). However, the metric that is derived in this chapter is designated as the "substance concentrating efficiency" of a process, albeit "substance collection efficiency" would be possible as well. This metric is the only existing indicator that measures concentrating and diluting effects. For the moment, the method is derived for "one-process" systems (taken from Rechberger and Brunner[119]). Chapter 3, Section 3.2.2 gives an example of how to apply SEA to an MFA system consisting of several processes. The discussion there also makes a case for the need to establish a concentrating waste-management system to achieve sustainable management for conservative substances.

As shown earlier in this chapter, a material balance is established by combining mass flows of all goods with the substance concentrations in those goods. In Figure 2.82, an exemplary MFA is displayed for a selected substance. For simplicity, the system is in an ideal steady state. There is neither an exchange with an internal stock ($\dot{m}_{storage} = 0$) nor a stock ($m_{stock} = 0$). The input and output goods of the system are defined by a set of elemental concentrations ($[c_I],[c_O]$) and a set of mass flows ($[\dot{m}_I],[\dot{m}_O]$). Hence, a system can be regarded as a procedure that transforms an input set of concentrations into an output set of concentrations; the same applies to the mass flows. Each system can be viewed as a unit that concentrates, dilutes, or leaves unchanged its throughput of a substance. In order to measure this transformation, an appropriate function that quantifies the various sets is required. The transformation can be defined as the difference between the quantities for the input (X) and the output (Y). This allows determination of whether a system concentrates (X − Y > 0) or dilutes (X − Y < 0) substances.

2.5.1.8.2 Information Theory

In order to calculate X and Y, a mathematical function from the field of information theory is used.[120] This function originates from Boltzmann's statistical description of entropy. It is formally and mathematically identical with Boltzmann's well-known H-theorem.[121] Information theory, developed by Shannon in the 1940s,[122] is used to measure the loss or gain of information within a system. In statistics, the so-called Shannon entropy is used to measure the variance of a probability distribution: the greater the variance, the less the information about the quantity of interest. Note that the thermodynamic entropy denoted "S" (J/mole/K), as introduced by Clausius[123] in 1865, is *formally* identical with the statistical entropy "H" (bit) of Shannon; however, there is no physical relationship between the two entropy terms. The following is based on Shannon's statistical entropy and not on thermodynamic entropy.

The statistical entropy H of a finite probability distribution is defined by Equations 2.21 and 2.22:

$$H(P_i) = -\lambda \cdot \sum_{i=1}^{k} P_i \cdot \ln(P_i) \geq 0 \qquad (2.21)$$

$$\sum_{i=1}^{k} P_i = 1 \qquad (2.22)$$

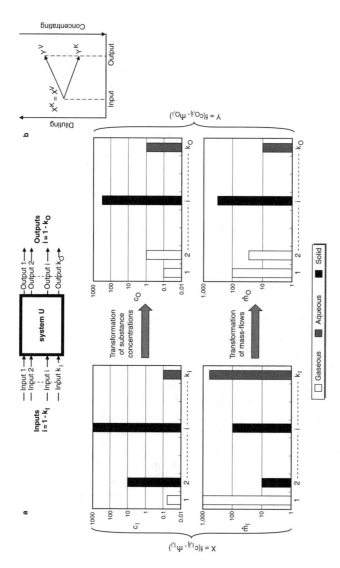

FIGURE 2.82 (a) A balance for substance j of a system U is determined by substance concentrations (c_{ij}) and mass-flows ($\dot{m}_i$) of all input and output goods. For all j and for steady-state conditions, the applicable equation is $\sum_{i=1}^{k_I} \dot{m}_{I,i} \cdot c_{I,ij} = \sum_{i=1}^{k_O} \dot{m}_{O,i} \cdot c_{O,ij}$. The system can be regarded as an algorithm that transforms sets of concentrations ($[c_I],[c_O]$) and mass-flows ($[\dot{m}_I],[\dot{m}_O]$) from the input (I) to the output (O). X and Y are functions that quantify the sets of concentrations and mass flows. (b) Comparison of a concentrating system K ($X^K - Y^K > 0$) and a diluting system V ($X^V - Y^V < 0$).[119] (From Rechberger, H. and Brunner, P.H., *Environ. Sci. Technol.*, 36, 809, 2002. With permission.)

Methodology of MFA

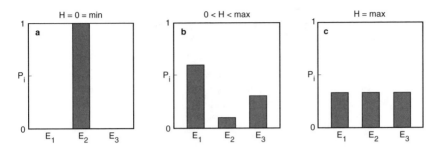

FIGURE 2.83 Probability distributions for a case where one of three events E_i can happen: extreme (a,c) and arbitrary (b) values of H. P_i probability for event i, $\Sigma P_i = 1$.[119]

where P_i is the probability that event i happens.[120]

In statistical mechanics, λ is defined by Boltzmann as the ratio of the gas constant per mole (R) to Avogadro's number (N_0): $k_B = R/N_0$, unit J/K. In information theory, λ is replaced by the term $1/(\ln 2)$. This converts the natural logarithm in Equation 2.21 to the logarithm to the base 2 (indicated as $ld(x)$ in the following equations). The unit of H then becomes 1 bit (short for "binary digit"). For two events with equal probability ($P_1 = P_2 = 1/2$), H is 1 bit, and the connection to coding and information theory becomes evident. The term $0 \times ld(0)$ is defined to be zero.[120] Figure 2.83 illustrates three different distributions with extreme as well as arbitrary values of H. A case is presented in which one of three events (E_i) can happen. In Figure 2.83a, the probability of event two is unity ($P_2 = 1$). The statistical entropy of such a distribution is zero. In Figure 2.83c, the probabilities for all three events are identical. The entropy of such a distribution becomes a maximum. This can be proven using the Lagrange multiplier theorem. Since H is a positive definite function, the distribution in Figure 2.83a must yield the minimum value of H. Hence all other possible combinations of probabilities (e.g., the distribution in Figure 2.83b) must yield a value of H in the range between zero and max.

2.5.1.8.3 Transformation of Statistical Entropy Function

In order to be applied to sets of concentrations and mass flows, the statistical entropy function is transformed in three steps.

2.5.1.8.3.1 First Transformation

The statistical entropy function is applied to both the input and the output of the investigated system (see Figure 2.82). During this first step, it is assumed that the mass flows of the investigated set of goods are identical and equal to unity ($\dot{m}_i = 1$). This simplification is necessary to understand the analogy between probability and concentration. To quantify the variance of the attribute "concentration" (instead of "probability"), Equation 2.21 with $\lambda = 1/(\ln 2)$ is transformed to Equation 2.23, replacing P_i by c_{ij}/c_j. The relative concentrations c_{ij}/c_j range between zero and one, and they can be regarded as a measure of the distribution of substance j among the goods. Equation 2.24 has no physical meaning and only serves to normalize the concentrations c_{ij}. (Since all $\dot{m}_i = 1$, the variable c_{ij} represents normalized substance flows.)

$$H^I(c_{ij}) = ld(c_j) - \frac{1}{c_j} \cdot \sum_{i=1}^{k} c_{ij} \cdot ld(c_{ij}) \tag{2.23}$$

$$c_j = \sum_{i=1}^{k} c_{ij} \tag{2.24}$$

where
i = 1, ..., k
j = 1, ..., n
c_{ij} = concentration of substance j in good i

Index k gives the number of goods in the set and n gives the number of investigated substances. As stated above, the maximum of H^I is found for $c_{1j} = c_{2j} = ... = c_{kj} = c_j/k$. Equation 2.23 consequently changes to

$$H^I_{max} = ld(k) \tag{2.25}$$

Thus, H^I ranges from zero to ld(k) for all possible sets of concentrations. Since the maximum of H^I is a function of the number of goods (k), the *relative* statistical entropy H^I_{rel} is used to compare sets with different numbers of goods. H^I_{rel} is defined as follows:

$$H^I_{rel}(c_{ij}) = \frac{H^I(c_{ij})}{H^I_{max}} = \frac{H^I(c_{ij})}{ld(k)} \tag{2.26}$$

The value of H^I_{rel} ranges between zero and one for all possible sets of concentrations; it is dimensionless.

2.5.1.8.3.2 Second Transformation

To further quantify the attribute "mass flow," Equation 2.23 is modified. The mean concentration in a good (c_{ij}) is weighted with the mass flow ($\dot{m}_i$) of this good; $\dot{m}_i$ can be regarded as the frequency of "occurrence" of the concentration c_{ij} (see Figure 2.84). The entropy H^{II} of a mass-weighted set of concentrations therefore is given by Equation 2.27 and Equation 2.28, which correspond to Equation 2.23 and Equation 2.24, respectively:

$$H^{II}(c_{ij}, \dot{m}_i) = ld(\dot{X}_j) - \frac{1}{\dot{X}_j} \sum_{i=1}^{k} \dot{m}_i \cdot c_{ij} \cdot ld(c_{ij}) \tag{2.27}$$

Methodology of MFA

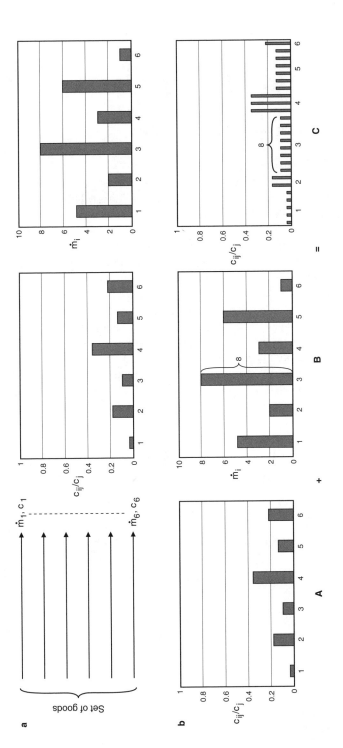

FIGURE 2.84 (a) Exemplary illustration of a set of $k = 6$ goods. The normalized concentrations c_{ij}/c_j and the mass flows $\dot{m}_i$ represent the distribution of substance j among the goods. (b) The mass flow of a good can be interpreted as the frequency of a concentration. The combination of the distribution of concentrations (A) and mass flows (B) can be defined as a weighted distribution of concentrations (C). Distribution C can be quantified by applying Equation 2.27.[126]

$$\dot{X}_j = \sum_{i=1}^{k} \dot{m}_i \cdot c_{ij} \qquad (2.28)$$

where

$\dot{m}_i$ = mass-flow of good i

$\dot{X}_j$ = total substance flow induced by the set of goods

2.5.1.8.3.3 Third Transformation

The last modification of the initial function concerns the gaseous and aqueous output goods (emissions). In contrast to the products (e.g., solid residues) of the investigated system, the emissions are diluted in air and receiving waters, which results in an increase in entropy. In Equation 2.30 and Equation 2.31, the index "geog" indicates the "natural" or geogenic concentrations of substances in the atmosphere and the hydrosphere. These concentrations serve as reference values to describe the dilution process. Factor "100" in Equation 2.30 and Equation 2.31 means that the emission (c_{ij}, $\dot{m}_i$; e.g., measured in a stack or wastewater pipe) is mixed with a geogenic flow ($c_{j,geog}, \dot{m}_{geog}$), so that the concentration of the resulting flow ($\dot{m}_i + \dot{m}_{geog}$) is 1% above $c_{j,geog}$. It has been demonstrated that this approximation reflects the actual (unlimited) dilution in the environmental compartment sufficiently for $c_{ij} \gg c_{j,geog}$.[118] For $c_{ij} = c_{j,geog}$ there is apparently no dilution. Hence, as a rule of thumb in cases where $c_{ij}/c_{j,geog} < 10$, Equations 2.30 and 2.31 have to be replaced by more complex terms that cover the total range $c_{j,geog} < c_{ij} < c = 1$ (g/g).[118] The applicability of Equation 2.30 and Equation 2.31 has to be checked in any case.

$$H^{III}(\underline{c}_{ij}, \underline{\dot{m}}_i) = \mathrm{ld}(\dot{X}_j) - \frac{1}{\dot{X}_j} \sum_{i=1}^{k} \underline{\dot{m}}_i \cdot \underline{c}_{ij} \cdot \mathrm{ld}(\underline{c}_{ij}) \qquad (2.29)$$

where $\underline{c}_{ij}$ is defined as

$$\underline{c}_{ij} = \begin{cases} c_{j,geog,g}/100 \\ c_{j,geog,a}/100 \\ c_{ij} \end{cases} \text{ for } \begin{cases} i = 1, \ldots, k_g \\ i = k_g + 1, \ldots, k_g + k_a \\ i = k_g + k_a + 1, \ldots, k \end{cases} \text{ for } \begin{cases} \text{gaseous} \\ \text{aqueous} \\ \text{solid} \end{cases} \text{outputs} \qquad (2.30)$$

where

k = number of total output goods
k_g = number of gaseous output goods
k_a = number of aqueous output goods
g = gaseous
a = aqueous

and where $\underline{\dot{m}}_i$ is defined as

Methodology of MFA

$$\dot{m}_i = \begin{cases} \dfrac{\dot{X}_{ij}}{c_{j,geog,g}} \cdot 100 \\ \dfrac{\dot{X}_{ij}}{c_{j,geog,a}} \cdot 100 \\ \dot{m}_i \end{cases} \text{ for } \begin{cases} i = 1,\ldots,k_g \\ i = k_g+1,\ldots,k_g+k_a \\ i = k_g+k_a+1,\ldots,k \end{cases} \text{ for } \begin{cases} \text{gaseous} \\ \text{aqueous} \\ \text{solid} \end{cases} \text{outputs} \quad (2.31)$$

where $\dot{X}_{ij}$ = substance flow of output i.

The geogenic concentrations in Equation 2.30 and Equation 2.31 can be replaced by more realistic background concentrations. In this case, the influence of the actual surrounding environment is better reflected in the evaluation process of the investigated system. For a simple comparison of options, this is usually not necessary.

Applied to the output of a system, Equation 2.29 quantifies the distribution of substance j. The maximum of H^{III} is reached when all of substance j is directed to the environmental compartment with the lowest geogenic concentration ($c_{j,geog,min}$). For heavy metals, this is usually the atmosphere, since $c_{j,geog,a} > c_{j,geog,g}$. The maximum of H^{III} is given by Equation 2.32.

$$H^{III}_{max,j} = ld\left(\dfrac{\dot{X}_j}{c_{j,geog,min}} \cdot 100\right) \quad (2.32)$$

Using Equation 2.26 and Equation 2.29 through Equation 2.32, the relative statistical entropy $RSE_{j,O} \equiv H^{III}_{rel}$ of the output (index O, measure Y in Figure 2.82) can be calculated. In the same way — using Equation 2.26, Equation 2.27, Equation 2.28, and Equation 2.32 — the $RSE_{j,I} \equiv H^{II}_{rel}$ for the input (index I, measure X in Figure 2.82) of a system is obtained. The difference in the RSE_j between the input and output of a system can be defined as the substance concentrating efficiency (SCE) of the system.[118]

$$SCE_j \equiv \dfrac{RSE_{j,I} - RSE_{j,O}}{RSE_{j,I}} \cdot 100 \quad (2.33)$$

The SCE_j is given as percentage and ranges between a negative value, which is a function of the input, and 100%. An SCE_j value of 100% for substance j means that substance j is transferred 100% into one pure output good. $SCE_j = 0$ is the result if RSE_j values of the input and output are identical. This means that the system neither concentrates nor dilutes substance j. However, this does not imply identical sets of mass flows and concentrations in inputs and outputs. The SCE_j equals a minimum if all of substance j is emitted into that environmental compartment that allows for maximum dilution (in general, the atmosphere). These relationships are illustrated in Figure 2.85.

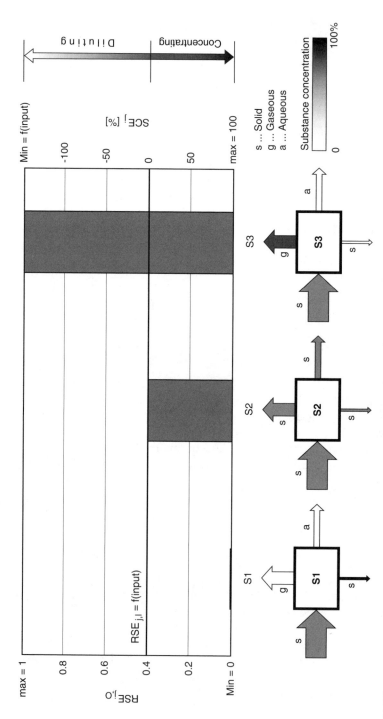

FIGURE 2.85 Relationship between relative statistical entropies RSE_j and substance concentrating efficiency SCE_j. S1 symbolizes a system that achieves maximum concentrating by producing a pure residue containing all of substance j. S2 is an example of a system that does not chemically discriminate between its outputs ($c_1 = c_0$). S3 transfers the substance entirely into the atmosphere, which means maximum dilution. (From Rechberger, H. and Brunner, P.H., *Environ. Sci. Technol.*, 36, 809, 2002. With permission.)

Methodology of MFA

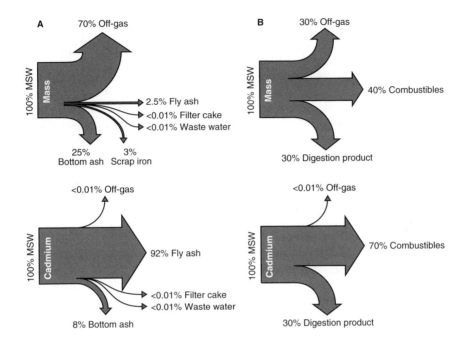

FIGURE 2.86 Partitioning of total mass and cadmium in two waste-treatment facilities: (A) MSW incineration[127] and (B) mechanical–biological treatment of MSW.[128] Incineration transfers 92% of cadmium into fly ash, a small fraction of the 2.5% of total MSW input that can be safely stored in appropriate underground disposal facilities. Mechanical–biological treatment does not produce residues with enhanced Cd concentrations with respect to the input, since the partitioning diagrams for mass and cadmium are essentially identical. With regard to management of Cd, solution A is superior to B. Statistical entropy analysis quantifies the advantage: $SCE_A = 42\%$, $SCE_B = 4.2\%$.

2.5.1.8.4 SEA Applications in MFA

SEA has been applied to MFA results of various waste-treatment facilities and multiprocess systems such as the European copper cycle (see Chapter 3, Section 3.2.2). SEA considers all information of an MFA except the magnitude of stocks. Only inputs into and outputs from a stock are included in SEA. The relevance of a stock has to be assessed by comparison with other geogenic or anthropogenic reference reservoirs (see Figure 2.80). Figure 2.86 gives an example how SEA quantifies the power of waste-treatment processes to concentrate cadmium.

REFERENCES

1. Baccini, P. and Brunner, P.H., *Metabolism of the Anthroposphere*, Springer, New York, 1991.
2. Guralnik, D.B. and Friend, J.H., Eds., *Webster's New World Dictionary of the American Language*, college ed., World Publishing, New York, 1968.
3. Gove, Ph.B., *Webster's Seventh New Collegiate Dictionary*, G&C Merriam, Springfield, MS, 1972.

4. Morris, W., Ed., *American Heritage Dictionary*, 2nd college ed., Houghton Mifflin, Boston, 1982.
5. Sax, N.I. and Lewis, R.J., Eds., *Hawley's Condensed Chemical Dictionary*, 11th ed., Van Nostrand Reinhold, New York, 1987.
6. Holleman, A.F. and Wiberg, E., *Lehrbuch der Anorganischen Chemie*, 101st ed., Walter de Gruyter, New York, 1995, chap. 1.
7. Brunner, P.H., Materials flow analysis: vision and reality, *J. Ind. Ecol.*, 5, 3, 2002.
8. Patten, B.C., A primer for ecological modeling and simulation with analog and digital computers, in *Systems Analysis and Simulation in Ecology*, Patten, B.C., Ed., Academic Press, New York, 1971, chap.1.
9. Ayres, R.U., Industrial metabolism: theory and policy, in *Industrial Metabolism: Restructuring for Sustainable Development*, Ayres, R.U. and Simonis, U.E., Eds., United Nations University Press, Tokyo, 1994, chap.1.
10. van der Voet, E., *Substances from Cradle to Grave: Development of a Methodology for the Analysis of Substance Flows through the Economy and the Environment of a Region*, Optima Druck, Amsterdam, 1996.
11. Graedel, T.E. and Allenby, B.R., *Industrial Ecology*, 2nd ed., Prentice Hall, Upper Saddle River, NJ, 2002, chap. 23.
12. van der Voet, E., Guinée, J.B., and Udo de Haes, H.A., Eds., *Heavy Metals: A Problem Solved?*, Kluwer Academic Publishers, Dordrecht, The Netherlands, 2000, chap. 1.
13. Patzel, N. and Baccini, P., Biological metaphors and system analysis for the investigation of urban systems, submitted to *Goeforum*.
14. U.S. Geological Survey, www.usgs.org.
15. U.S. Department of Energy, www.doe.gov.
16. Landner, L. and Lindeström, L., Zinc in society and in the environment, Swedish Environmental Research Group (MFG), Kil, Sweden, 1998; serg@mfg.se.
17. Schachermayer, E., Bauer, G., Ritter, E., and Brunner, P.H., Messung der Güter- und Stoffbilanz einer Müllverbrennungsanlage, Projekt MAPE, Monographie No. 56, Umweltbundesamt GmbH, Vienna, 1995.
18. Morf, L., Ritter, E., and Brunner, P.H., Güter- und Stoffbilanz der MVA Wels, Projekt Mape III, TU-Wien, Institut für Wassergüte und Abfallwirtschaft, Wien, 1997.
19. Morf, L.S. and Brunner, P.H., The MSW incinerator as a monitoring tool for waste management, *Environ. Sci. Technol.*, 32, 1825, 1998.
20. Hedbrant, J. and Sörme, L., Data vagueness and uncertainty in urban heavy-metal data collection, *Water Air Soil Pollut.: Focus*, 1, 43–53, 2001.
21. Morgan, M.G., Henrion, M., and Small, M., *Uncertainty — A Guide to Dealing with Uncertainty in Quantitative Risk and Policy Analysis*, Cambridge University Press, Cambridge, U.K., 1990, p. 332.
22. Baccini, P. and Bader, H.-P., *Regionaler Stoffhaushalt — Erfassung, Bewertung und Steuerung*, Spektrum Akademischer Verlag, Heidelberg, 1996.
23. Hebbeker, T., Statistische Methoden der Datenanalyse, www-eep.physik.hu-berlin.de/~hebbeker/lectures/st78_l03.htm (accessed Feb. 2003).
24. Asperl, A. and Paukowitsch, P., Eine geometrische Interpretation der Ausgleichsrechnung, www.geometrie.tuwien.ac.at/asperl/projekt/start.htm (accessed Feb. 2003).
25. Adamek, T., Numerische Methoden für Ingenieure, www.uni-stuttgart.de/bio/adamek/numerik/NuMe02.pdf (accessed Feb. 2003).
26. Reichert, P., Concepts Underlying a Computer-Program for the Identification and Simulation of Aquatic Systems, Schriftenreihe der EAWAG, No. 7, EAWAG, Duebendorf, Switzerland, 1994.

27. Glenck, E. and Wall, J., State of the Art Computer Programs for Material Balancing and Accounting — SAMBA, Institute for Water Quality and Waste Management, Vienna University of Technology, Vienna, 1996.
28. Brunner, P.H. et al., Die Stoffflussanalyse als Grundlage von UVP und Öko-Auditing, von der Theorie zur praktischen Anwendung, ÖWAV, Vienna, 1996.
29. Fehringer, R. and Brunner, P.H., Kunststoffflüsse und die Möglichkeit der Verwertung von Kunststoffen in Österreich, Monographic Bd. 80, Federal Environmental Agency, Vienna, 1997.
30. Möller, A., Grundlagen stoffstrombasierter Betrieblicher Umweltinformationssysteme, Projekt-Verlag, Bochum, Germany, 2000.
31. Umberto 4 Manual, Institute for Energy and Environmental Research Heidelberg Ltd. (IFEU) in cooperation with the Institute for Environmental Informatics Hamburg Ltd. (IFU), Heidelburg, 2001.
32. GaBi 4 Manual, IKP, University of Stuttgart and PE Europe GmbH, Stuttgart, YEAR.
33. Goedkoop, M., The Eco-Indicator 95, Pré Consultants, Amersfoort, The Netherlands, 1995.
34. Organization for Economic Cooperation and Development (OECD), Environmental Indicators, OECD core set, OECD, Paris, 1994.
35. Hammond, A., Adriaanse, A., Rodenburg, E., Bryant, D., and Woodward, R., Environmental Indicators: A Systematic Approach to Measuring and Reporting on Environmental Policy Performance in the Context of Sustainable Development, World Resource Institute, Washington, D.C., 1995.
36. Schmidt-Bleek, F., Wieviel Umwelt braucht der Mensch? MIPS – Das Mass für ökologisches Wirtschaften, Birkhäuser, Berlin, 1994.
37. Hinterberger, F., Luks, F., and Schmidt-Bleek, F., Material flows vs. "natural capital": what makes an economy sustainable? *Ecol. Econ.*, 23, 1, 1997.
38. Schmidt-Bleek, F., Bringezu, S., Hinterberger, F., Liedtke, C., Spangenberg, J., Stiller, H., and Welfen, M.J., MAIA-Einführung in die Material-Intensitäts-Analyse nach dem MIPS-Konzept, Birkhäuser, Berlin, 1998.
39. Schmidt-Bleek, F., *Das MIPS-Konzept*, Droemer Knaur, Munich, 1998.
40. Schmidt-Bleek, F., MIPS: a universal ecological measure? *Fresenius Environ. Bull.*, 2, 306, 1993.
41. Ritthoff, M., Rohn, H., and Liedtke, C., MIPS berechnen — Ressourcenproduktivität von Produkten und Dienstleistungen, Wuppertal Spezial 27, Wuppertal, Germany, 2002.
42. Schmidt-Bleek, F., MIPS revisited, *Fresenius Environ. Bull.*, 2, 407, 1993.
43. Cleveland, C.J. and Ruth, M., Indicators of dematerialization and the materials intensity of use, *J. Ind. Ecol.*, 2, 3, 15, 1999.
44. Reijnders, L., The factor X debate: setting targets for eco-efficiency, *J. Ind. Ecol.*, 2, 1, 13, 1998.
45. Krotscheck, C. and Narodoslawsky, M., The sustainable process index: a new dimension in ecological evaluation, *Ecol. Eng.*, 6, 241, 1996.
46. Narodoslawsky, M. and Krotscheck, C., The sustainable process index (SPI): evaluating processes according to environmental compatibility, *J. Hazardous Mater.*, 41, 383, 1995.
47. Krozer, J., Operational Indicators for Progress towards Sustainability, EU project final report, No. EV-5V-CT94-0374, The Hague, The Netherlands, 1996.
48. Krotscheck, C., König, F., and Obernberger, I., Ecological assessment of integrated bioenergy systems using the sustainable process index, *Biomass Bioenergy*, 18, 341, 2000.

49. Eder, P. and Narodoslawsky, M., Input-output based valuation of the compatibility of regional activities with environmental assimilation capacities, in Proceedings of the Inaugural Conference of the European Branch of the International Society for Ecological Economics, Paris, May 1996.
50. Steinmüller, H. and Krotscheck, C., Regional assessment with the sustainable process index (SPI) concept, in Proceedings of the 7th International Conference on Environ-Metrics, Innsbruck, 1997.
51. Consoli, F., Boustead, I., Fava, J., Franklin, W., Jensen, A., de Oude, N., Parish, R., Postlethwaite, D., Quay, B., Seguin, J., and Vignon, B., Guidelines for life-cycle assessment: a "code of practice," SETAC, Brussels, 1993.
52. Udo de Haes, H.A., Ed., Towards a Methodology for Life Cycle Impact Assessment, SETAC, Brussels, 1996.
53. Guinée, J.B. et al., Life Cycle Assessment: An Operational Guide to the ISO Standards, Ministry of Housing, Spatial Planning and the Environment, The Hague, The Netherlands, 2001.
54. Rydh, C.J. and Karlström, M., Life cycle inventory of recycling portable nickel-cadmium batteries, *Resour. Conserv. Recycling*, 34, 289, 2002.
55. Song, H.-S. and Hyun, J.C., A study on the comparison of the various waste management scenarios for PET bottles using the life-cycle assessment (LCA) methodology, *Resour. Conserv. Recycling*, 27, 267, 1999.
56. Finnveden, G. and Ekvall, T., Life-cycle assessment as a decision-support tool — the case of recycling versus incineration of paper, *Resour. Conserv. Recycling*, 24, 235, 1998.
57. Andersson, K., Ohlsson, T., and Olsson, P., Screening life-cycle assessment (LCA) of tomato ketchup: a case study, *J. Cleaner Prod.*, 6, 277, 1998.
58. Amatayakul, W. and Ramnäs, O., Life-cycle assessment of a catalytic converter for passenger cars, *J. Cleaner Prod.*, 9, 395, 2001.
59. Furuholt, E., Life-cycle assessment of gasoline and diesel, *Resour. Conserv. Recycling*, 14, 251, 1995.
60. Potting, J. and Blok, K., Life-cycle assessment of four types of floor covering, *J. Cleaner Prod.*, 3, 201, 1995.
61. Landfield, A.H. and Karra, V., Life-cycle assessment of a rock crusher, *Resour. Conserv. Recycling*, 28, 207, 2000.
62. Widman, J., Environmental impact assessment of steel bridges, *J. Construct. Steel Res.*, 46, 291, 1998.
63. Bundesamt für Umweltschutz, Ökobilanzen von Packstoffen, Zusammenfassender Übersichtsbericht, Schriftenreihe Umweltschutz, No. 24, Bundesamt für Umweltschutz, Bern, 1984.
64. Burgess, A.A. and Brennan, D.J., Application of life-cycle assessment to chemical processes, *Chem. Eng. Sci.*, 56, 2589, 2001.
65. Lenzen, M. and Munksgaard, J., Energy and CO_2 life-cycle analyses of wind turbines — review and applications, *Renewable Energy*, 26, 339, 2002.
66. Schleisner, L., Life-cycle assessment of a wind farm and related externalities, *Renewable Energy*, 20, 279, 2000.
67. Seppälä, J., Koskela, S., Melanen, M., and Palperi, M., The Finnish metals industry and the environment, *Resour. Conserv. Recycling*, 35, 61, 2002.
68. Brentrup, F., Küsters, J., Kuhlmann, H., and Lammel, J., Application of the life-cycle assessment methodology to agricultural production: an example of sugar beet production with different forms of nitrogen fertilisers, *Eur. J. Agron.*, 14, 221, 2001.

69. Haas, G., Wetterich, F., and Köpke, U., Comparing intensive, extensified and organic grassland farming in southern Germany by process life-cycle assessment, *Agric. Ecosystem Environ.*, 83, 43, 2001.
70. Finnveden, G., Methodological aspects of life-cycle assessment of integrated solid waste management systems, *Resour. Conserv. Recycling*, 26, 173, 1999.
71. Seppälä, J., Melanen, M., Jouttijärvi, T., Kauppi, L., and Leikola, N., Forest industry and the environment: a life-cycle assessment study from Finland, *Resour. Conserv. Recycling*, 23, 87, 1998.
72. Myer, A. and Chaffee, C., Life-cycle analysis for design of the Sydney Olympic Stadium, *Renewable Energy*, 10, 169, 1997.
73. Krozer, J. and Vis, J.C., How to get LCA in the right direction? *J. Cleaner Prod.*, 6, 53, 1998.
74. Ayres, R.U., Life cycle analysis: a critique, *Resour. Conserv. Recycling*, 14, 199, 1995.
75. Graedel, T.E., *Streamlined Life-Cycle Assessment*, Prentice-Hall, Englewood Cliffs, NJ, 1998.
76. Udo de Haes, H.A., Ed., *Towards a Methodology for Life Cycle Impact Assessment*, SETAC Europe, Brussels, 1996.
77. Heiljungs, R., Guinée, J.B., Huppes, G., Lankreijer, R.M., Udo de Haes, H.A., and Wegener-Sleeswijk, A., Environmental Life-Cycle Assessment of Products: Guide and Backgrounds, Centre of Environmental Science, Leiden, 1992.
78. Müller-Wenk, R., Die ökologische Buchhaltung: ein Informations- und Steuerungsinstrument für umweltkonforme Unternehmenspolitik, Campus Verlag, Frankfurt, 1978.
79. Braunschweig, A. and Müller-Wenk, R., *Ökobilanzen für Unternehmungen: Eine Wegleitung für die Praxis*, Haupt, Bern, Switzerland, 1993.
80. Ahbe, S., Braunschweig, A., and Müller-Wenk, R., Methodik für Ökobilanzen auf der Basis ökologischer Optimierung, Schriftenreihe Umwelt No. 133, BUWAL, Bern, Switzerland, 1990.
81. Brand, G., Braunschweig, A., Scheidegger, A., and Schwank, O., Bewertung in Ökobilanzen mit der Methode der ökologischen Knappheit — Ökofaktoren 1997, Schriftenreihe Umwelt No. 297, BUWAL, Bern, Switzerland, 1998.
82. Hertwich, E.G., Pease, W.S., and Koshland, C.P., Evaluating the environmental impact of products and production processes: a comparison of six methods, *Sci. Total Environ.*, 196, 13, 1997.
83. Habersatter, K., Ökobilanzen von Packstoffen — Stand 1990, Schriftenreihe Umwelt No. 132, BUWAL, Bern, Switzerland, 1991.
84. Hellweg, S., Time- and Site-Dependent Life-Cycle Assessment of Thermal Waste Treatment Processes, Ph.D. thesis, Swiss Federal Institute of Technology, Zurich, 2000.
85. Szargut, J., Morris, D., and Steward, F., *Exergy Analysis of Thermal, Chemical, and Metallurgical Processes*, Hemisphere Publishing, New York, 1988.
86. Rant, Z., Exergie, ein neues Wort für technische Arbeitsfähigkeit, *Forsch. Ing. Wesen*, 22, 36, 1956.
87. Barin, I., *Thermochemical Data of Pure Substances*, VCH, Weinheim, Germany, 1989.
88. Ayres, R.U. and Ayres, L.W., *Accounting for Resources*, Vol. 2, The Life Cycle of Materials, Edward Elgar, Cheltenham, U.K., 1999.
89. Costa, M.M., Schaeffer, R., and Worrell, E., Exergy accounting of energy and materials flows in steel production systems, *Energy*, 26, 363, 2001.
90. Michaelis, P. and Jackson, T., Material and energy flow through the UK iron and steel sector, part 1: 1954–1994, *Resour. Conserv. Recycling*, 29, 131, 2000.

91. Michaelis, P. and Jackson, T., Material and energy flow through the UK iron and steel sector, part 2: 1994–2019, *Resour. Conserv. Recy.*, 29, 209, 2000.
92. Ertesvag, I.S. and Mielnik, M., Exergy analysis of the Norwegian society, *Energy*, 25, 957, 2000.
93. Wall, G., Exergy — A Useful Concept within Resource Accounting, Report No. 77-42, Institute of Theoretical Physics, Chalmers University of Technology and University of Göteborg, Göteborg, Sweden, 1977; www.exergy.se/ftp/paper1.pdf.
94. Wall, G., Sciubba, E., and Naso, V., Exergy use in the Italian society, *Energy*, 19, 1267, 1994.
95. Wall, G., Exergy, ecology and democracy — concepts of a vital society, in *ENSEC '93 International Conference on Energy Systems and Ecology*, Szargut, J. et al., Eds., ENSEC, Krakow, Poland, 1993, p. 111.
96. Rosen, M.A. and Dincer, I., Exergy as the confluence of energy, environment and sustainable development, *Exergy Int. J.*, 1, 3, 2001.
97. Connelly, L. and Koshland, C.P., Exergy and industrial ecology — part 1: an exergy-based definition of consumption and a thermodynamic interpretation of ecosystem evolution, *Exergy Int. J.*, 1, 146, 2001.
98. European Union, Directive 2000/76/EC of the European Parliament and of the Council of 4 December 2000 on the Incineration of Waste, Official Journal of the European Communities, Brussels, 2000.
99. Fiedler, H. and Hutzinger, O., Sources and sinks of dioxins: Germany, *Chemosphere*, 25, 1487, 1992.
100. Wall, G., Exergy — A Useful Concept, Ph.D. thesis, Chalmers University of Technology, Göteborg, Sweden, 1986.
101. Baehr, H.D., *Thermodynamik*, 7th ed., Springer, Berlin, 1989.
102. Johansson, P.-O., *Cost-Benefit Analysis of Environmental Change*, Cambridge University Press, New York, 1993.
103. Hanley, N. and Spash, C.L., *Cost-Benefit Analysis and the Environment*, Edward Elgar, Brookfield, VT, 1995, p. 5.
104. Schönbäck, W., Kosz, M., and Madreiter, T., *Nationalpark Donauauen: Kosten-Nutzen-Analyse*, Springer, New York, 1997, p. 3.
105. Hutterer, H., Pilz, H., Angst, G., and Musial-Mencik, M., Stoffliche Verwertung von Nichtverpackungs-Kunststoffabfällen, Kosten-Nutzen-Analyse von Massnahmen auf dem Weg zur Realisierung einer umfassenden Stoffbewurtschaftung von Kunststoffabfällen, Monographien, Band 124, Austrian Federal Environmental Agency, Vienna, 2000.
106. Döberl, G., Huber, R., Brunner, P.H., Eder, M., Pierrard, R., Schönbäck, W., Frühwirth, W., and Hutterer, H., Long-term assessment of different waste management options — a new, integrated and goal-oriented approach, *Waste Manage. Res.*, 20, 322, 2002.
107. Hileman, B., Precautionary principle, *Chem. Eng. News*, 76 (Feb. 9), 16, 1998.
108. Martin, P.H., If you don't know how to fix it, please stop breaking it! The precautionary principle and climate change, *Found. Sci.*, 2, 263, 1997.
109. Haigh, N., The introduction of the precautionary principle into the UK, in *Interpreting the Precautionary Principle*, O'Riordan, T. and Cameron, J., Eds., Earthscan Publications, London, 1994, chap. 13.
110. Sand, P.H., The precautionary principle: a European perspective, *Hum. Ecol. Risk Assess.*, 6, 445, 2000.

111. United Nations Conference on Environment and Development, Report of the Conference, Rio de Janeiro, A/CONF.151/26/Rev.1.1, United Nations, New York, 1992.
112. deFur, P.L. and Kaszuba, M., Implementing the precautionary principle, *Sci. Total Environ.*, 288, 155, 2002.
113. Sandin, P., Dimensions of the precautionary principle, *Hum. Ecol. Risk Assess.*, 5, 889, 1999.
114. Applegate, J.S., The precautionary preference: an American perspective on the precautionary principle, *Hum. Ecol. Risk Assess.*, 6, 413, 2000.
115. Güttinger, H. and Stumm, W., Ökotoxikologie am Beispiel der Rheinverschmutzung durch den Chemie-Unfall bei Sandoz in Basel, *Naturwiss.*, 77, 253, 1990.
116. SUSTAIN, Forschungs- und Entwicklungsbedarf für den Übergang zu einer nachhaltigen Wirtschaftsweise in Österreich, Institute of Chemical Engineering, Graz University of Technology, Austria, 1994.
117. Daly, H.E., Sustainable growth: an impossibility theorem, in *Valuing the Earth*, Daly, H.E. and Townsend, K.N., Eds., MIT Press, Cambridge, MS, 1993, p. 271.
118. Rechberger, H., Entwicklung einer Mehtode zur Bewertung von Stoffbilanzen in der Abfallwirtschaft, Ph.D. thesis, Vienna University of Technology, Vienna, 1999.
119. Rechberger, H. and Brunner, P.H., A new, entropy-based method to support waste and resource management decisions, *Environ. Sci. Technol.*, 36, 809, 2002.
120. Vogel, F., *Beschreibende und schließende Statistik*, R. Oldenbourg, Vienna, 1996.
121. Boltzmann, L., *Vorlesungen über Gastheorie*, Barth, Leipzig, 1923.
122. Shannon, C.E., A mathematical theory of communications, *Bell System Tech. J.* (formerly *AT&T Tech. J.*), 27, 379, 1948.
123. Clausius, R., Über verschiedene für die Anwendungen bequeme Formen der Hauptgleichungen der mechanischen Wärmetheorie, *Poggendorff's Annalen*, 125, 353, 1856.
124. Obernosterer, R., Brunner, P.H., Daxbeck, H., Gagan, T., Glenck, E., Hendriks, C., Morf, L., Paumann, R., and Reiner, I., Materials accounting as a tool for decision making in environmental policy — Mac TEmPo case study report: urban metabolism, the city of Vienna, Institute for Water Quality and Waste Management, Vienna University of Technology, Vienna, 1998.
125. Brunner, P.H., Daxbeck, H., and Baccini, P., Industrial metabolism at the regional and local level: a case study on a Swiss region, in *Industrial Metabolism — Restructuring for Sustainable Development*, Ayres, R.U. and Simonis, U.E., Eds., United Nations University Press, New York, 1994, p. 163.
126. Rechberger, H. and Brunner, P.H., Die Methode der Stoffkonzentrierungseffizienz (SKE) zur Bewertung von Stoffbilanzen in der Abfallwirtschaft, in *Müll-Handbuch*, Hösel, G., Bilitewski, B., Schenkel, W., and Schnurer, H., Eds., Erich Schmidt Verlag, Berlin, 2002, chap. 8506.1.
127. Schachermayer, E., Bauer, G., Ritter, E., and Brunner, P.H., Messung der Güter- und Stoffbilanz einer Müllverbrennungsanlage (Projekt MAPE), Monographie No. 56, Environmental Protection Agency, Vienna, 1995.
128. Fehringer, R., Rechberger, H., Pesonen, H.-L., and Brunner, P.H., Auswirkungen unterschiedlicher Szenarien der thermischen Verwertung von Abfällen in Österreich (Projekt ASTRA), Vienna University of Technology, Institute for Water Quality and Waste Management, Vienna, 1997.
129. Paumann, R., Obernosterer, R., and Brunner, P.H., Wechselwirkungen zwischen anthropogenem und natürlichem Stoffhaushalt der Stadt Wien am Beispiel von Kohlenstoff, Stickstoff und Blei, Vienna University of Technology, Institute for Water Quality and Waste Management, Vienna, 1997.

130. Krauskopf, K.B., *Introduction to Geochemistry*, 2nd ed., McGraw-Hill, New York, 1979.
131. Legarth, J.B., Alting, L., Danzer, B., Tartler, D., Brodersen, K., Scheller, H., and Feldmann, K., A new strategy in the recycling of printed circuit boards, *Circuit World*, 21, 10, 1995.
132. Tauber, C., *Spurenelemente in Flugaschen*, Verlag TÜV Rheinland GmbH, Köln, 1988.
133. Heyde, M. and Kremer, M., *Recycling and Recovery of Plastics from Packagings in Domestic Waste*, Eco-Informa Press, Bayreuth, Germany, 1999.
134. Stahel, U., Fecker, I., Förster, R., Maillefer, C., and Reusser, L., Bewertung von Ökoinventaren für Verpackungen, Schriftenreihe Umwelt No. 300, BUWAL, Bern, Switzerland, 1998.
135. Huijbregts, M.A.J., Priority assessment of toxic substances in the frame of LCA: time horizon dependency of toxicity potentials calculated with the multi-media fate, exposure and effects model USES-LCA, Institute for Biodiversity and Ecosystem Dynamics, University of Amsterdam, 2000; www.leidenuniv.nl/interfac/cml/lca2/ (accessed April 2002).
136. Ayres, R.U., Martinás, K., and Ayres, L.W., Eco-thermodynamics: exergy and life cycle analysis, Working Paper 96/04/EPS, INSEAD, Fontainebleau, France, 1996.

3 Case Studies

Looking at the result of a material flow analysis (MFA), it seems easy and straightforward to define the system, collect the data, calculate the results, and draw conclusions. In practice, one starts not with the result but, quite often with a poorly defined, highly complex problem that must first be simplified and structured. After the goals of an MFA have been clearly defined, the real art consists of skillfully designing a system of boundaries, processes, flows, and stocks that facilitates solution of a given problem at the least cost. As in any other art, a precondition for mastering the art is repetitive use of the basic tools. The more experience one gains, the easier it becomes to set up an appropriate and cost-effective system. An expert skilled in MFA will be able to define a system in any new field quite efficiently with only a few alterations of the initial draft. Beginners will often have to revise their systems many times to cope with incomplete information about the important processes, stocks, and flows within the system; inappropriate systems boundaries; missing, bad, or incompatible data; etc.

MFA is usually a multidisciplinary task. Materials flow through many branches of an economy, and they cross boundaries such as the interfaces "anthroposphere–environment" or "water–air–soil." Hence, it is of prime importance to seek advice from experts of those disciplines that are relevant to a particular MFA. For example, if regional nutrient balances are investigated by MFA, it is necessary to include the knowledge of partners from agriculture, nutrition, sewage treatment, water quality, and hydrology, either by forming a project team or by engaging the experts as consultants when needed. Sometimes this cooperation leads to new research questions, since research within individual disciplines may be stimulated by specialized knowledge from other fields of study (see Section 3.1.2).

An MFA can be a time-consuming and costly task. This is especially true if an MFA is performed for the first time in a new field, such as a study of regional heavy metal flows (see Sections 3.1.1 and 3.4.1). It may well be that the basic data of the region — hydrological data on precipitation, evaporation, flows of surface water and groundwater, and groundwater stock — have not been previously assessed. Without a requisite set of base information, an MFA cannot succeed. In such cases, additional human and financial resources will be needed.

Performing an MFA in a particular field for the first time is a distinctly different and more difficult task than repeating the analysis for additional materials (heavy metals → nutrients) or for further time periods. The latter two tasks require much less effort because the system has been set up and basic data have already been collected. If the costs for an initial MFA seem to be high, one should take into account the fact that the basic data can be used for future MFA and similar consecutive studies, such as annual environmental reporting.

The following 14 case studies demonstrate how MFA can be applied in the fields of environmental management, resource management, and waste management. The final case study — an example of regional materials management (lead) — is provided to demonstrate that MFA is especially well suited to address problems related to multiple fields, such as the three mentioned here. The regional lead study by MFA was not undertaken to address a specific problem, but it revealed conclusions important for the three fields of environmental management, resource management, and waste management. The case studies are intended to increase the reader's experience of MFA applications. The reader is also directed to review some of the original literature cited in the case studies. Nevertheless, for those who want to master the fine art[1] of MFA, it will be indispensable to gain additional experience by performing as many MFA studies as possible on their own.

3.1 ENVIRONMENTAL MANAGEMENT

Most material flow analysis studies have been undertaken to solve problems related to environmental management. A recent overview of the potential of MFA in this field is given in MAcTEmPo.[2] In general, MFA is a tool well suited for:

- Early recognition of environmental loadings
- Linking of emissions to sources and vice versa
- Setting of priorities for management measures
- Designing new processes, goods, and systems in view of environmental constraints

As seen in Chapter 2, an MFA is usually the starting point of any life-cycle assessment (LCA) and environmental impact statement. It is also useful as a base for an environmental management and audit system (EMAS) at the company level (see Chapter 3.1.4). If a company's financial accounting system is linked to a material input–output flow and stock analysis, it can be efficiently used to measure the company's environmental performance. The following case studies demonstrate that MFA can be used to investigate

- Single-substance issues (e.g., emissions of heavy metals or nutrients)
- Multisubstance problems (e.g., environmental impact statement of a coal-fired power plant)

They also show the wide scale of spatial application: A single power plant, a small region of 66 km², and a large watershed such as the entire River Danube with 820,000 km² can all be investigated using the same MFA approach.

3.1.1 CASE STUDY 1: REGIONAL LEAD POLLUTION

Heavy metals are important substances for both economic as well as environmental reasons. Due to their physical-chemical properties, they can withstand weathering (zinc coatings of steel), improve the properties of other materials (chromium in steel,

Case Studies

cadmium as an additive in polyvinyl chloride [PVC]), or serve to improve the efficiency of energy systems (lead in gasoline, mercury in batteries). Some heavy metals are not essential for the biosphere, but many are toxic for humans, animals, plants, and microorganisms. It is thus important to control the flows and stocks of heavy metals to avoid harmful flows and accumulations and to make the best use of heavy metals as resources.

This case study is taken from a comprehensive study (RESUB[3]) on the flows and stocks of 12 elements in a Swiss region (Bunz Valley) of 66 km^2 and 28,000 inhabitants. The purpose was to develop a methodology to assess material flows and stocks within, into, and out of a region in a thorough and integrated way. In addition, the significance of the findings for the management of resources and the environment was to be investigated. There was no given goal in view of environmental management. The case study portrayed in this chapter represents merely a small fraction of the entire RESUB project. Only the flows and stocks of lead relevant to environmental management are discussed. The implications of these flows and stocks for resource management are examined in Chapter 3.4.1. The detailed procedure described below confirms that an MFA is a multidisciplinary task that requires knowledge, information, and support from many fields.

3.1.1.1 Procedures

In a first step, the region is defined according to Figure 3.1. For the spatial boundary, the administrative boundary of the region "Bunz Valley" is chosen because, by chance, this border coincides well with the hydrological boundary. (This is often the case in mountainous or hilly regions, where the watershed serves well to delineate an administrative boundary.) Because water flow is fundamental for many material flows, it is important that a reliable regional water balance be established. If the spatial boundary does not coincide with the hydrological boundary, it may be difficult to establish a water balance. Hence, it is often crucial to find a good compromise between regional boundaries that match the administrative region, thus allowing the use of data collected by the regional administration, and hydrological boundaries that yield a consistent water balance.

As a boundary in time, a period of one year is selected because existing data about the anthroposphere (e.g., tax revenues, population data, and fuel consumption) and the environment (e.g., data on precipitation, surface water flow, and concentrations in soil and groundwater) show that a sampling period of one year is representative for the region during the period of 1985 to 1990.

In this chapter, each process is labeled with a letter and each flow with a number. These letters and numbers help to identify the corresponding processes and flows in Figure 3.1, Tables 3.1 to 3.10, and in the calculations at the end of this chapter. The system is defined by the following 10 processes and 20 flows of goods.

3.1.1.1.1 Private Households

The process "private households" (PHH) summarizes the flows and stocks of materials through the 9300 private households of the region. Import goods (1) relevant for lead comprise leaded gasoline (in the process of being phased out) and consumer

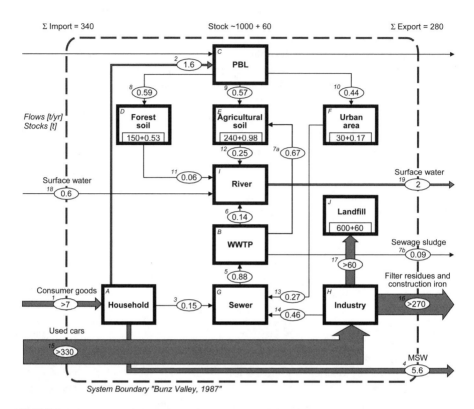

FIGURE 3.1 Results of the MFA of lead flows (t/year) and stocks (t) through the Bunz valley.[3] (From Brunner, P.H., Daxbeck, H., and Baccini, P., Industrial metabolism at the regional and local level: a case study on a Swiss region, in *Industrial Metabolism — Restructuring for Sustainable Development*, Ayres, R.U. and Simonis, U.E., Eds., United Nations University Press, Tokyo, 1994. With permission.)

goods such as lead in stabilizers, caps topping wine bottles, etc. Output goods are exhaust gas from cars (2), sewage (3), and municipal solid waste (MSW) (4).

The lead flows through private households are calculated as follows. Lead input in consumer goods is calculated based on the flows of sewage and MSW. This is a major shortcoming, since neither stocks nor flows of construction materials and appliances in private households are taken into account. To measure flows of lead in such goods is an extremely laborious and costly task. Thus, for this study, it is assumed that all lead that enters households leaves them within one year. This hypothesis is incorrect, since it does not account for the lead stock in households. Nevertheless, it is estimated that this error is of little relevance for the conclusions and the overall lead balance of the region.

Figures for lead in sewage are calculated as follows. The number of inhabitants (capita) connected to the sewer system (percentage of the regional population) is multiplied by per capita lead-emission factors (g/capita/year) determined elsewhere in similar regions.

Case Studies

Lead in MSW is similarly calculated. The number of inhabitants (capita) times MSW generation rate (kg/capita/year) times lead concentration in MSW (g/kg) yields the lead flow in MSW. MSW generation rate is available from regional waste-management companies. Lead concentration in MSW is taken from measurements of the residues of waste incineration (see Section 3.3.1).

It is assumed that all lead emitted by car exhausts stems from leaded gasoline. Fuel consumed for room heating contains less than 0.05 t of lead (calculated as the amount of fuel consumed times lead concentration in fuel) and is not taken into further consideration. Lead in car exhaust is calculated as follows. The number of cars licensed in the region is multiplied by the average mileage (in km/year) of a car (taken from national statistics), the average consumption of gasoline per km (l/km, from car manufacturers' statistics) and the mean lead content of the gasoline (mg/l, from gasoline producers and federal statistics). The results are crosschecked by figures from regional traffic monitoring and a model that takes into account the road network (fractions of highways, urban roads, roads outside of settlement areas) and speed-dependent emission factors (figures for lead concentration in gasoline are kept constant). Lead emissions of trucks are considered to be small and are neglected because, in this region, trucks are operated on diesel only, and diesel does not contain significant amounts of lead.

Figures about the total input into households are rounded because they do not include lead-containing goods that contribute to the stock and thus have little effect on accuracy. For the overall results and conclusions, this accuracy is sufficient. If it turns out that, for the conclusions, the difference between the calculated and the rounded value is decisive, a more thorough investigation into the lead flows through private households would become necessary.

For detailed information about the calculation, see Table 3.1.

3.1.1.1.2 Wastewater Treatment Plant

In the process "wastewater treatment plant" (WWTP), wastewater (5) is treated, resulting in cleaned wastewater (6), sewage sludge (7a and b), off-gas, and small amounts of sievings and sandy sediments. Due to the chemical species of lead in wastewater, off-gas is of no quantitative relevance for this heavy metal. Preliminary sampling and chemical analysis of the concentrations in sievings and sediments show that the amount of lead in these fractions is small. Hence, most lead leaves the WWTP in sewage sludge and purified wastewater. The flows of wastewater, purified wastewater, and sludge are measured during one year (m^3/year) and are sampled and analyzed for lead (g/m^3). The flow of wastewater is determined by a venturi device at the inflow of the WWTP. Samples are taken continuously from wastewater and purified wastewater by a so-called Q/s sampler that samples proportional to the water flow. The flow of sewage sludge is measured as the total volume of sludge transferred to sludge transport vehicles during one year. Samples are taken whenever sludge is transferred to transport vehicles. There are three WWTPs in the region: one large plant and two small plants. Only the large and one of the small plants are included in the measuring campaign. For the third plant, the same material flows and balances are anticipated as for the other small treatment plant. The data of the three plants are summarized as a single WWTP process.

TABLE 3.1
Calculation of Lead Flows through Process "Private Household"

Flow No.	Operator	Description	Units	Value
		Stock		
		initial value	kg	not considered
		rate of change	kg Pb/year	not considered
		Inputs		
1		Consumer goods and leaded gasoline (balanced):		
		exhaust gas (2)	kg Pb/year	1,596
	+	sewage (3)	kg Pb/year	151
	+	MSW (4)	kg Pb/year	5,600
	=	Total lead flow	kg Pb/year	7,347
		Outputs		
2		Exhaust gas:		
		number of cars	cars	14,000
	×	mileage	km/car/year	15,000
	×	consumption of gasoline	l/km	0.08
	×	lead content of gasoline	mg Pb/l	95
	=	Total lead flow	mg Pb/year	1.6×10^9
	=		kg Pb/year	1,596
3		Household sewage:		
		number of inhabitants	capita	28,000
	×	connected to sewer system	—	1
	×	lead emission per capita	g Pb/capita/year	5.4
	=	Total lead flow	g Pb/year	1.51×10^5
	=		kg Pb/year	151
4		Municipal solid waste (MSW):		
		number of inhabitants	capita	28,000
	×	MSW generation rate	kg/capita/year	400
	×	lead concentration in MSW	g Pb/kg MSW	0.5
	=	Total lead flow	g Pb/year	5.6×10^6
			kg Pb/year	5,600

Note: Process A in Figure 3.1.

For detailed information about the calculation, see Table 3.2.

3.1.1.1.3 Planetary Boundary Layer

The process "planetary boundary layer" (PBL) denotes the lowest layer of the atmosphere. It is about 500 m high and is well suited as a "distribution" process for the RESUB case study. In regional studies, it is usually not possible to measure a material balance of the PBL because it is a daunting analytical and modeling task. However, it is possible to make certain assumptions and simplifications that allow

TABLE 3.2
Calculation of Lead Flows through Process "Wastewater Treatment Plant"

Flow No.	Operator	Description	Units	Value 1	Value 2	Total Value
			Stock			
		initial value	kg	not considered		
		rate of change	kg Pb/year	−16		
			Inputs			
5		WWTP input:		TP 1	TP 2 + 3	TP 1 + 2 + 3
		wastewater flow	l/year	6.1×10^9	2.26×10^9	8.36×10^9
	×	lead concentration	µg Pb/l	121.7	59.2	
	=	Total lead flow	µg Pb/year	7.42×10^{11}	1.34×10^{11}	
	=		kg Pb/year	742	134	876
			Outputs			
6		WWTP output:		TP 1	TP 2 + 3	TP 1 + 2 + 3
		purified water flow	l/year	6.1×10^9	2.26×10^9	8.36×10^9
	×	lead concentration	µg Pb/l	20.7	6.3	
	=	Total lead flow	µg Pb/year	1.26×10^{11}	1.42×10^{10}	
	=		kg Pb/year	126	14	140
7a		Sewage sludge (used):		TP 1	TP 2 + 3	TP 1 + 2 + 3
		sludge flow	kg dry/year	8.06×10^5	2.14×10^5	1.02×10^6
	×	lead concentration	mg Pb/kg dry	875	216	
	×	used inside of the region	%	92	36	
	=	Total lead flow	mg Pb/year	6.49×10^8	1.66×10^7	
	=		kg Pb/year	649	17	665[a]
7b		Sewage sludge (exported):		TP 1	TP 2 + 3	TP 1 + 2 + 3
		sludge flow	kg dry/year	8.06×10^5	2.14×10^5	1.02×10^6
	×	lead concentration	mg Pb/kg dry	875	216	
	×	exported out of the region	%	8	64	
	=	Total lead flow	mg Pb/year	5.64×10^7	2.96×10^7	
	=		kg Pb/year	56	30	86

Note: Process B in Figure 3.1.

[a] Total value does not equal the sum of the value 1 and 2 due to rounded values.

using the PBL as a suitable process to account for flows from the atmosphere to the soil and vice versa.

Because the case study region is surrounded by regions that have basically the same metabolic characteristics, it is reasonable to suppose that the emissions into air are similar for all surrounding regions. Thus, it can be assumed that imported lead corresponds to exported lead, i.e., that the amount of lead imported and deposited in the region equals the amount of domestic lead exported and deposited outside of the region. Note that the total flow of lead through the PBL (not given

TABLE 3.3
Calculation of Lead Flows through Process "Planetary Boundary Layer"

Flow No.	Operator	Description	Units	Value 1	Value 2	Value 3
			Stock			
		initial value	kg	not considered		
		rate of change	kg Pb/year	0		
			Inputs			
2		Exhaust gas:[a]				
		Total lead flow	kg Pb/year	1596		
			Outputs			
8, 9, 10		Deposition:		forest	agriculture	urban
		deposition ratio	—	3	1	5
		deposition rate	kg Pb/ha/year	0.294	0.098	0.490
	×	surface	ha	2000	3700	900
	+	additional lead (roads)	kg Pb/year	—	200	—
	=	Total lead flow	kg Pb/year	588	563	441

Note: Process C in Figure 3.1.

[a] Data from PHH (see Table 3.1).

in Figure 3.1) is about three to four times larger than the flow of lead deposited. The flows from the PBL to the soil consist of wet and dry depositions to forest (8), agricultural land (9), and urban (10) soils. The lead flows to the soil are determined by two methods.[4]

First, based on the assumption of uniform metabolism of all neighboring regions, the total regional emission to PBL is divided among the land areas, taking into account differences in vegetation and surface. Second, preliminary measurements at 11 sampling stations throughout the region show little significant differences for long measuring periods. Therefore, for a period of one year, wet and dry deposition of lead is measured only at two sampling stations in the region. Based on the wet and dry deposition results, and on models given in Beer,[4] lead flows are calculated for the corresponding soils. The deposition on urban soils takes into account that most lead is emitted in the proximity of roads, thus the load per hectare of urban soils is comparatively larger than that of agricultural and forest soils. Results from the two approaches agree fairly well. The method based on actual measurements of dry and wet depositions yields values about 30% higher than the distribution of the total emissions over the region.

For detailed information about the calculation, see Table 3.3.

3.1.1.1.4 Lead Flows and Stocks in Soils

The region consists of 3700 ha soil used for agriculture, 2000 ha forest soil, and 900 ha settlement area. The area actually covered with buildings, roads, and other

Case Studies 175

TABLE 3.4
Calculation of Lead Flows through Process "Forest Soil"

Flow No.	Operator	Description	Units	Value
		Stock		
		initial value	kg	150,000
		rate of change	kg Pb/year	529
		Inputs		
8		Deposition:[a]		
		Total lead flow	kg Pb/year	588
		Outputs		
11		Runoff:		
		deposition	kg Pb/year	588
	×	runoff factor	—	0.1
	=	Total lead flow	kg Pb/year	59

Note: Process D in Figure 3.1.

[a] Data from PBL (see Table 3.3).

constructions is much smaller. The hydrological balance reveals the water flow to and from the forest soil and the agricultural soil. Precipitation (measured by continuous automatic rain measurement) minus evaporation (estimated with various models and finally calculated according to Primault[5]) yields the net water flow to the soil. This amount of water is divided among the fractions of agriculture and forest soils, taking into account differences in evapotranspiration of forests and agricultural crops. The water reaching the soil is divided into surface runoff and interflow (both reaching surface waters) and the fraction seeping to groundwater. The concentrations of lead in soil leachate are estimated based on another research project on heavy metal mobility in soils.[6] Erosion is approximated according to Von Steiger and Baccini.[7] Both assessments are individually tailored for forest and agricultural soils. The calculations show that 10% of the deposited lead on forest soil (11), 20% of agricultural soil (12), and up to 60% of built-up areas (13) can be found in the runoff. This lead is transported to receiving waters.

For detailed information about the calculation, see Table 3.4, Table 3.5, and Table 3.6.

3.1.1.1.5 Sewer System

The mixed sewer system receives wastewater from private households (3), urban surface runoff (13), and industry (14) and transports the resulting sewage (5) to the WWTP. Lead in wastewater from industry can be estimated by balancing the process "sewer," taking into account lead flows in wastewater from private households, in surface runoff from urban areas, and in the resulting sewage.

TABLE 3.5
Calculation of Lead Flows through Process "Agricultural Soil"

Flow No.	Operator	Description	Units	Value
		Stock		
		initial value	kg	240,000
		rate of change	kg Pb/year	982
		Inputs		
9		Deposition:		
	+	from PBL	kg Pb/year	563
	+	from WWTP	kg Pb/year	665
	=	Total lead flow	kg Pb/year	1,228
		Outputs		
12		Runoff:		
		deposition	kg Pb/year	1,228
	×	runoff factor	—	0.2
	=	Total lead flow	kg Pb/year	246

Note: Process E in Figure 3.1.

TABLE 3.6
Calculation of Lead Flows through Process "Urban Areas"

Flow No.	Operator	Description	Units	Value 1	Value 2	Total Value
		Stock				
		initial value	kg	30,000		
		rate of change	kg Pb/year	176		
		Inputs				
10		Deposition:[a]				
		Total lead flow	kg Pb/year	441		
		Outputs				
13		Runoff:		buildings	green	buildings + green
		deposition	kg Pb/year	221	221	
	×	runoff factor	—	1.00	0.20	
	=	Total lead flow	kg Pb/year	221	44	265

Note: Process F in Figure 3.1.

[a] Data from PBL (see Table 3.3).

TABLE 3.7
Calculation of Lead Flows through Process "Sewer"

Flow No.	Operator	Description	Units	Value
		Stock		
		initial value	kg	n.d.[a]
		rate of change	kg Pb/year	0
		Inputs		
3		Household sewage:[b]		
		Total lead flow	kg Pb/year	151
13		Urban area runoff:[c]		
		Total lead flow	kg Pb/year	265
14		Industry sewage (balanced):		
		WWTP input (5)	kg Pb/year	876
	−	PHH sewage (3)	kg Pb/year	151
	−	UA runoff (13)	kg Pb/year	265
	=	Total lead flow	kg Pb/year	460
		Outputs		
5		WWTP input:[d]		
		Total lead flow	kg Pb/year	876

Note: Process G in Figure 3.1.

[a] n.d. = not determined.
[b] Data from PHH (see Table 3.1).
[c] Data from UA (see Table 3.6).
[d] Data from WWTP (see Table 3.2).

For detailed information about the calculation, see Table 3.7.

3.1.1.1.6 Industry

The process "industry" proves to be a real challenge. Despite the region's small size and low population, there are 1,300 companies with 11,000 employees active in the region. The main task is to find within this large number those companies that play an important role in the regional lead flow. As a first step, all sectors except the production sector are removed from the list. Of the remaining 323 companies, those with less than 20 employees are eliminated. The remaining 102 businesses are included in the investigation, which consists of an interview and a questionnaire about the material flows and stocks of each company. Of these, 61 companies cooperate actively and supply comprehensive data about their material turnover. It appears that only a few are handling lead-containing goods. A car shredder and an adjacent iron smelter using shredded cars to produce construction rods are dominating the process "industry." Hence, the inputs into the process "industry" are used cars (15). The outputs comprise, on the one hand, construction rods (16a) and filter residues (16b) of the smelter that are exported. On the other hand, the car shredder produces organic shredder residues (17) consisting of plastics, textiles, and biomass (wood, paper, leather, and hair) mixed with residual metals of every kind. These so-called automotive shredder residues (ASR) are landfilled within the region.

The lead flows through "industry" are assessed as follows. Input in "industry" is calculated as the number of used cars times the concentration of lead in a car. This yields a minimal figure. It may well be that other goods are shredded and treated in the shredder as well. (Note that this uncertainty is not important for the final conclusion. It would be the same even if double the amount of lead were used in "industry.") The number of cars processed by the shredder is supplied by the shredder operator. Figures for lead concentration in used cars are found in the literature or can be received from car manufacturers. The smelter operator supplies figures about the amount of construction steel produced, the amount of filter residue exported, as well as the lead concentration in the steel and the filter residue. The latter figure is confirmed by local authorities, who periodically monitor emissions of the smelter. Note that there is no emission flow given for the smelter in Figure 3.1. This is due to the excellent air-pollution control (APC) device of the smelter, which keeps annual emissions at a level that is orders of magnitude below lead emissions from gasoline. Thus the smelter is of no relevance for lead emissions in the region anymore. The amount of lead in industry sewage is similar to other regional lead flows.

For detailed information about the calculation, see Table 3.8 and Table 3.9.

3.1.1.1.7 Surface Waters

The process "river" consists of a river flowing through the region and a small tributary originating predominantly within the region. Groundwater entering or leaving the region does not play an important role. The process "river" receives water from the river inflow (10), the agricultural soil (11), the forest soil (12), and the WWTP (8). The river outflow (13) leaves the region. The direct lead flow from settlement areas to surface waters has not been taken into account. First, the region has a mixed sewer system, and most urban runoff is collected and treated in the WWTP. Second, the settlement area is small (<10%) in comparison with the agricultural and forest soils; hence neglecting this flow may be justified. The surface water flow has been determined in the course of a complete water balance that is measured for the comprehensive RESUB project. Existing measuring stations continuously record the flow of river water in and out of the region. Since these stations are not located at the exact systems boundaries, the differences are compensated by taking into account the area contributing water to the river. The river water is sampled continuously at the same stations with Q/s samplers. It turns out that these state-of-the-art samplers are well suited to collect dissolved substances and suspended particles, but they cannot catch aliquots of large particles. In the course of a rainstorm, when the river transports larger particles, debris, and chunks of biomass, the samplers are not working appropriately. In addition to the limits in sampling technology, there are practical problems. On several occasions, large water flows during heavy storms have destroyed or carried off the sampling equipment. Hence, it is advisable to have short sampling periods (e.g., one week). In case of invalid sampling, the missing data cover a smaller fraction of the total measuring period and thus are of less weight.

Case Studies

TABLE 3.8
Calculation of Lead Flows through Process "Industry"

Flow No.	Operator	Description	Units	Value
		Stock		
		initial value	kg	not considered
		rate of change	kg Pb/year	0
		Inputs		
15a		Used cars:		
		number of used cars	cars/year	120,000
	×	lead per car (excl. battery)	kg Pb/car	2.5
	=	Total lead flow	kg Pb/year	300,000
15b		Scrap metal:		
		scrap metal	kg/year	6.50×10^7
	×	lead content	kg Pb/kg	0.0005
	=	Total lead flow	kg Pb/year	32,500
15		Industry input:		
		used cars (15a)	kg Pb/year	300,000
	+	scrap metal (15b)	kg Pb/year	32,500
	=	Total lead flow	kg Pb/year	332,500
		Outputs		
16a		Construction iron:		
		construction iron	kg/year	1.45×10^8
	×	lead content	kg Pb/kg	0.0005
	=	Total lead flow	kg Pb/year	72,500
16b		Filter residues:		
		filter residues	kg/year	1.50×10^7
	×	lead content	kg Pb/kg	0.0133
	=	Total lead flow	kg Pb/year	200,000
16		Industry output:		
		construction iron (16a)	kg Pb/year	72,500
	+	filter residues (16b)	kg Pb/year	200,000
	=	Total lead flow	kg Pb/year	272,500
14		Industry sewage:[a]		
		Total lead flow	kg Pb/year	460
17		Automotive shredder residues (balanced):		
		industry input (15)	kg Pb/year	332,500
	−	industry output (16)	kg Pb/year	272,500
	−	industry sewage (14)	kg Pb/year	460
	=	Total lead flow	kg Pb/year	59,540

Note: Process H in Figure 3.1.

[a] Data from Sewer (see Table 3.7).

TABLE 3.9
Calculation of Lead Flows through Process "Landfill"

Flow No.	Description	Units	Value
	Stock		
	initial value	kg	600,000
	rate of change	kg Pb/year	59,540
	Inputs		
17	ASR:[a]		
	Total lead flow	kg Pb/year	59,540

Note: Process J in Figure 3.1.

[a] Data from Industry (see Table 3.8).

Given these shortcomings, values for lead flows in the river must be regarded as minimum flows.

For detailed information about the calculation, see Table 3.10.

3.1.1.2 Results

In this chapter, conclusions are drawn regarding the use of MFA for environmental management. In particular, the results are used to show how MFA serves to provide early recognition of environmental hazards, how it can be used to establish priorities for environmental measures, and how MFA can be used for efficient environmental monitoring. In Section 3.4.1, the same case study is further used to point out the potential for regional materials management and resource conservation. The numerical results of the MFA analysis of lead in the region are given in Figure 3.1.

3.1.1.2.1 Early Recognition of Environmental Hazards

The difference between lead import and lead export amounts to approximately 60 t/year. Hence, lead is accumulated in the region. The existing stock of lead totals about 1000 t. A "doubling time" for the lead stock of 17 years can be calculated. In other words, if the regional flows of lead remain the same for the next 100 years, the stock will have increased from 1000 t to 7000 t! (Note that according to Chapter 1, Section 1.4.5.1, there is no indication yet that lead flows will decrease; on the contrary, based on past developments, they are likely to increase further.) Without the present study, this buildup of lead occurs unnoticed. As shown in Chapter 1, Section 1.4.5.4, such accumulations of substances are a rule for all urban regions. What makes this case study special is the huge extent of the accumulation. About one-sixth of the lead imported stays "forever" within the region. Thus, it is highly important to investigate the fate of potentially toxic lead in the region. Does the accumulation in the soil result in an increase of lead in plants up to a level of concern

TABLE 3.10
Calculation of Lead Flows through Process "River"

Flow No.	Operator	Description	Units	Value 1	Value 2	Total Value
			Stock			
		initial value	kg	not considered		
		rate of change	kg Pb/year	−948		
			Inputs			
18		Surface water (import):		Holzbach	Bunz	sum
		water flow	l/year	3.69×10^9	3.19×10^{10}	3.56×10^{10}
	×	lead concentration	µg Pb/l	4.6	18.4	
	=	Total lead flow	µg/year	1.70×10^{10}	5.87×10^{11}	
			kg Pb/year	17	587	604
11		Forest runoff:[a]				
		Total lead flow	kg Pb/year	59		
12		Agricultural runoff:[b]				
		Total lead flow	kg Pb/year	246		
6		WWTP output:[c]				
		Total lead flow	kg Pb/year	140		
			Outputs			
19		Surface water (export):				
		water flow	l/year	6.70×10^{10}		
	×	lead concentration	µg Pb/l	29.8		
	=	Total lead flow	µg/year	2.00×10^{12}		
	=		kg Pb/year	1,997		

Note: Process I in Figure 3.1.

[a] Data from Forest Soil (see Table 3.4).
[b] Data from Agricultural Soil (see Table 3.5).
[c] Data from WWTP (see Table 3.2).

---

for animal or human food? Will there be a steady increase in lead flows from the soil to the surface water? When will lead concentrations reach a level that endangers the standards for surface water or drinking water? What about the concentration of lead in dust, will it increase, too?

While MFA is helpful in identifying the problem and formulating relevant questions, the questions cannot be answered by simple MFA alone. It is necessary to engage specific experts, e.g., in the field of metal transfer between soil and plants, between soil and surface water and groundwater, and between soil and air. The merit of MFA is the ability to identify a future environmental problem that has neither been on the agenda nor even known before the study is begun.

From an environmental point of view, the largest flow, stock, and accumulation in stock is caused by the lead imported in used cars, shredded, and landfilled as automotive shredder residue (ASR). The landfill is by far the main regional "accumulator" for lead. Assuming similar practice over the past 10 years, it can be estimated that approximately 600 t of lead are buried in the landfill. This is due to the fact that the separation of lead by the car shredder is incomplete. Some elemental lead as well as lead compounds in plastic additives are transferred to the ASR. The hypothesis that ASR may contribute to the pollution of the regional hydrosphere is discussed below.

Lead is accumulated in the soil, too. The doubling periods are between 170 (urban soil) and 280 (agricultural soil) years. If the use of lead continues in the same way, standards for lead concentrations in soils will be exceeded in the future. It is a matter of soil-protection strategy (and, in a wider perspective, environmental protection) whether such a slow approach to a limit needs to be controlled or not, and if so, when. The "filling up" strategy raises the question of what options future generations will have when they inherit the "full" soil. If soil inputs and outputs can be kept in equilibrium, the concentrations in the soil will remain constant. The case-study region will come close to this condition if leaded gasoline is phased out. (Note that since the residence time of aerosol-borne lead in the atmosphere is several days, which facilitates the transport of lead over long distances, this measure will be effective only if neighboring regions adopt the same strategy.) Lead concentration increases fastest in urban soils. MFA results suggest that persons eating food grown in urban areas, such as home gardeners, consume the highest amount of heavy metals such as lead. Hence, material balances and analysis of urban soils and gardens are needed in order to protect consumers of home-grown products in cities.

3.1.1.2.2 Establish Priorities for Environmental Measures

The landfill is the most important stock of lead, so it must be investigated and controlled first. Based on the following calculation, it can be hypothesized that lead is leaching into the groundwater and surface waters. The balance of the process "river" reveals a deficit of 0.95 t/year. Compared with the lead flows from soils and WWTP, this is a large figure. The most likely stock that can lead to such a large flow is the landfill stock. At present, the fate of lead in the landfill is not known. Since much of the ASR has been landfilled without bottom and top liners, it may be that lead is leaching into ground and surface waters. Even if the landfills are constructed according to the state of the art, with impermeable bottom and top liners, it still has to be expected that the liners will become permeable over the long run (>100 years), thus polluting ground and surface waters for long time periods.[8]

The following issues are crucial for possible mobilization and emission of lead from the landfill: the interaction of the landfill body with the surroundings (atmosphere, precipitation), the transformations of ASR within the landfill due to biochemical and chemical reactions, and the chemical speciation of the lead in ASR. Investigations into these aspects are specialized tasks that cannot be performed by MFA. Such investigations require in-depth analysis by experts in the field of transformation and leaching processes in soils and landfills. It is not possible within the case study to follow up this hypothesis (e.g., by collecting leachates and analyzing it for lead). In any case, it is of first priority to ensure that the ASR is landfilled in a manner that ensures long-term immobilization of heavy metals. If disposing of

# Case Studies

raw ASR leads to leaching of heavy metals and significant water pollution, pretreatment of ASR before landfilling will be mandatory.

The second largest regional flow of lead is due to MSW, which is exported from the region and incinerated. This flow is an order of magnitude larger than the lead flow in sewage. Thus, compost from MSW is less suited for application to land than sewage sludge because it will overload the soil with lead in a comparatively short time. For this region, it is recommended that sewage sludge not be applied to soils as well, since it loads the soil with additional lead. It is clear that a decision to use compost or sludge in agriculture cannot be based on lead alone. The approach taken here is exemplary and also has to be applied to a full range of substances such as other heavy metals, nutrients, and organic substances.

As mentioned above, MSW is incinerated outside the region. In order to protect the soil from the large flow of lead in MSW, thermal treatment of MSW has to be combined with efficient air-pollution control. If incineration transfers less than 0.01% of lead into air, the resulting increase of lead in soil will be below 1% in 8000 years (assuming uniform deposition within the region). State-of-the-art MSW incinerators exhibit such transfer coefficients for lead to the atmosphere. Regarding the smelter, the MFA supports the conclusion that the emissions to air are of no priority ever since the furnace has been equipped with a high-efficiency fabric filter system, reducing the lead emissions to less than 50 kg/year.

3.1.1.2.3 Environmental Monitoring

Once an MFA of a region is established, many opportunities for monitoring substance flows and stocks arise:

MFA can replace soil monitoring programs. Such programs are costly and are limited in their forecasting capabilities. If statistically significant changes in soil concentrations are to be detected by traditional soil monitoring, then (1) either very intensive sampling programs with large numbers of samples or (2) long time periods are required. Because the funds for such intense sampling are not usually available, it takes decades until significant changes in the soil become visible. However, with a single measuring campaign, MFA can predict how the soil concentration will evolve over time. The results indicate whether there is a danger before high concentrations are reached. If inputs to the soil are changed, e.g., through the addition of sewage sludge or a ban on leaded gasoline, the effect of such measures can be evaluated by an MFA before they are implemented. In contrast, traditional soil monitoring would take years to confirm accumulation or depletion of soil pollutants in a statistically significant way.

Combining MFA with the analysis of sewage sludge allows monitoring of the process "industry." For example, before the smelter was equipped with high-efficiency fabric filters, a wet scrubbing system removed metals from the off-gas stream. By accident, it happened that some of the lead-loaded scrubber effluent reached the sewer system and severely contaminated the activated sludge in the treatment plant. While wastewater has a short residence time in the treatment plant, sewage sludge in a digester or storage tank represents a "memory" of several weeks. Thus sludge samples from the digester can show a prolonged increase in lead concentrations. In combination with concentration data for other metals, it may be possible to identify the source of pollution by assigning metal "fingerprints" (concentration ratios) of sewage sludge to those of scrubber liquid.

Likewise, the combination of MFA and monitoring of MSW incineration residue allows one to assess the flow of lead, as well as other substances, through private households (see Chapter 3.3.1).

Finally, MFA facilitates an additional type of monitoring. If there is no information available about a process, it may be possible to estimate the missing material flows by mass-balancing the adjacent processes. In the above case study, data about substance flows through the shredder are not available. Rough data about lead in new cars from the last decade, combined with data supplied by the smelter on lead in the two outputs (filter dust and construction rods), allows the flow of lead through the car shredder to be assessed without analyzing the shredder itself.

3.1.1.3 Basic Data for Calculation of Lead Flows and Stocks in the Bunz Valley

In the calculations presented in Tables 3.1 through 3.10, each process is labeled with a letter and each flow with a number. These letters and numbers help to identify the corresponding processes and flows in Figure 3.1. The description of each process is structured as follows:

- Name of the process
- Lead stock inside the process
- Rate of change of the lead stock
- Name of input flows (including a list of quantities that are used to calculate the lead flows)
- Name of output flows (including a list of quantities that are used to calculate the lead flows)

3.1.2 CASE STUDY 2: REGIONAL PHOSPHOROUS MANAGEMENT

Nutrients such as nitrogen, phosphorous, potassium, and carbon are essential for the biosphere. They are the key factors controlling growth and enabling species and populations to develop or causing them to vanish. They are especially crucial for the production of food for humans and animals. Because of limitations inherent to the soil–plant system, not all nutrients delivered to the soil can be taken up by plants.[9] Thus, agricultural losses of nutrients are common, and they cannot be avoided. Yet, they can be reduced by farming practices that are directed toward minimizing losses to the environment. Nutrients in surface waters enhance the growth of algae (eutrophication). As a consequence, the oxygen content in surface water is reduced due to the increased plankton mass, mass death, and decomposition of organisms. As the oxygen concentration decreases, fish and other organisms find it increasingly difficult to survive. Due to transformations in soil and groundwater, nitrogen can also be lost as NO_x or NH_3 to the atmosphere, contributing to the formation of tropospheric ozone and particulate matter, respectively. Hence, the control of nutrients is of prime importance for the management of resources as well as of the environment.

Case Studies

Case studies 2 and 3 both relate to nutrient pollution. The difference between the two is the scale: a small region of 66 km² and 28,000 inhabitants in Case Study 2 vs. the entire River Danube with a watershed of 820,000 km² and 85 million inhabitants in Case Study 3. It is noteworthy that for both scales, the same MFA approach can be taken. Nevertheless, there are focal differences in the task of balancing nutrients on these two scales. The challenge on the large scale is to put together a team (often international) that uses the same approach along the entire stream, allowing true comparison and combination of the individual results.

3.1.2.1 Procedures

Like the lead example in Section 3.1.1, Case Study 2 is a part of the comprehensive RESUB project; it focuses on flows and stocks of phosphorous (P). The procedure is the same as for lead, with some small changes due to the way phosphorous is used. The systems boundaries in space and time are identical. Only agricultural soil is taken into account, since the flow of P on forest and urban soils is comparatively small. Two additional processes for animal breeding and plant production are introduced. Hence, again 10 processes and 19 flows of goods are taken into account (Figure 3.2).

As a first step, the water balance is estimated. Water is important for the flow of phosphorous because P can be transported both in dissolved state (leaching) and as a particle (runoff and erosion). Hence, a comprehensive water balance for the region is needed (Figure 3.3). To minimize the costs for an annual water balance, the relevant hydrological flows and processes must be identified by a systems analysis (Figure 3.4). By a provisional semiquantitative water balance, the main water flows and stocks are identified in order to set priorities for the following costly assessment and measurement program. The main purpose is to achieve sufficient accuracy with the least number of expensive measurements. A potential problem for water balancing is the mismatch between the regional (administrative) and hydrological boundaries. In this study, the two definitions of the region coincide well. The small deviations are compensated for assuming the same net precipitation for areas within and outside the administrative region. Determining the flows and stocks of groundwater is a necessary but usually difficult and resource-consuming task. It is therefore beyond the possibility of most regional MFAs. If groundwater data are not available, and if there are major groundwater inflows, outflows, or changes in stock, a hydrological balance might not be possible. In such cases, MFA has to be limited to a specific regional problem not related to the hydrosphere, or it fails altogether. Data for evapotranspiration can be calculated using various formulas (according to Penman or Primault[5]) and regional data on climate and vegetation. The path of water from precipitation to groundwater and surface water can only roughly be assessed, too. In the present study, there is sufficient information about groundwater outflow from the region to neighboring regions.

The following equation is used for the hydrological balance:

$$\text{Precipitation} + \text{surface water import} + \text{groundwater import} + \text{drinking water import} = \text{evapotranspiration} + \text{surface water export} + \text{groundwater export} + \text{drinking water export} + \text{change in stock}$$

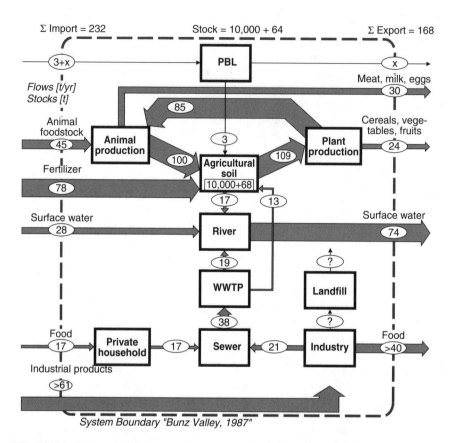

FIGURE 3.2 Regional phosphorous flows and stocks.[3] (From Brunner, P. H., Daxbeck, H., and Baccini, P., Industrial metabolism at the regional and local level: a case study on a Swiss region, in *Industrial Metabolism — Restructuring for Sustainable Development*, Ayres, R.U. and Simonis, U.E., Eds., United Nations University Press, New York, 1994. With permission.)

The flows and stocks of water in eight goods, listed in Table 3.11, are measured for a period of one year. Samples are taken for the same time period for most of these goods. Since drinking water is produced from groundwater, it is assumed that drinking and groundwater have the same concentrations. Measurement and sampling methods, frequencies, and locations are given in Table 3.11 and Figure 3.4. For more information about establishing regional water balances, refer to Henseler et al.[11] After analyzing the water balance, the next step is to measure the flows and stocks of phosphorus. For each good investigated, the flow is multiplied by the concentration of P within that good to determine the phosphorus fluxes. In the following, it will be explained how the data presented in Figure 3.2 are assessed.

3.1.2.1.1 Private Household

The flow of food-derived P into private households is established using data about household food consumption[12] and the nutrient content of food.[13] Phosphorus in household detergents and cleaners is not taken into account, since federal legislation

Case Studies

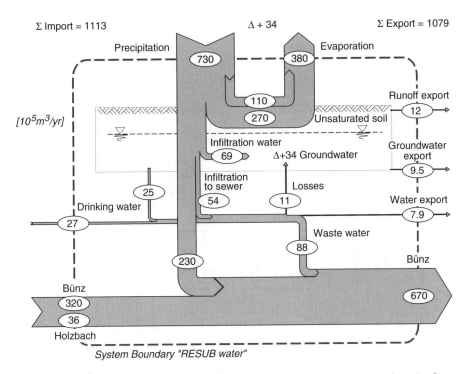

FIGURE 3.3 Results of regional water balance: while the river passes the region, the flow of surface water is doubled by the net precipitation input (precipitation minus evapotranspiration). Bunz and Holzbach are two rivers in the valley.[11] (From Henseler, G., Scheidegger, R., and Brunner, P.H., *Vom Wasser*, 78, 91, 1992. With permission.)

banned P for these purposes. A rough estimation of other P flows showed that they are so small that they do not have to be taken into account (<1% of total regional flow, <10% of flow through private households). The P output of households is not measured but instead is calculated according to Figure 3.5 and the conservation of mass. Of the P entering private households, 90% is assumed to leave by means of wastewater and the remaining 10% by MSW. MSW is not considered further, since it is treated in a MSW incinerator outside the region. It is worth mentioning that composting of MSW can hardly be justified on the grounds of nutrient conservation, since its maximum contribution to regional P management is marginal and about 1% of the total nutrient use in agriculture.

3.1.2.1.2 River

The good "surface water" is flowing in and out of the region at rates of 35×10^6 and 67×10^6 m^3/year, respectively (Figure 3.3). The P concentrations in the input and output of the region are measured (0.8 mg/l and 1.1 mg/l, respectively) and multiplied by the corresponding water flow, resulting in 28 t P/year and 74 t P/year, respectively (Figure 3.2). Some phosphorus flows such as those in precipitation, drinking water import and export, evapotranspiration, and groundwater are less than

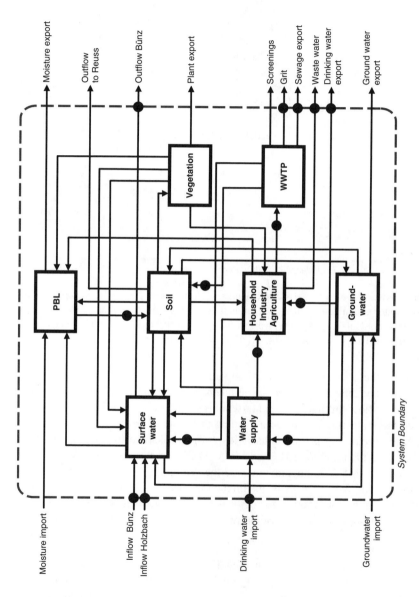

FIGURE 3.4 Determination of regional water balance.[11] (● = flow measurements and sampling points) (From Henseler, G., Scheidegger, R., and Brunner, P.H., *Vom Wasser*, 78, 91, 1992. With permission.)

TABLE 3.11
Measuring and Sampling Procedure To Establish a Regional Water Balance

Good	Number of Measuring Stations	Method of Flow Measurement	Measuring Period	Method of Substance Sampling	Sampling Period
Precipitation	3	rain gauge	1 year (365 × 24 h)	composite sample	27 × 2 weeks
Surface water	3	river gauge	1 year (365 × 24 h)	composite sample	27 × 2 weeks
Wastewater	2	venturi	1 year (365 × 24 h)	composite sample	27 × 2 weeks
Sewage sludge	2	container	1 year	composite sample	10 per year
Sievings from WWT[a]	1	balance	1 year	grab sample	3 per year
Sand from WWT[b]	1	volume	1 year	grab sample	3 per year
Drinking water	9	meter	1 year	grab sample	9 per year
Groundwater	5	water table	1 year	no sampling [c]	—

[a] Screenings from wastewater pretreatment.
[b] Sediment of wastewater pretreatment.
[c] Groundwater identical to drinking water.[11]

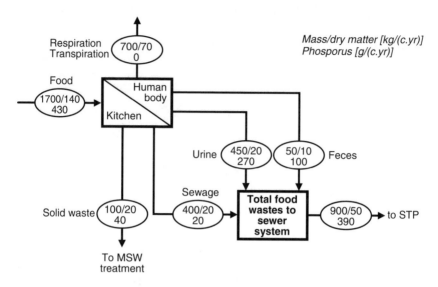

FIGURE 3.5 Flow of food, food dry matter, and phosphorous contained in food through private households.[90] STP: sewage treatment plant. (From Baccini, P. and Brunner, P.H., *Metabolism of the Anthroposphere*, Springer, New York, 1991. With permission.)

1% of the total regional phosphorus flow. Therefore, they are not taken into account for the phosphorus balance.

3.1.2.1.3 Sewer System

The flows of P through the sewer system are calculated using data from household outputs and measured WWTP inputs. The figure for P in "industry" wastewater is calculated as the difference between WWTP input and household wastewater (38 − 17 = 21 t P/year). In the WWTP, the output to the surface waters is measured by multiplying the volume of treated wastewater by the concentration of P measured in 52 biweekly samples of treated wastewater (19 t P/year). P contained in sludge and applied in agriculture (13 t P/year) is measured in metering the total flow of sludge when transferred to transport vehicles, and samples of P are taken for analysis during this transfer.

3.1.2.1.4 Industry

Two aspects are important for the process "industry": Food is stored temporarily in a large stock of interregional importance, and polyphosphates are used in some amounts in regional chemical and other companies. The P contained in food leaves the region unchanged as an export good. P in polyphosphates is transferred to the sewer and is a large source for P in the WWTP.

3.1.2.1.5 Landfill

The process "landfill" is not investigated, although it is possible that phosphorous-containing wastes (biomass, detergents) have been landfilled in the past and that P is leaching to ground and surface waters.

Case Studies

3.1.2.1.6 Agriculture

Three types of agricultural production systems — "animal breeding," "crop raising," and "miscellaneous" — are defined based on their different managerial characteristics. These agricultural practices are investigated under three processes: "animal production" (production of animals and diary products), "plant production" (production of wheat, corn, vegetables, etc.), and "agricultural soil," as shown in Figure 3.2. For each production system, the use of mineral fertilizer, manure use, animal products, and harvested goods are measured per unit of agricultural area and monitored for two years. Phosphorus contents of all goods are determined analytically to estimate the annual entry of phosphorus to the soil. All data are double-checked against the values taken from agricultural information sources. Input through manure and fertilizer and output through harvest are then extrapolated, taking the values of the three production systems described above and considering actual farming practice in the region (e.g., number of animals, amount of produce, area for crop production, etc.). Flows of goods in the process "animal production," such as animals, fodder, and dairy products, are checked through field accounts. Figures for sewage sludge are collected from WWTPs and those for deposition, erosion, and runoff are taken from the literature. Figure 3.2 displays the amount of P in the output goods, namely harvested plants like cereals, vegetables, and fruits (export of 24 t P/year), and animal feedstock cycled within the region (85 t P/year).

The flow of P to plant production including agricultural soil (X) is calculated as follows:

$$X = \text{fertilizer} + \text{manure} + \text{atmospheric deposition} +$$
$$\text{sewage sludge} - (\text{animal feed produced} + \text{cereals, vegetables, fruits})$$
$$= 194 - 109 = 85 \text{ t P/year}$$

The amount of P stored in the agricultural soil (S) is calculated as follows:

$$S = X - (\text{erosion} + \text{runoff and leaching to surface and groundwater})$$
$$= 85 - 17 = 68 \text{ t P/year}$$

The groundwater inflow into the region is close to zero, and the groundwater outflow from the region is small. Therefore it has been assumed that all P running off and leaching from soils is eventually reaching the regional river.

3.1.2.2 Results

The results of this case study include the regional water balance as well as the P flows and stocks of the region. The water balance is summarized in Figure 3.3. There are two large imports (precipitation and river inflow) and exports (river outflow and evapotranspiration) of water. The main water flow through the region consists of humidity contained in air; this flow has not been considered here because it is not relevant for the P case study. On its way through the region, the surface water flow (river water) is doubled by the input of net precipitation (precipitation minus evapotranspiration). Of the surface water produced within the region, 28% is treated wastewaters from one major and two small sewage treatment plants. Thus, the ratio of wastewater to surface water flow is relatively high, and accordingly, the regional potential for dilution of wastewaters is rather small. Hence, efficient wastewater treatment is highly important for the quality of the river water.

An increase in groundwater stock of ≈10% of net precipitation is observed during the assessment campaign. It reveals that the year of measurement is a rather "wet" year; it distinctly deviates from the 10-year average of the hydrological balance, which shows a tendency toward decreasing groundwater stock.

The results of the analysis of flows and stocks of phosphorous are presented in Figure 3.2. As in the case of lead, the P imports (232 t/year) outweigh the P exports (168 t/year) by far, resulting in an accumulation of 64 t of P per year. The main sink for P is the soil. It contains already 10,000 t, and 68 t/year are added. (Note: the difference between 68 t/year and 64 t/year given above for regional accumulation stems from the uncertainty of the processes "WWTP" and "sewer" that are not in balance.) In MFA, it is often the case that inputs, outputs, and changes in stocks of processes do not match, and hence uncertainties remain (see Chapter 2.3). The largest amount of P is imported for agricultural activities (45 t/year of fodder for animals and 78 t/year of fertilizer for plant production). By manure (100 t/year), fertilizer (78 t/year), sewage sludge (13 t/year), and atmospheric deposition (3 t/year), 194 t/year of P are applied to the regional soil. Plants take up 109 t/year, and 17 t/year passes to the surface waters by leaching and erosion. A regional silo for food holds large amounts of P that are continuously replenished, accounting for a P flow of 40 t/year. The use of P for industrial water treatment amounts to another 21 t/year. Possibly, more P is used but not accounted for in "industry."

3.1.2.2.1 Environmental Protection

Besides the unknown amount of P in landfills, there are two main issues concerning P management in the region. First, P is accumulating in the soil (+68 t/c, corresponding to an increase of the P stock in the soil of +0.68% per annum) and eroding/leaching to the surface waters (17 t/year). Second, P is directly discharged with purified sewage into the receiving waters. The load of P in the river increases from 28 t/year at the inflow to 74 t/year at the outflow. The flow of river water is doubled by the addition of regional net precipitation. Hence, the P concentration in the river increases by ≈30%. If all upstream and downstream regions contribute to the P load in surface waters in the same way, eutrophication is likely to take place in downstream lakes and reservoirs. Thus, the maximum allowable P load to the river must be assessed by also taking into account the potential and limitations for P dilution of the surface waters outside of the region.

3.1.2.2.2 Early Recognition and Monitoring

The MFA of P facilitates early recognition of P accumulation in the soil before it actually happens. This is important for water-pollution control. If the load of P into the river needs to be limited, there are two theoretical options (the uncontrolled landfills are not discussed here as a potential source because they have not been investigated):

Case Studies 193

1. The removal efficiency for P in the sewage treatment plant WWTP can be increased from about 30 to 50% to >90% in a relatively short time (months).
2. The flow of P from the soil to the surface waters can be reduced.

The second option does not allow quick reduction of P flows: For a given agricultural practice, the amount of P eroded is mainly a function of the P stock in the soil. Hence, it is either necessary to change agricultural practices or to reduce the P stock in the soil, which takes a long period of time (decades to centuries). MFA makes it possible to forecast accumulation (as well as depletion) of P in the soil long before it actually happens. Taking into account the current amount of P in the soil ($\approx$10,000 t) and the annual accumulation of 68 t/year, it can be assessed that the P concentration in soils will be doubled in about one and a half centuries, if present agricultural practice is maintained. This will lead to a large increase in eroded P, offsetting the reduction of P in the river due to improved elimination of P in sewage treatment.

Direct soil monitoring yields results with large standard deviations. Thus, even an intensive soil sampling and analysis campaign will not identify P accumulation within a few years because mean values will not be significantly different within one decade. MFA provides timely predictions of the change in soil stocks with one single measuring campaign of soil concentrations and P inputs into the soil. Of course, if agricultural practice is changed, the data and calculations have to be adapted to the new situation, too.

3.1.2.2.3 Priorities

Comparing the various flows of P to the soil, it becomes clear that sewage sludge is a comparatively small source of P, supplying less than 10% of the total soil input. Thus, in terms of resource conservation, the application of sewage sludge on farm land is not important and is of low priority. The dominant flows are due to the cycling of a large stock of P within agricultural production by the soil-plant-animal-soil system. The ratio "input to product output" of the different processes is noteworthy: The process "animal" (production of animals and diary produce) consumes 130 t P/year to produce 30 t P/year; "plant production including soil" requires 194 t P/year to produce 109 t P/year in plants harvested. This translates to efficiencies of 23% for the use of P in "animal production" and 56% for "plant production including soil." Clearly, if P becomes a limited resource, then either the efficiency of P in animal farming has to be increased, or the dietary intake has to be shifted toward less meat and more vegetarian foods.

It is also noteworthy that composting of household garbage is insignificant and of low priority regarding the P flow within the region. Assuming that less than 20% of food bought by private households is discarded as municipal solid wastes (80% being eaten, eventually transformed to urine and feces, and collected with sewage), composting of separately collected garbage would supply only about 3 t P/year for agricultural production, equaling $\approx$2% of total agricultural input.

3.1.3 CASE STUDY 3: NUTRIENT POLLUTION IN LARGE WATERSHEDS

Case Study 3 is part of an in-depth investigation into water-quality management of the Danube Basin and is described in the report "Nutrient Balances for Danube Countries."[14] High nutrient loads are recognized as one of the most severe problems of the River Danube, the Danube Delta, and the "final sink" Black Sea. The ecosystem of the Danube Delta is severely endangered, and a large part of the Black Sea is critically eutrophic.[15] The main objective of this case study is to prepare a basis for decisions regarding the protection of the water quality of the Danube, its delta, and the Black Sea. In particular, the goal is to use MFA to establish reliable information about sources, flows, stocks, and sinks of phosphorous and nitrogen in the Danube Basin.[16,17] The main difference between case studies 2 and 3 is the scale: Instead of an area of 66 km^2 and a population of 28,000 inhabitants (Case Study 2), the "Danube" case study covers 820,000 km^2 and includes 12 countries with a total population of 85 million. Despite the large difference in scale, the same MFA methodology is applied.

This case study shows that the application of MFA is independent of scale and that the same methodology can be applied to small and very large systems. Nevertheless, there are clear differences in the focus and procedures according to the scale. In general, multinational MFAs on such a large scale require the joint effort of several research groups from each of the participating countries. It is necessary to acquire an adequate and comprehensive data set of the flows and stocks of nutrients from agriculture, industry and trade, private households, water and wastewater management, and waste management. And it is essential to use a uniform methodology in collecting, calculating, and evaluating the data. Hence, capacity building and know-how transfer is important to ensure that all groups and participants are applying the same methodology.

The results of such a study directed toward decision making in environmental protection can have different consequences for the partners engaged. While one country may not be a large factor for the pollution of the watershed, another may turn out to be a major contributor. Hence, if a common level of water protection is established and corresponding remediation measures are taken, the financial consequences may be quite severe for the latter and only marginal for the former. It is thus of utmost importance to use a rigid, consistent, and transparent methodology that is accepted by all partners. MFA represents such a methodology well suited for multinational teams collaborating in delicate situations.

Key questions of the case study are: What are the main sources of nutrients, and what measures are appropriate to reduce the nutrient flows to environmentally acceptable levels? Traditionally, emission inventories and ambient water-quality measurements are used to answer these questions. As a novel approach, comprehensive material flow analysis is applied to the entire watershed. The main advantage is that all processes in a region are looked at and that the total inputs, outputs, and stocks are investigated. Nutrient flows are tracked from their very beginning (fertilizer, animal feedstock, agricultural production) to the consumer (private households),

Case Studies

to waste management, to surface and groundwaters, and finally to the River Danube. Since the balance principle is applied to all processes, cross-checking of flows and stocks becomes possible at many points within the system investigated.

3.1.3.1 Procedures

As mentioned before, one of the main tasks when exploring such a large system is to set up a broad international group that learns and uses the same MFA methodology. Ten national teams from Austria, Bulgaria, Czech Republic, Germany, Hungary, Moldavia, Romania, Slovakia, Slovenia, and Ukraine, each consisting of several experts, are participating in the study. In a first step, the common MFA methodology as well as water-quality goals and principles are established. The system boundaries are defined in space and time. The least number of processes is selected that allows full description of all necessary flows and stocks and still does not result in excessive work. To facilitate assembly of the individual data and results, the system is defined uniformly for all teams (see Figure 3.6).

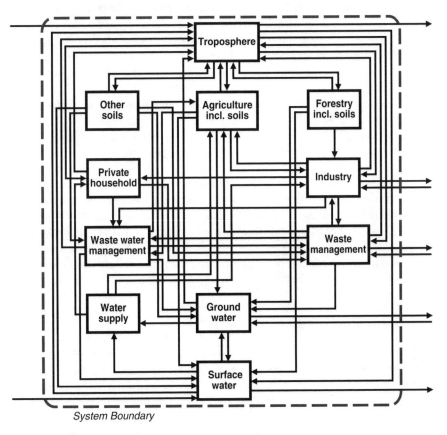

FIGURE 3.6 System definition for nutrient balancing in the Danube River watershed. The same system is used for all national balances and for the balance of the total catchment area.[14]

Next, data are collected to balance each of the processes of Figure 3.6. Existing measurements, regional statistics, literature data, expert advice, and sometimes additional measurements are used to assemble a data set as comprehensive as possible. For example, for the process "agriculture including soils," this means finding information about all inputs such as mineral fertilizer, atmospheric deposition, nitrogen fixation, sewage sludge, compost, seedlings, and outputs such as crops harvested, animal products, eroded soil, gaseous losses, leachate, and percolation. Manure is recycled within the process and thus can be looked at as both an output and input at the same time if no export or import of manure takes place. Stocks comprise nutrients in the soil, in stored manure, animal biomass, and stockpiles of fertilizer. The procedure is similar to the one described for Case Study 2 in Section 3.1.2. All other processes of Figure 3.6 are balanced similarly.

Some processes are not easy to balance: erosion from forest and agricultural soils in alpine areas can only be roughly estimated. Denitrification in natural systems (e.g., soil, aquifer) is not well known. The fate of intermediate stocks in the River Danube and in the soil over time is not sufficiently understood yet. Data about the efficiency of wastewater treatment and about corresponding nutrient removal are not available in all of the eastern European countries. During the time of centrally planned economies, much information on agriculture, water-quality management, and waste management was collected and stored on a large scale. However, since the transition of these economies to a free-market economy, much less information is available. In part this is because it is too costly to gather comprehensive data. On the other hand, the price to access existing information increased dramatically after the economic transition.

It is important that all partners exchange information during collection of data. They also must make sure that they use compatible figures. For instance, it is likely that the balance of a cow is similar in most countries of the Danube Basin, and thus that the input and output figures are comparable for most teams. If there are differences, like the significant variation between the nutrient metabolism of a Ukrainian and an Austrian cow, explanations must be available. Often, the balance principle brings such differences to light and allows cross-checking and verifying such differences. Thus, the balance principle can be highly valuable in the "negotiation" process within a group comprising teams from many countries. It ensures transparency, enables data verification, and results in acceptance of others' results.

3.1.3.2 Results

A large-scale multinational MFA proves to be a time- and resource-consuming task. It takes time until all know-how is transferred, incorporated, and well applied in practice. It takes even more time to find all the necessary data. The exchange of information, iterations, and adaptations of the individual work of the different participating groups again takes time. It may be that a partner is not able to perform its task and that a new team has to be engaged in the middle of the project. Given these factors, it is clear that a large-scale MFA cannot be undertaken in a couple of weeks. It is likely to take one year or more to complete such a comprehensive task.

Case Studies

The Danube case study produces a lot of data and many results that can be used to support decisions regarding water quality management. For results regarding wastewater management and water pollution control, see Zessner et al.[18] The results presented here identify the most important sources and pathways of nutrients in the Danube River basin. The purpose of this presentation is to demonstrate how a very large data set can be compressed to present a few relevant results. In Figure 3.7, the system given in Figure 3.6 is transformed and presented in a format that identifies the major imports and exports of nutrients into the surface waters of the Danube catchment area. This format still shows the same 11 processes, but it centers on the surface waters. Imports and exports are quantified, and conclusions regarding the importance of all flows can be taken according to their mass flow. Note that mass flows alone do not permit one to evaluate the effects of nutrients in surface waters. It is necessary first to transform nutrient flows into nutrient concentrations by dividing nutrient flows through water flows.

The same case study data are even more condensed in Figure 3.8. It becomes clear that the more aggregated the data are, the easier it is to get a message across: Figure 3.8 clearly identifies agriculture as the dominant source of nutrients in the Danube Basin. In addition, it suggests the hypothesis that improving existing wastewater treatment plants (WWTP) is probably more important for reducing emissions than connecting all households and industries to sewers. This hypothesis of course has to be verified with data about the fraction of people and companies connected to sewer systems, and on nutrient removal in wastewater treatment within the catchment area. In addition, the costs of upgrading WWTP and of connecting households and industries to sewers need to be known. The advantage of an MFA approach is that such hypotheses can be set up, making further investigations more straightforward.

Another way of presenting data is displayed in Table 3.12 and Table 3.13, which combine results about sources (agriculture, household, industry, etc.) with results about pathways. Hence, they are well suited to serve as a base to set priorities for decisions regarding nutrient emission reductions. The following conclusions can be drawn: Agriculture is the main source of nutrient inputs into the River Danube. Erosion and runoff are the main pathway of nutrients from agriculture to surface waters for both P and N. The direct inputs of liquid manure are high and amount to about 12% of total N and 20% of total P loads. Private households are the second largest source of nutrients, contributing around 20% of total N and P. Approximately 10% of both N and P originates from industry. MFA yields the following results regarding pathways: for surface waters, the main nitrogen load (35%) is due to exfiltration of groundwater. Erosion/runoff; direct inputs from agriculture, households, and industry; and effluents from WWTP contribute each about 20 to 25% of the total load to the Danube River. About 60% of N and 40% of P originate from nonpoint sources. Retention (sedimentation and denitrification) amounts to 15% of N and close to 50% for P.

The detailed results of all teams (presented in PHARE[14]) make it possible to identify the nutrient contribution of each country. Almost half of the nutrient input into the Danube catchment area comes from Rumania, while Austria, Germany, and Hungary contribute together about one-third of the total load. An interesting and not yet resolved question concerns the allocation of the dilution potential of the River Danube: what is the amount of nutrients a country may release to the Danube?

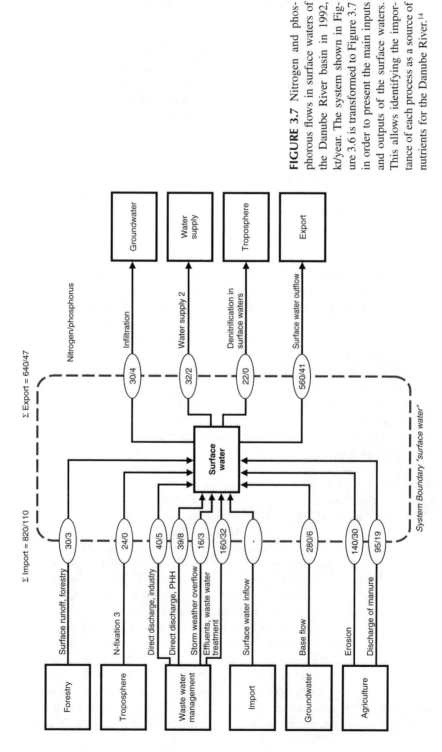

FIGURE 3.7 Nitrogen and phosphorous flows in surface waters of the Danube River basin in 1992, kt/year. The system shown in Figure 3.6 is transformed to Figure 3.7 in order to present the main inputs and outputs of the surface waters. This allows identifying the importance of each process as a source of nutrients for the Danube River.[14]

Case Studies

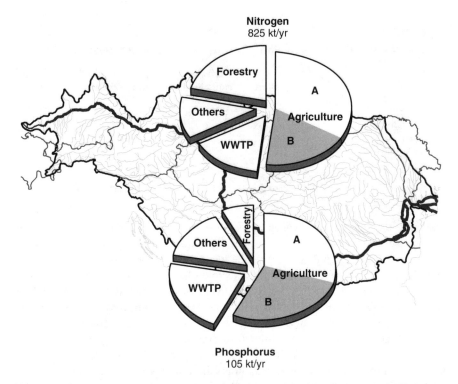

FIGURE 3.8 Sources of nutrients in the catchment area of the Danube River in 1992, kt/year. The charts show clearly the importance of agriculture for P and N emissions. Erosion and leaching from agricultural fields dominates all sources (A). Direct discharges and discharges via treatment plants of animal wastes are the second most important path of nutrients to the Danube (B). The direct inflows from private households and industry (others) are smaller than the effluents from wastewater treatment plants (WWTP). Diffusive inputs from forestry are small for P but more significant for N.[17] (From Brunner, P.H. and Lampert, Ch., *EAWAG News*, 43E, June 1997, p. 15–17. With permission.)

TABLE 3.12
Sources and Pathways of Nitrogen in the Danube River (1992)

N, %	Agriculture	Private Household	Industry	Others	Total
Erosion/runoff	17	0	0	4	21
Direct discharges	12	4	6	2	24
Base flow	17	4	0	13	35
Sewage treatment plant	6	10	5	0	20
Total	51	19	10	19	100

Note: Total input equals 100%. Base flow represents flows to the Danube via groundwater.[17]

TABLE 3.13
Sources and Pathways of Phosphorous in the Danube River (1992)

P, %	Agriculture	Private Household	Industry	Others	Total
Erosion/runoff	28	0	0	3	31
Direct discharges	18	6	6	3	33
Base flow	2	2	0	2	6
Sewage treatment plant	9	14	7	0	30
Total	57	22	13	8	100

Note: Total input equals 100%. Base flow stands for flows to the Danube via groundwater.[17]

Assuming that the carrying capacity for nutrients of the Danube basin is known, there are several ways to answer the above question. A per capita load limit favors countries with large populations; it can be justified on the grounds that every human being has a similar metabolism and thus should have an equal share. This method of allocation neglects the fact that some countries are better suited for agriculture than others and, because of their agricultural activities, will have a larger nutrient turnover. A per area load limit favors large countries but does not consider population density. A per net-precipitation limit takes into account the regional dilution of nutrients: if regional nutrient emissions are heavily diluted by a large amount of net precipitation (precipitation minus evapotranspiration), the resulting concentration in the River Danube may still be low and below carrying capacity. However, this argument does not hold for the Black Sea, where the total flow is important.

The River Danube, like many of the large river systems in the world, has become an important path for wastes such as nutrients from countries within its catchment area. The main question is how future loads will develop. At present, eastern European countries are experiencing a low standard of living. It can be assumed that the per capita turnover will rise rapidly in the future and that population will grow, too. Both factors will increase total nutrient turnover as well as waste generation. If no actions are taken, the capacity of the River Danube, the delta, and the "final sink" Black Sea will be overloaded, with serious ecological and economic consequences. It is not the "classical" resource problem (lack of nutrients) that will limit the development of the region. Rather, it will be the lack of appropriate sinks that restricts progress. Strategies to limit nutrient loads need to be discussed and developed. Crucial issues will be: type of agriculture; population density, lifestyle, and consumption; and standards and enforcement for emissions from industry, households, and WWTP. In any case, transregional and international agreements will be required to solve the allocation problem.

3.1.4 CASE STUDY 4: MFA AS A SUPPORT TOOL FOR ENVIRONMENTAL IMPACT ASSESSMENT

In an environmental impact assessment (EIA), potential impacts of a project such as a new plant (power plant, municipal incinerator) or system (road, harbor) on the

Case Studies

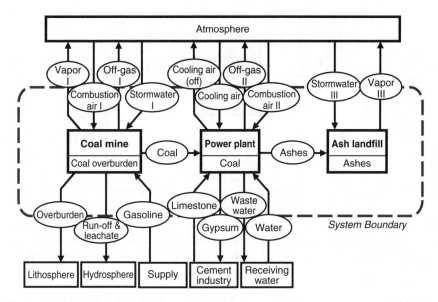

FIGURE 3.9 Systems definition of electricity production in a coal-fired power plant including the processes "coal mining" and "ash landfilling." The figure includes all exports and imports that are necessary to establish an EIA and EIS.

environment are identified, quantified, evaluated, predicted, and monitored. In the event that a significant impact is acknowledged, more detailed studies have to be carried out, finally leading to the preparation of an environmental impact statement (EIS). Meanwhile, more than half of the nations around the world require EIA for certain projects. The U.S. National Environmental Policy Act of 1969 (NEPA) provided one of the earliest sets of EIA requirements. Many countries followed and modeled their requirements after NEPA.[19] In Europe in the 1970s, some federal environmental legislation began to require mandatory EIAs. However, the final breakthrough was not achieved until the European Directive 85/337/EEC on the "assessment of the effects of certain public and private projects on the environment"[20] became effective. MFA that is based on a sound methodological framework is considered a useful tool to support both EIA and EIS.[21]

In the case study SYSTOK,[22] the impact of electricity production from coal on the local and regional environment is investigated. For this purpose, a three-process system comprising coal mining, the coal-fired power plant, and landfilling of the ash is defined (Figure 3.9). The contribution of this system to the anthropogenic and geogenic metabolism of the region is determined (Figure 3.10). The case study focuses on particular technologies of mining, power generation, air-pollution control, and landfilling. It is clear that the results cannot be generalized to other coal-fired power plants. If the technology is changed, if the coal composition and heating value is different, or if the landfill leaks to the groundwater, the impact on the environment will also be different. The reason for including this case study is to show that MFA serves well as a base for EIS and EIA. MFA can be applied independently of the technology used or the input composition.

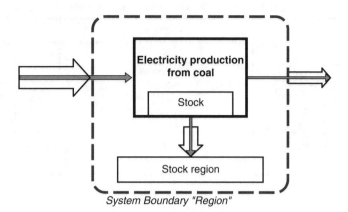

FIGURE 3.10 The contribution of a coal-fired power plant to a region's metabolism. The system consists of the three processes "coal mining," "power generation," and "landfilling of the ash" that can be summarized in a single process, "electricity production from coal."

The findings of SYSTOK serve as a basis for the operator to optimize the plant and to prepare an environmental impact statement. One of the features of SYSTOK is that a system, in this case "electricity production from coal," is to be embedded comprehensively into a region. However, the definition of "the region" is not clear *a priori*, so appropriate approaches for the demarcation of spatial and temporal systems boundaries have to be developed. SYSTOK exemplifies how a comparatively simple MFA can lead to valuable new conclusions regarding system boundaries.

3.1.4.1 Description of the Power Plant and Its Periphery

All coal used in the power plant (1 million t/year) is produced in a nearby open-pit coal mine with a surface area of 2 km². Temporary interruptions in mining or power generation are buffered by interim coal storage of 2 million tons, located at the premises of the power plant. In order to gain 1 ton of coal, an average of 6.7 tons of overburden have to be removed. Of this mining waste, 70% is transported to locations outside of the mine and the rest is filled back into the mine. For extraction, transportation (trucks and conveyors), and processing of the coal and overburden, 1400 t/year of gasoline and 25 million kWh/year of electricity are needed. The reservoir of coal still in the mine is estimated to be 11 million tons.

Brown coal has a low heating value of 8.4 to 13 MJ/kg$_C$ (index$_C$ stands for coal) and bears much noncombustible ash, making transportation expensive. Thus, the power plant is situated near the coal mine in order to avoid long transport distances. Because of high specific costs of power generation at this plant, it operates only during peak demand. The average time of operation is 4000 h/year, and the maximum electric capacity is 330 MW. The coal is pulverized to coal dust and injected into the combustion chamber at a feed rate of 300 t$_C$/h at full load. The average coal consumption (ca. 265 t$_C$/h) is somewhat lower due to periods of partial loading. Approximately 33 kg/t$_C$ of wet bottom ash (40% water content) are removed from the water basin that serves as an air seal toward the combustion chamber. The power

Case Studies 203

plant is equipped with an electrostatic precipitator (ESP) to collect particulates (ESP ash) with an efficiency of 99.85%. The ratio of dry bottom ash to dry ESP residue (or fly ash) is ca. 1/9. The ESP residue is humidified (20% water content) and, together with bottom ash, transported by conveyor belts to the landfill. Sulfur dioxide (SO_2) is removed from the flue gas in a wet scrubber using limestone ($CaCO_3$, 20 kg/t_C). Absorption of SO_2 occurs with an efficiency of more than 90%, producing ca. 37 kg of gypsum per t_C (water content is 12%). Finally, nitrogen oxides (NO_x) are reduced to molecular nitrogen (N_2) in a catalyst by injecting ca. 1.2 kg of ammonia solution (NH_3, 33%) per t_C. Other chemicals used include hydrochloric acid (HCl, 0.02 kg/t_C, 33%) and sodium hydroxide (NaOH, 0.007 kg/t_C, 50%) for conditioning of feeder and perspiration water as well as chemicals used to stabilize water hardness, inhibit corrosion, etc. (<0.002 kg/t_C). Combustion requires 4000 kg/t_C of air, resulting in ca. 5000 kg/t_C of off-gas. Water is used to feed the steam cycle (20 kg/t_C) and for cooling (2300 kg/t_C). Waste heat is dissipated in a cooling tower having an air-exchange rate of 8500 kg/t_C. Part of the cooling water (25%) is discharged to surface water; the rest evaporates in the cooling tower.

The landfill for the ashes consists of a natural vale made of quartzite with a surface of ca. 0.35 km². This basin, which is a former mine area of coal, has a low permeability for water. Hence, precipitation (ca. 1000 l/m²) creates a lake in the landfill. The water is used during dry periods to wet the ashes and is supposed to finally evaporate. No leachate leaves the ash landfill as long as the landfill is maintained by the operator.

3.1.4.2 System Definition

The system displayed in Figure 3.9 is divided further into three processes: "coal mining," "power plant" including air-pollution control, and "ash landfill." For EIA and EIS, a black-box approach is appropriate. It is not necessary to take into account more detailed subprocesses that would require much additional information. The substances are selected based on knowledge about combustion processes and the main product "coal." Carbon is the priority substance in any combustion process, since the content of organic carbon in ashes and flue gas is a measure of combustion efficiency and of the formation of organic pollutants. Past experience with coal-fired power plants has shown that they are major sources of emissions of SO_2, NO_x, HCl, and heavy metals such as arsenic and selenium.[23] Modern legislation (Clean Air Act) sets stringent standards for these emissions. The procedure to select substances relevant for environmental impact assessments is given in Table 3.14. The substance concentrations in coal are compared with concentrations in the Earth's crust. The following six elements are significantly more concentrated in coal than in the crust: As, Se, Hg, S, Cl, and N. The relevant goods are determined by mass-balancing each process and by assuming that goods inducing a substance flow <1% of the total throughput of a substance can be neglected. This requires an iterative approach among the steps "establishing substance balances," "selection of goods," and "establishing total mass balances." For a first step, the premises of mining, power station, and landfilling are selected as a spatial system boundary. Materials balances are

TABLE 3.14
Determination of Relevant Substances for the Analysis of a Coal-Fired Power Plant by Relating Average Concentrations of Selected Substances in Coal and Ash to Concentrations in the Earth's Crust

Substance	Concentration in Coal (A), mg/kg	Concentration in Ash (B), mg/kg	Concentration in Earth's Crust (C), mg/kg	Ratio A/C	Ratio B/C
Arsenic	12	64	1	12	64
Lead	6	37	13	0.5	2.8
Cadmium	0.1	0.64	0.2	0.5	3.2
Chromium	30	170	100	0.3	1.7
Copper	13	84	55	0.24	1.5
Selenium	0.9	2.2	0.05	18	44
Zinc	27	190	70	0.4	2.7
Nickel	27	130	75	0.36	1.7
Mercury	0.3	0.45	0.08	3.8	5.6
Sulfur	6500	3000	260	25	12
Chlorine	1000	—	130	7.7	—
Nitrogen	12,000	—	20	600	—

determined for an average operational year (temporal system boundary). Figure 3.9 shows the system defined.

3.1.4.3 Results of Mass Flows and Substance Balances

The mass flows of all goods are listed in Table 3.15. Except for mining, where it is not known whether and how much runoff and leachate is draining to surface and groundwaters, all processes are mass-balanced. The main quantitative features of the system are: (1) The coal mine will be exhausted in about 10 years. (2) Compared with the emissions of the power station, the off-gas from mining is negligible. (3) The power plant is actually a giant fan moving huge amounts of air. The mass of air required for cooling dominates the mass flows of the system, exceeding combustion air flow by more than an order of magnitude. Additionally, the power plant consumes more than 1 t of water per ton of coal. (4) The main flow of solid waste is generated during mining (overburden). The power plant produces a large net "hole," since coal is extracted and the volume of backfilled ash is much smaller.

The substance balances as displayed in Figure 3.11 provide an overview of the qualitative behavior of the system. These balances are calculated based on data for substance concentrations in overburden ("soil" in Table 3.19), gasoline, coal (Table 3.14), and the transfer coefficients for the power plant that have been measured on site (see Table 3.16).

A comparison of sulfur concentration in coal and in the Earth's crust shows that a large amount of sulfur is extracted from the crust via coal. The power plant transfers sulfur quite efficiently into the product gypsum (86%). Before desulfurization

TABLE 3.15
Mass Balance for Goods of the System "Electricity Production by a Coal-Fired Power Plant," 1000 t/year

Input		Output	
\multicolumn{4}{c}{Process "Coal Mining"}			
Combustion air I	29	Off-gas I	31
Storm water I	2,000	Vapor	2,000–n.d.
Gasoline	1.4	Overburden	4,700
		Runoff and leachate	n.d.
		Coal	1,000
Total	2,000		7,700
Stock (coal)	11,000		1,000
Stock (total)	145,000[a]	Stock change	–5,700[a]
		Process "Power Plant"	
Coal	1,000	Ashes	280
Cooling air (input)	85,000	Cooling air (output)	87,000
Combustion air II	4,000	Off-gas II	5,000
Water	2,500	Wastewater	760
Limestone	20	Gypsum	390
Total	93,000		93,000
Stock (coal)	2,000	Stock change	0
		Ash Landfill	
Storm water III	350	Vapor III	350
Ashes	280		
Total	630		350
Stock (ashes)	4,100	Stock change	+280

Note: Values are rounded; n.d. = not determined.

[a] Estimated, includes coal and overburden.

became a part of the air-pollution control system of the plant, this sulfur was emitted, too, resulting in a sulfur transfer to the atmosphere of >90%.

The foremost flow of mercury is associated with the good "overburden" or mining waste. During combustion, the atmophilic mercury is evaporated and leaves the plant evenly distributed between ESP residue and off-gas. A small part is precipitated with gypsum. Electricity production in coal-fired power plants extracts significant amounts of mercury from the Earth's crust and disperses a substantial amount via the stack. So far, 2.2 t of mercury have been deposited in the ash landfill and more than 10 t in the overburden deposit. It is interesting to compare these stocks with other mercury stocks. The consumption of mercury has been assessed to range between 0.66 g/capita/year in Stockholm and 1.5 g/capita/year in the U.S. For stock, 10 g/capita have been determined in Stockholm.[24,25] This means that the landfill contains the same amount of mercury as is stored in buildings, infrastructure, and

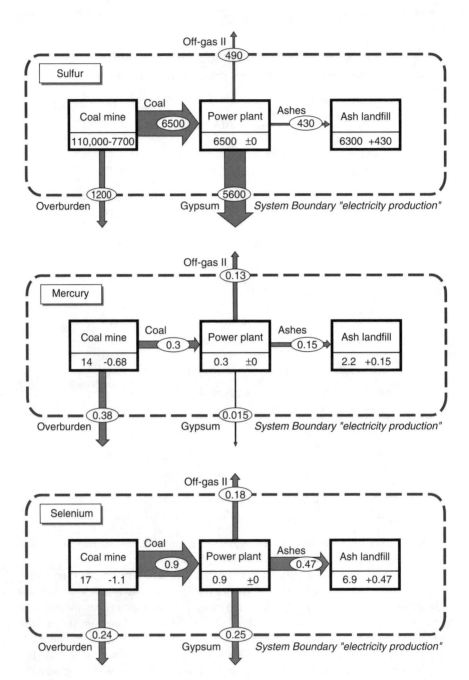

FIGURE 3.11 Substance balances for sulfur, mercury, and selenium for the system "electricity production" (flows are in t/year and stocks in t). The stock in the process "coal mine" includes coal and overburden.

TABLE 3.16
Partitioning of Selected Substances in a Coal-Fired Power Plant, % of Input

Substance	ESP Fly Ash	Bottom Ash	Gypsum	Off-Gas II
Carbon	2	1.3	<0.05	97
Sulfur	5.7	0.93	86	7.6
Mercury	50	0	5	45
Arsenic	99	0.4	0.4	<0.1
Selenium	52	0.6	28	20

long-lasting goods associated with a region of 220,000 inhabitants. In the overburden deposit, a mercury stock corresponding to more than 1 million persons is contained. While this mercury is dispersed over a large region (2000 km^2), the landfill's mercury is located in a comparatively small area (0.35 km^2) and therefore is easier to control.

About half the amount of selenium that enters the power plant is transferred to the ashes and landfilled, and 20% is emitted into the atmosphere. The relevance of this path for the environment will be discussed below. Note that gypsum holds about 30% of the selenium contained in coal.

The relation between the power plant and the surrounding region is discussed in the following sections.

3.1.4.4 Definition of Regions of Impact

The mine, power plant, and ash landfill have several impacts. In the first place, they supply power to consumers. Thus, a region can be defined in a product-related way. Second, they create jobs and income and thus serve an economic region. Third, the mine, power plant, and landfill are situated in an administratively defined region such as a community or a province. Fourth, they have an impact on the environment. Depending on the "conveyor belt" that transports an emission, substances released by the coal-fired power plant may affect a small or large area. The ash landfill has only a local impact during the transfer of the ash to the landfill site. Sulfur and mercury emitted by off-gases are distributed over a large (global) area through atmospheric transport. Thus, the size of the region is determined by the distance an emitted substance travels from the plant and by the effect of this substance on the environment. In each of these four regions of impact, there are specific problems, benefits, and stakeholders with regard to the power plant.

In the SYSTOK study, three regions of impact are defined.

3.1.4.4.1 Product-Related Region

The power plant supplies electricity to a certain area. In principle, this area is defined by the amount of electricity the plant supplies, by the demand per customer, and by the population (or customer) density. Due to the liberalization of the electricity market, this area can only be defined as a virtual region, since customers may be served far away from the plant. The product region is changing constantly in response to the market situation. Thus, for SYSTOK, the size of this virtual region is calculated by the average electricity production of the plant, the average consumption per capita, and the national population density.

$$A_{PR} = \frac{P \cdot h \cdot f}{e \cdot \rho_P} \approx 2100 \text{ km}^2$$

where
- P = output of the power plant (330 MW)
- h = operating hours per year (4000 h)
- f = factor considering partial-load operation (0.85)
- e = specific demand for electricity in Austria (including private households, industry, service, administration, traffic, agriculture) (5.6 MW·h/capita/year)
- ρ_p = population density in Austria (95 capita/km²)

3.1.4.4.2 Administratively Defined Region

The region is defined by the borders of the administrative unit, i.e., the district that represents the legislative and administrative authority for the plant operator. The advantage of this definition is twofold:

1. The region as a spatial unit is well accepted and known and is governed by an authority supervising the plant.
2. Data are usually collected on the level of administrative regions, thus facilitating the allocation of data.

3.1.4.4.3 Region Defined by Potential Environmental Impacts

As mentioned before, this area is different for particulate, gaseous, and aqueous emissions, and it is also substance specific. For gaseous emissions, dispersion models help to determine the region. Criteria for selecting the boundaries may be:

1. Concentration limit (ambient standard) for a substance ($c_{crit} = c_{lim}$)
2. A fraction of the concentration limit, since the limit should not be used up by the power plant alone ($c_{crit} = c_{lim}/10$)
3. A proviso that the power plant does not change the current ambient concentrations in a significant way ($c_{crit} = c_{background} \times 1.1$)
4. A proviso that the power plant does not change the geogenic (or "natural," without present anthropogenic influences) concentrations in a significant way ($c_{crit} = c_{geog} \times 1.1$; see Figure 3.12)

Figure 3.13 exemplifies the differently defined regions, and Table 3.17 presents sizes of regions calculated according to different definitions for SYSTOK.

Case Studies

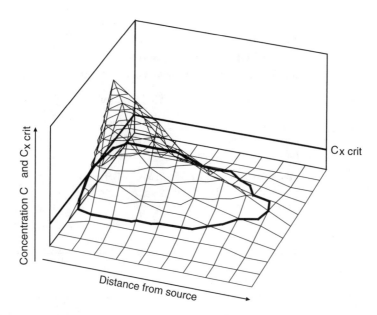

FIGURE 3.12 Application of a dispersion model to determine the border of a substance-specific, environmentally relevant region. The regional boundary with regard to substance x is defined as the area within $c_x > c_{x\,crit}$.

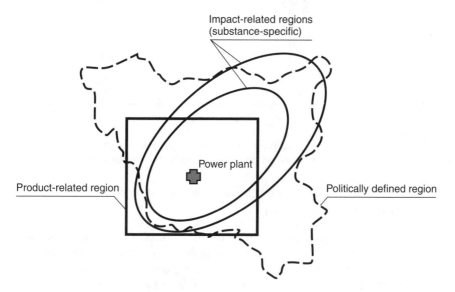

FIGURE 3.13 Regions defined according to three criteria — "consumers of electricity" (product-related region), "administrative designation," and "environmental impacts" — overlap but are not identical.

TABLE 3.17
Sizes of Differently Defined Regions for a 330-MW Coal-Fired Power Plant

Region Definition	Size, km²
Administrative (district)	678
Product (electricity)	2100
Environmental (example SO_2)	620

Note: The boundary for the environmentally defined region of SO_2 is determined by $c_{crit} = c_{geog} \times 1.1$, where $c_{geog} \approx 10 \, \mu g/m^3$.

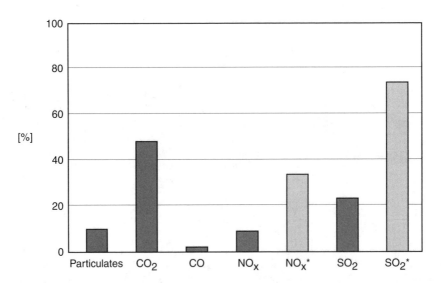

FIGURE 3.14 Contribution of the power plant to the total emissions of the product-related region (=100%). (* Indicates emissions before introduction of advanced air-pollution control.)

3.1.4.5 Significance of the Coal Mine, Power Plant, and Landfill for the Region

In addition to the procedure given in Figure 3.12, there are other means of determining the significance of emissions for the region. In Figure 3.14, emissions from the power plant are compared with the total regional emissions of various air pollutants. As a base for comparison, the product-related region is chosen. The emissions of the region (R_i) are assessed by the following equation:

$$R_i = X_i - P_i \cdot \frac{A_{PR}}{A_{AU}} + P_i$$

TABLE 3.18
Comparison of Substance Flows Induced by a Coal-Fired Power Plant and by MSW in the Product-Related Region

Substance	Flow via Coal, g/capita/year	Flow via MSW, g/capita/year	Coal/MSW
Arsenic	18	0.8	21
Copper	19	100	0.2
Lead	9	170	0.1
Cadmium	0.15	2.3	0.1
Mercury	0.45	0.4	1.2
Selenium	1.3	0.2	8
Chromium	45	53	0.9
Nickel	40	18	2.3
Zinc	40	230	0.2
Carbon	420,000	930,000.0[a]	0.4
Sulfur	10,500	2000.0[a]	5

[a] Figures for carbon and sulfur include the contribution by fossil fuels, which is much larger than MSW.

with X_i as the average emissions of an appropriate administrative (AU) unit for which data are available (state, district). P_i stands for the emissions of the power plant, and A_{PR} and A_{AU} are the respective areas of product-related region and administrative unit.

The power plant is responsible for about half of the region's CO_2 emissions. Removal of SO_2 and catalytic reduction of NO_x significantly reduced the plant's contribution to the regional emissions. A further decrease in particulates and NO_x will result in modest improvements of air quality only (<10%). For CO, the power plant's emissions are not relevant at all.

The throughput of heavy metals by the power plant is set in relation to the product-related region, too. Table 3.18 shows the annual flows of selected metals and nonmetals through the power plant. Comparing these flows to the corresponding total flows through the region is time consuming. Much information, which is usually not available, is required. Therefore, only the materials flows through private households are taken into account. The contributions of industry and the service sector are not considered. Since the consumption of heavy metals in private households is not well known either, the amount of metals in MSW, which is available from measurements (see Section 3.3.1), is taken as a reference. Table 3.18 gives the ratio of substance flows via coal and flows via MSW on a mass-per-capita and year basis. For comparison, the flows of carbon and sulfur in fossil fuels are also shown. The coal-fired power plant is responsible for a high turnover of arsenic, selenium, and sulfur when compared with the generation of MSW in private households.

The relevance of heavy metal emissions can be assessed by the "anthropogenic vs. geogenic flows" approach described in Chapter 2, Section 2.5.1.1. Applied to

SYSTOK, the following question has to be answered: Does the power plant change substance concentrations in any of the environmental compartments? A simplified model is used to identify the effect of power plant emissions on the soil concentrations in the product-related region of 2100 km². By deposition, metals such as lead and cadmium may be accumulated in the top 30-cm layer of soil, the soil depth that is turned over by plowing. Assuming an average soil density of 1.5 kg/m³, the regional compartment "soil" holds a mass of 2100 · 10⁶ m² · 0.3 m · 1.5 kg/m³ = 950 · 10⁶ kg.

The cumulative emissions of the power plant can be estimated based on the total coal throughput of 33 million t, the mean substance concentrations in coal, and the transfer coefficients (TC) for off-gas.

$$E_i t = 33 \cdot 10^6 t \cdot c_i \text{ mg/kg} \cdot 10^{-6} \cdot TC$$

Table 3.19 shows that only emissions of selenium and mercury are of relevance. Note that the model draws on two major simplifications. First, it is understood that the product-related region is identical to the substance-specific impact-related regions. Second, it is assumed that deposition is evenly distributed over the region. Concerning the first simplification, one has to consider that particulate removal takes place in an efficient electrostatic precipitator. Hence, the emitted particulates are small, most with diameters <2 μm. Such particles (aerosols) have a long residence time in the atmosphere (≈1 week) and do not sediment in the vicinity of the power plant. They are washed out of the atmosphere by precipitation. Thus, the average frequency of rainfall determines the travel distance of these particles. In Central Europe, this is around 1 week and results in a significantly larger region

TABLE 3.19
Accumulation of Metals in Soils due to Emissions of a Coal-Fired Power Plant

Substance	Coal, mg/kg	TC	Emission, t/τ[a]	Soil,[96] mg/kg	Soil Reservoir,[b] t	Enrichment, %/τ[a]
As	12	0.001	0.4	8	7600	0
Pb	6	0.005	0.99	25	23,800	0
Cd	0.1	0.041	0.14	0.2	190	0.1
Cr	30	0.001	0.99	40	38,000	0
Cu	13	0.014	6	15	14,300	0
Se	0.9	0.2	5.9	0.1	95	6.3
Zn	27	0.009	8	30	28,500	0
Ni	27	0.003	2.7	20	19,000	0
Hg	0.3	0.45	4.5	0.2	190	2.3

[a] τ = total time of operation (ca. 33 years).
[b] Soil reservoir is calculated based on the product-related region.

Case Studies

than the product-related region (at least ten times larger). Hence, the product-related region overestimates substance accumulation in the soil by more than one order of magnitude.

The relevance of the second simplification can be assessed when dispersion models for particulates are analyzed. In most cases, the ratio between maximum concentration and mean concentration is less than ten. This means that the assumption of substances being evenly distributed over the region underestimates the actual accumulation by a maximum factor of 10. Considering both simplifications, it can be concluded that the chosen model rather overestimates the enrichment in soils caused by the power plant, and the emissions can be rated as nonrelevant with the exception of selenium and mercury, where more detailed investigations are necessary.

3.1.4.6 Conclusions

The case study shows that the power generation based on coal is relevant for enhanced flows of arsenic, selenium, mercury, sulfur, and carbon within the region's metabolism. The landfill represents a considerable reservoir for certain metals within the region. It can be considered as a point source that is comparatively easy to control. On the other hand, in the event of insufficient immobilization or leaching of the containment, the landfill can substantially contribute to the pollution of the region's environment. The landfill requires constant water management. Solutions have to be developed for the future, when the power plant is not in operation anymore and when funds are no longer available for landfill aftercare. Since leaching will be a constant threat, it is necessary to investigate whether immobilization of filter ash is more economic than aftercare of the landfill for several thousand years:

The emissions of the power plant are of little relevance for the region, with the exceptions of CO_2 (greenhouse gas) and, of minor importance, SO_2. The comparatively low retention capacity for mercury and selenium should be a focus for future improvement efforts of the operators after more detailed investigations and measurements. Generally, the study shows that the power plant actually does not pose a severe burden to the region's environment. The operators use these results for their environmental impact statement and for communication with concerned local people.

3.1.5 PROBLEMS — SECTION 3.1

Problem 3.1: Assess the effects of the following measures on the regional lead flows and stocks given in Figure 3.1. Show quantitatively and discuss the following effects of reductions in lead concentrations in soil, surface waters, and landfill: (a) ban on leaded gasoline, (b) ban on application of sewage sludge to land, and (c) construction of a new MSW incinerator in the region with an air-pollution control efficiency for lead of 99.99% treating the waste of 280,000 persons from the Bunz Valley and neighboring regions.

Problem 3.2: Consider a region of 2500 km² and 1 million inhabitants. Only one river flows through this region. At the inflow, the river has a flow rate of 1 billion m³/year, and the concentration of phosphorous (P) is 0.01 mg/l. The river discharges into a lake that represents a reservoir of 2.8 billion m³; residence time of water in the lake is 1 year. (Precipitation, evaporation, etc. are not considered; assume that the river is unchanged when flowing through the lake.) The anthroposphere of the region comprises the following processes: agriculture, food industry, private households, composting of biomass wastes from private households, and wastewater treatment. Per capita consumption of P for nutrition is 0.4 kg/capita/year; 20% of this demand is supplied by food industry within the region. For cleaning purposes, 1 kg/capita/year of P is used, with 70% contained in detergents for textiles; all detergents are imported. Assume that 90% of total nutritional P and 100% of total detergent-based P are directed to the wastewater treatment plant (WWTP). The transfer coefficient (TC) for P into sewage sludge is 0.85. The remaining 10% of nutritional P is contained in biomass waste from households that is composted without loss of P and applied to the soil. The stock of P is assessed at ca. 380,000 t. Agriculture imports 2400 t of P in fertilizers and 1200 t of P in animal feed; 80% of this P flow goes to the soil as manure, dung, and residues from harvesting. The balance is input to the regional food industry. The TC for P into food production wastes is 0.6. Food products that are not consumed within the region are exported. Approximately 1% of P that is annually applied to soil escapes to the surface water (river) as a result of erosion.

1. Draw a qualitative flowchart for the region described (system boundary, flows, processes).
2. Quantify the P flows and stocks (t/year and t) of the system. What is the accumulation of P in the soil?
3. Assume that P in detergents for textiles is phased out. Is this measure sufficient to prevent eutrophication (limit for eutrophication = 0.03 mg/l) of the lake?
4. What further measures do you suggest to prevent eutrophication?

Problem 3.3: Discuss and quantify the "reaction time" of different measures to control phosphorous flows in the Danube River basin. "Reaction time" is defined as the time span in days, weeks, months, years, etc. between the decision to take an action and a measurable effect in the Danube River. Note that "reaction time" also includes planning and implementation (construction, startup).

1. Reduction of phosphorous fertilizer input to soils by a resource tax
2. Connecting 95% of all private households to sewer systems
3. Increasing the removal efficiency for P in sewage treatment from 50% to >80%
4. Banning direct discharges from agriculture
5. Assessment of "reaction time" if P is banned in all detergents (assuming that one-third of the P flow through private households in the Danube basin stems from phosphorous-containing detergents)

Case Studies   215

TABLE 3.20
Example of Average Substance Concentrations in MSW, mg/kg

As	Pb	Cd	Cu	Se	Zn	Hg	S	Cl	N
10	500	10	1000	1	1200	1	3000	7000	5000

Draw a general conclusion regarding the reduction of P flows to the river Danube when you evaluate the effectiveness of the measures discussed above.

Problem 3.4: Compare the materials turnover of a coal-fired power plant and an MSW incinerator. The feed rate for coal is 300 t/h; for an MSW it is 30 t/h. Select the substances As, Pb, Cd, Cu, Se, Zn, Hg, S, Cl, N. Substance concentrations for coal are given in Table 3.2; for MSW, see Table 3.20. Discuss your findings with respect to air-pollution control.

The solutions to the problems are given on the Web site, www.iwa.tuwien.ac.at/MFA-handbook.

3.2 RESOURCE CONSERVATION

The main advantage of the application of MFA for resource conservation is the comprehensive information about sources, flows, and sinks of materials. This makes it possible to set priorities in resource conservation, to recognize early the benefit of material accumulations (e.g., in urban stocks), and to design new processes and systems for better control and management of resources. In this chapter, two groups of substances (nutrients and metals) and two groups of goods (plastic materials and construction materials) are discussed in view of resource conservation.

3.2.1 CASE STUDY 5: NUTRIENT MANAGEMENT

Nutrients are essential resources for the biosphere. Life without nitrogen and phosphorous is not possible. The atmosphere represents an unlimited reservoir for nitrogen. The industrial transformation of N_2 to chemical compounds such as ammonium and nitrate that can be taken up by plants requires energy. Hence, the amount of nitrogen available within the anthroposphere is limited mainly by energy supply. In contrast, phosphorous is taken from concentrated phosphate minerals that are limited in extent. It is assessed that at present consumption rates, concentrated phosphate deposits will be used up in about 100 years.[26] Thus, in order to conserve energy and resources, both nitrogen and phosphorous have to be managed with care.

The purpose of the following case study is to show how MFA can be used to set priorities in resource conservation. Measures are analyzed in view of their effectiveness regarding nutrient recycling. The total flows of phosphorous and nitrogen are investigated. The entire activity "to nourish" is analyzed from agriculture to food processing to private households. Losses and wastes are identified and quantified along the process chain. Because MFA of nutrients has been discussed before (see Case Studies 2 and 3 in Sections 3.1.2 and 3.1.3, respectively), the main emphasis is focused on the interpretation of the results. The procedure for establishing nutrient flow analysis of the activity "to nourish" is not given in detail. For further information, see Baccini and Brunner.[90]

3.2.1.1 Procedures

The activity "to nourish" is investigated on a national level. A system comprising five processes for the production, processing, distribution, and consumption of food is defined (Figure 3.15) and investigated. For each process, information about inputs and outputs are obtained from available sources, including national statistics about import, export, and production of fertilizer, agricultural produce, and food; agricultural information databases about the use of fertilizer and production of agricultural products; reports from food-processing companies, wholesale companies, and distributors of food; medical literature about human consumption and excretion of nutrients; and databases about concentrations and loadings of nutrients in wastewater, municipal solid wastes, and compost.

It is important to start with reliable data about the structure of the national agricultural sector: What are the main agricultural products; how are they produced; what is the nutrient input required for the production; and how large is the amount of nutrients actually harvested? Internal cycles of agriculture are to be investigated, such as the soil–plant–animal–manure–soil nutrient cycle. Figures for total losses of nutrients in agricultural practice are usually not available. Farmers use different definitions for wastes and losses. Shortfalls have to be calculated as the difference between total input and total output of the agricultural sector. The same method can be applied to the processes "industrial processing and distribution" to calculate or cross-check figures for losses, wastes, and wastewaters.

Using the sources pointed out above, the process "household" can be balanced as shown in Figure 3.5. The average amount of food consumed per capita and per year is taken from national statistics. Note that if such statistics are based on bookkeeping of individual households, they usually do not contain out-of-house consumption; in such cases, it will be necessary to increase the figure for food consumption by 20 to 30%. Waste-analysis data yield the amount of food residues in MSW. If such data are not available, it can be assumed that 5 to 10% of food purchased is discarded with MSW. Information about kitchen wastewater is taken from studies about sewage production in households. It can also be estimated that about 20 to 25% of food entering a household is discarded via the kitchen sink. Note that cooking water may contain considerable amounts of dry matter and (dissolved) salts. Of course, the partitioning of food in households between MSW, wastewater, and human consumption is a function of cultural aspects, too: in societies that are traditionally scarce in resources, the amount of kitchen wastes is considerably smaller. If grinders are installed in kitchen sinks, the food fraction in wastewater will be higher. If fast food plays a major nutritional role, food wastes in kitchens will be smaller because most food entering households has already been processed. In such cases, packaging wastes may be larger.

Case Studies

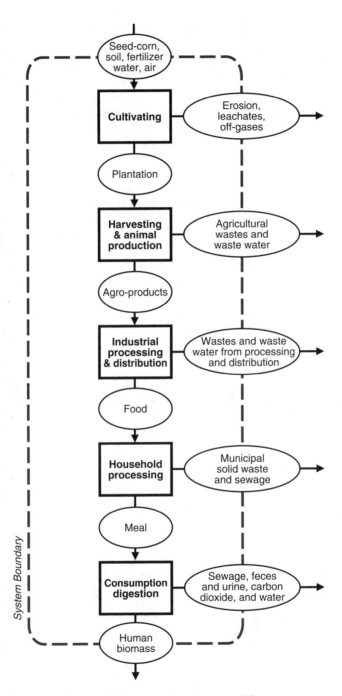

FIGURE 3.15 Activity "to nourish," presented as a system of five processes.

Data about respiration, urine, and feces is found in the medical literature on human metabolism. This information contains figures about N and P in food, urine, and feces, too. It is important to cross-check all data. The output of agricultural production can be compared with the input into food industry, the output of food industry to the consumption of the total population, and the output of the total population to the input into wastewater treatment and waste management. If the data for balancing the individual processes have been collected independently for each process, the redundancy of such cross-checking will be high, and the accuracy of the total nutrient balance can be improved significantly.

3.2.1.2 Results

To demonstrate the relevant results, the five processes in Figure 3.15 are combined to the three processes presented in Figure 3.16 and Figure 3.17. Food-related flows of P and N through agriculture, industrial processing and distribution, and consumption are displayed on a per capita base. In view of resource conservation, agriculture is the most important process, where 80% of P and close to 60% of N are lost during agricultural production. "Losses" are flows to groundwater, surface water, and air for N, and erosion/surface runoff and accumulation in soils for P. In order to optimize nutrient management, agricultural practice has to be changed as a first priority. Since nutrients are still comparatively cheap, there is no economic incentive yet for such a change. It seems timely to investigate how new or other technologies

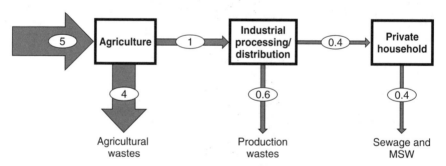

FIGURE 3.16 Phosphorous flow through the activity "to nourish," kg/capita/year.

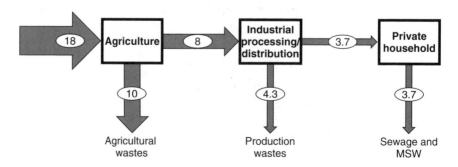

FIGURE 3.17 Nitrogen flow through the activity "to nourish," kg/capita/year.

Case Studies

TABLE 3.21
Partitioning of Food-Derived P and N in Private Households

	P, g/capita/year	P,[a] %	P,[b] %	N, g/capita/year	N,[a] %	N,[b] %
Food input	430	100	8.6	3700	100	20.5
Output						
MSW	40	9	0.8	300	8	1.7
Kitchen wastewater	20	5	0.4	200	6	1.1
Respiration	0	0	0.0	110	3	0.6
Urine	270	63	5.4	2600	70	14.4
Feces	100	23	2.0	490	13	2.7
Total food-related output	430	100	8.6	3700	100	20.5

[a] Percent of food nutrient input into household.
[b] Percent of total nutrient import into activity "to nourish" from Figure 3.16 and Figure 3.17. Data from Baccini and Brunner.[90]

can make better use of nutrients in agriculture. While the primary objective today is to prevent nutrient losses in order to protect the environment, it is likely that within a century, resource scarcity of phosphorous may become a driving force for changes in agriculture.

A key factor for nutrient losses in agriculture is consumer lifestyle. During the change from a resource-scarce society to an affluent society, the dietary tradition usually changes from low meat consumption to a diet that is rich in animal protein. The production of meat and poultry requires a much larger nutrient turnover than cereals and vegetables. Hence, the shift in dietary habits causes an increase in nutrient losses, too.

Losses of nutrients in industrial processing and distribution are much smaller than in agriculture, similar to those in households. The main difference between industrial processing/distribution and households is the number of sources: There are about 1000 times more households. Thus, from a reuse point of view, it is much more efficient to collect and recycle wastes from industrial sources than from consumers. The results summarized in Table 3.21 demonstrate clearly the limited contribution of individual households to the overall nutrient flows. If all food-derived nutrients from households are recycled, less than 10% of P and about 20% of N requirements of agriculture can be satisfied. Table 3.21 also shows the contribution of each household output to nutrient conservation, thus serving as a base for decisions regarding nutrient conservation and waste management. Composting of MSW is an inefficient measure to recycle nutrients. If all MSW were turned into compost, the contribution to agriculture would only be 1 to 2%. The fraction of nutrients in wastewater from households is about ten times larger. Thus, priority in nutrient recycling should be on wastewater and not on solid waste.

Another interesting fact is revealed by MFA and presented in Table 3.21: The amount of nutrients in urine is three (P) to five (N) times larger than in feces. This opens up new possibilities. Separate collection of urine could allow more than half of all nutrients entering a household to be accumulated in a relatively pure, concentrated, and homogeneous form. Several concepts have been proposed to manage this so-called anthropogenic nutrient solution (ANS).[27] They are all based on a new type of toilet that is designed to collect urine separately from feces. The sewer system would be used after midnight to collect ANS stored in households during the daytime, thus permitting specific treatment and recycling of N, P, and K. Or ANS could be stored in households for longer time periods and collected separately with mobile collection systems. In any case, in order to prepare a fertilizer of high value, hazards such as endocrine substances and pharmaceuticals would have to be removed before ANS could be be applied in agriculture.

Note that MFA of the activity "to nourish" is the basis for identifying the relevant nutrient flows and for developing alternative scenarios. In order to test the feasibility of the scenarios, technological, economic, and social aspects have to be investigated. New ways of managing urine and feces will only be successful if the same or greater convenience for the consumer is guaranteed.

3.2.2 CASE STUDY 6: COPPER MANAGEMENT

In Chapter 1, Section 1.4.5.1, it has been documented that modern economies are characterized by unprecedented material growth. Consumption of metals has increased while metal prices have decreased due to more efficient mining and refining technologies.[28] Up to 80 to 90% of all resources consumed by mankind have been used in the second half of the 20th century (Figure 3.18).

Within the anthropogenic metabolism, heavy metals are comparatively unimportant from a *mass* point of view, since they represent less than 10% of all inorganic

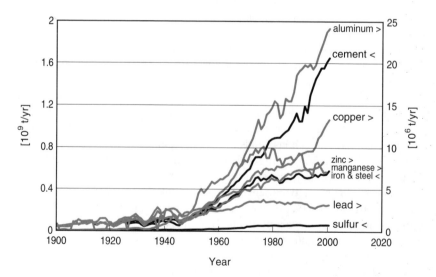

FIGURE 3.18 World use of selected resources, t/year. About 80 to 90% have been used since 1950.[91]

Case Studies

goods* consumed.[29] However, heavy metals play an important role in the production and manufacture of many goods. They can improve the quality and function of goods and are often crucial in extending the lifetime and range of application of goods. Their importance is based on their specific chemical and physical properties, e.g., corrosion resistance, electrical conductivity, ductility, strength, heat conductivity, brightness, etc.

In 1972, the Club of Rome was among the first to point out the scarcity of resources in the book *The Limits to Growth*.[30] Meadows et al. predicted that resources such as copper will be depleted within a short time of only a few decades. Prognoses about the depletion time** of metals have been constantly revised and extended as a result of newly found reserves and advanced exploitation technologies. For certain metals essential for modern technology — lead, zinc, copper, molybdenum, manganese, etc. — some authors expect shortages within the next several decades.[31] There is controversy about whether this limitation will restrict future growth (see Becker-Boost and Fiala[32]). Up to now, some but not all functions of metals can be mimicked by other materials.

Current metals management cannot be considered sustainable. During and after use, large fractions of metals are lost as emissions and wastes. Consequently, in many areas, concentrations of metals in soils as well as in surface and groundwaters are increasing. As discussed in Chapter 1, Section 1.4.5.2, human-induced flows of many metals surpass natural flows. Figure 1.7 displays the example of cadmium.[90] While geogenic processes mobilize roughly 5.4 kt/year of cadmium, human activities extract about 17 kt/year from the Earth's crust. Comparatively large anthropogenic emissions into the atmosphere are causing a significant accumulation of cadmium in the soil. Global emissions of cadmium should be reduced by an order of magnitude to achieve similar deposition rates as those determined for natural cadmium deposition. On a regional basis, the reduction goal should be even higher. Since most of the anthropogenic activities are concentrated in the Northern Hemisphere, the cadmium flows in this region have to be reduced further in order to protect the environment properly. The stock of anthropogenic cadmium grows by 3 to 4% per year. It needs to be managed, disposed of, and recycled carefully in order to avoid short- and long-term environmental impacts.

Heavy metals are limited valuable resources, but they are also potential environmental pollutants. New strategies and methods are needed for the management of heavy metals. A first prerequisite for efficient resource management is appropriate information about the use, location, and fate of these substances in the anthroposphere. Based on such information, measures to control heavy metals in view of resource optimization and environmental protection have to be designed. This case study discusses sustainable management of copper using information about copper flows and stocks in Europe as determined by Spatari and colleagues.[33]

---

* Excluding water.
** The number of years left until a resource is exhausted under a constant use rate.

3.2.2.1 Procedures

The copper household is evaluated by statistical entropy analysis (SEA). In Chapter 2, Section 2.5.1.8, the SEA method was introduced for single-process systems. In this case study, a system consisting of multiple processes is analyzed, requiring additional definitions and procedures. SEA can be directly applied to copper databases with no further data collection and little computational effort. The procedure has been developed and described by Rechberger and Graedel.[34]

3.2.2.1.1 Terms and Definitions

A set of "material flows" consists of a finite number of material flows. The "distribution" of a substance represents the partitioning of a substance among a defined set of materials. The distribution (or distribution pattern) is described by any two of the three properties $\dot{M}_i$, $\dot{X}_i$, c_i for all materials of the set (see Figure 3.19).

3.2.2.1.2 Calculations

The following equations are used to calculate the statistical entropy H of a set of solid materials*. The number of materials in the set is k, and the flow-rates $(\dot{m}_1,\ldots,\dot{m}_k)$ and substance concentrations $(c_1, \ldots, c_k)$ are known.

$$\dot{X}_i = \dot{m}_i \cdot c_i \qquad (3.1)$$

$$\tilde{m}_i = \frac{\dot{m}_i}{\sum_{i=1}^{k} \dot{X}_i} \qquad (3.2)$$

$$H(c_i, \tilde{m}_i) = -\sum_{i=1}^{k} \tilde{m}_i \cdot c_i \cdot ld(c_i) \geq 0 \qquad (3.3)$$

The concentrations in Equation 3.1 and Equation 3.3 are expressed on a mass-per-mass basis in equivalent units (e.g., $g_{substance}/g_{product}$ or $kg_{substance}/kg_{product}$, etc.) so that $c_i \leq 1$. If other units are used (e.g., %, mg/kg), Equation 3.3 must be replaced by a corresponding function.[35] The variable $\tilde{m}_i$ represents standardized mass fractions of a material set. If the c_i and $\tilde{m}_i$ are calculated as described, the extreme values for H are found for the following distributions (see Figure 3.20):

1. The substance is only contained in one of the k material flows (i = b) and appears in pure form $\Sigma \dot{X}_i = \dot{X}_b = \dot{m}_b$. Such a material set represents the substance in its highest possible concentration. The statistical entropy H

---

* If gaseous and aqueous flows (emissions) are also to be considered, more complex equations such as given in Chapter 2, Section 2.5.1.8 have to be applied. The system analyzed in this chapter contains solid materials/copper flows only.

Case Studies

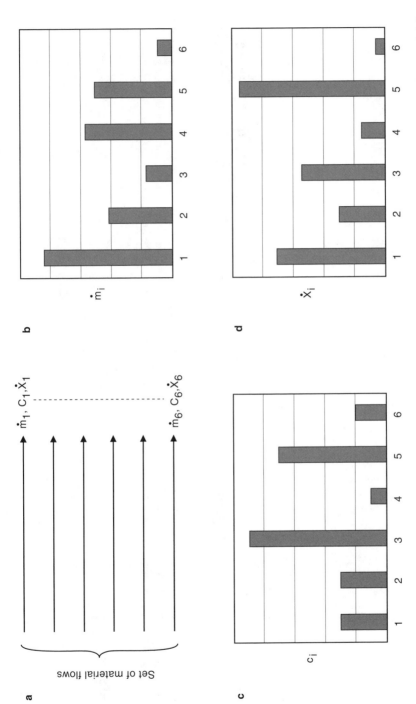

FIGURE 3.19 (a) Exemplary set of six material flows. (b) Mass flows of the set, mass/time. (c) Concentrations of the substance in the material flows, mass/mass. (d) Distribution of the substance among material flows (fraction).

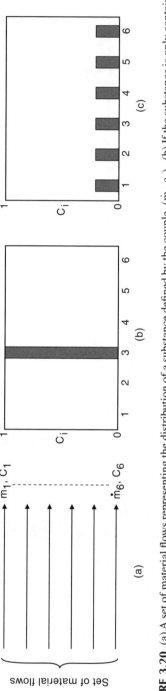

FIGURE 3.20 (a) A set of material flows representing the distribution of a substance defined by the couple $(\dot{m}_i, c_i)$. (b) If the substance is only contained in one material flow, the statistical entropy H is 0. If the substance is equally distributed among the material flows, H reaches the maximum. (c) Any other distribution yields an H-value between zero and max. (From Rechberger, H. and Graedel, T.E., *Ecol. Econ.*, 42, 59, 2002. With permission.)

Case Studies

of such a distribution is zero, which is also a minimum, since H is a positive definite function for $c_i \leq 1$ (Figure 3.20b).

2. The other extreme is when all material flows have the same concentration ($c_1 = c_2 = \ldots = c_k$). Such a material set represents the substance in its highest possible diluted form. For such a distribution, the statistical entropy is a maximum. Any other possible distribution produces an H value between these extremes (Figure 3.20c).

The maximum of H is expressed as

$$H_{max} = \text{ld}\left(\sum_{i=1}^{k} \tilde{m}_i\right) \quad (3.4)$$

Finally, the relative statistical entropy (RSE) is defined as

$$\text{RSE} \equiv H/H_{max} \quad (3.5)$$

A material flow system usually comprises several processes that are often organized in process chains. Figure 3.21 displays such a system comprising four processes (P) linked by ten material flows (F), including one loop (recycling flow F9).

The procedure for evaluating a system by statistical entropy analysis depends on the structure of the system. For the system investigated in this chapter, the statistical entropy development can be calculated as described in the following two sections.

3.2.2.1.2.1 Determination of Number and Formation of Stages

If the number of processes in the system is n_P, then the number of stages is $n_S = n_P + 1$, where the stage index $j = 1, 2, \ldots, n_S$. The system as a whole can be seen as a process that transfers the input step by step, with each step designated as a "stage." Stages are represented by a set of material flows (see Figure 3.21b). The first stage is defined by the input into the first process of the process chain. The following stages are defined by the outputs of processes 1 to n_P. So stage j (j > 1) receives (1) the outputs of process $j - 1$ and (2) all outputs of preceding processes that are not transformed by the system (export flows and flows into a stock). Flows out of a stock are treated as input flows into the process. Flows into a stock are regarded as output flows of the process (see process P3, flow F7 in Figure 3.21a). This means that the stock is actually treated as an independent external process. However, for the sake of clarity, stocks are presented as smaller boxes within process boxes (see Figure 2.1). Finally, recycling flows are treated as export flows. The allocation of material flows to stages is displayed in Figure 3.21b. The diagram shows how substance flows through the system become increasingly branched from stage to stage, resulting in different distribution patterns of substances.

3.2.2.1.2.2 Modification of Basic Data and Calculation of RSE for Each Stage

The basic data, flow rates of materials, and substance concentrations ($\dot{m}_i, c_i$) of the investigated system are determined by MFA. Normalized mass fractions $\tilde{m}_i$ are

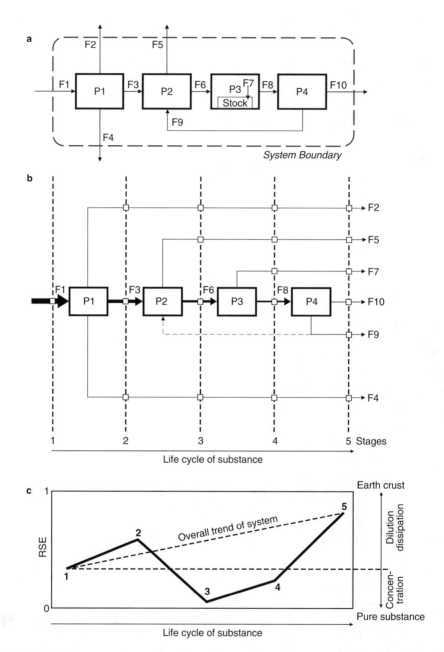

FIGURE 3.21 (a) Basic structure of a system made up of a process chain including one recycling loop. (b) Allocation of the system's material flows to five stages. For example, stage 3 is represented and defined by flows F2, F5, F6, and F4. Stages 2 to 5 represent the transformations of the input (stage 1) caused by processes 1 to 4. (c) The partitioning of the investigated substance in each stage corresponds to a relative statistical entropy (RSE) value between maximal concentration (0) and maximal dilution (1). (From Rechberger, H. and Graedel, T.E., *Ecol. Econ.*, 42, 59, 2002. With permission.)

derived using Equation 3.1 and Equation 3.2. Application of Equation 3.3 to the couple $(c_i, \dot{m}_i)$ or to each stage yields the statistical entropy H for that stage. H_{max} is a function of the total normalized mass flow represented by a stage (see Equation 3.4). This normalized mass flow grows with subsequent stages if the concentrations of the materials decrease, since $\Sigma c_i \cdot \dot{m}_i = 1$ (combine Equation 3.1 and Equation 3.2). One can assume maximum entropy when materials of a stage have the same concentration as the Earth's crust (c_{EC}) for the substance under study. H_{max} is then given by

$$H_{max} = ld\left(\frac{1}{c_{EC}}\right) \qquad (3.6)$$

The reason for this definition of H_{max} is related to resource utilization. If, for example, copper is used to produce a good that has a copper concentration of 0.00006 kg/kg (the average copper content of the Earth's crust[36]), this product has the same resource potential for copper as the average crustal rock. Thus, a stage with entropy $H = H_{max}$ defines a point at which enhanced copper resources no longer exist. Using Equation 3.5 and Equation 3.6, the RSE for each stage can be calculated. Figure 3.21c demonstrates that a system as a whole can be either concentrating, "neutral" (balanced), or diluting, depending on whether the RSE for the final stage is lower than, equal to, or higher than the first stage.

3.2.2.1.3 Copper Data and Copper System of Study

Figure 3.22 illustrates copper flows and stocks in Europe in 1994, developed as part of a comprehensive project carried out at the Center for Industrial Ecology, Yale University. For a discussion of the quality, accuracy, and reliability of the data, see Graedel et al.[37] and Spatari et al.[33] Evaluating copper management practices on the basis of these data poses a challenge. At present, Europe is an "open system" for copper and depends heavily on imports. The total copper import (2000 kt/year) is more than three times higher than the domestic copper production from ore (≈590 kt/year; ore minus tailings and slag). Large amounts of production residues result from the use of copper, but with the present system boundaries, they are located outside of the system and therefore are not considered in an evaluation of European copper management. For a "true" evaluation, exports of goods containing copper and imports such as old scrap have to be taken into account, too. Thus, it is necessary to define a virtual "autonomous system" that (1) is independent of import and export of copper products and wastes and (2) incorporates all external flows into the system. Hence, in this virtual system, the copper necessary to support domestic demand is produced entirely within the system, depleting resources and producing residues. The estimated data for this supply-independent scenario are given in parentheses in Figure 3.22, which represents a closed system that includes all material flows relevant for today's copper management.

Table 3.22 gives the data that are used to calculate the entropy trends. The flow-rates for copper ($\dot{X}_i$) are from Spatari et al.[33] The concentrations for copper (c_i) and their ranges are either from literature references or best estimates. The ranges

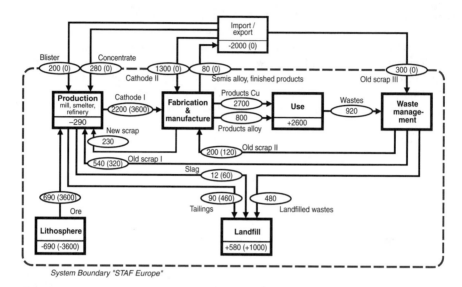

FIGURE 3.22 Copper flows and stocks for Europe in 1994 (values rounded, kt/year). The values given in parentheses represent a virtual and autonomous copper system with the same consumption level but no copper imports and exports. (From Rechberg, H. and Graedel, T.E., *Ecol. Econ.*, 42, 59, 2002. With permission.)

provide the basis for assessing the uncertainty of the final entropy trends. The flow rates for materials ($\dot{m}_i$) are calculated using Equation 3.7:

$$\dot{m}_i = \dot{X}_i / c_i \cdot 100 \qquad (3.7)$$

3.2.2.2 Results

3.2.2.2.1 Relative Statistical Entropy of Copper Management and of Alternative Systems — Status Quo and Virtual Supply-Independent Europe

The entropy trends are calculated using Equation 3.3 to Equation 3.7, the data given in Table 3.22, and the appropriate flow charts. Figure 3.23 shows the trend of the relative statistical entropy along the life cycle of copper for two systems: (1) the status quo of 1994 and (2) the supply-independent Europe (both displayed in Figure 3.22). The assignment of material flows to stages is illustrated in Figure 3.24.

Both systems behave similarly, with the production process reducing the RSE from stage 1 to stage 2, since ore (copper content 1 g/100 g) is refined to plain copper (content >99.9 g/100 g). Note that the RSE for stage 2 is not zero, since mining ores and the smelting concentrates produce residues (tailings and slag). The more efficient a production process is (efficiency being measured by its ability to transform copper-containing material), the more closely the RSE of stage 2 approaches zero, meaning that the total amount of copper appears in increasingly

TABLE 3.22
Data on Material Flows of European Copper Management

Material	Material Flow ($\dot{m}_i$), kt/year	Copper Concentration (c_i), g/100 g	Copper Flow (X_i),[33] kt/year
Ore	69,000	1 (0.3–3)[97–99]	690
Concentrate	930	25 (20–35)[97,98,99]	280
Blister	205	98 (96–99)[97]	200
Cathode I	2200	100	2200
Flow out of stock (production)	290	100	290
Cathode II	1300	100	1300
Tailings	90,000	0.1 (0.1–0.75[a])[99]	90
Slag	1700	0.7 (0.3[b]–0.7)[99]	12
New scrap	260	90 (80–99)[c]	230
Old scrap I	680	80 (20–99)[97]	540
Old scrap II	250	80 (20–99)[97]	200
Old scrap III	380	80 (20–99)[97]	300
Semi alloy and finished products	110	70 (7–80)[c]	80
Products (pure Cu)	27,000	10 (1–50)[c]	2700
Products (Cu alloy)	11,000	7 (1–40)[c]	800
Flow into stock (use)	1,200,000	0.2 (0.1–0.3)[c,38]	2600
Wastes	460,000	0.2 (0.1–0.3)[c,38]	920
Landfilled wastes	460,000	0.10[d]	480

Note: Values are rounded.

[a] Higher value for period around 1900.
[b] Lower value for period around 1925.
[c] Informed estimate.
[d] Calculated by mass balance on waste-management process.

From Rechberg, H. and Graedel, T.E., *Ecol. Econ.*, 42, 59, 2002. With permission.

---

purer form.* Producing semiproducts and consumer goods from refined copper increases the RSE from stage 2 to stage 3 because of the dilution of copper that occurs in manufacturing processes. It is obvious that dilution takes place when copper alloys are produced. Similarly, installing copper products into consumer goods (e.g., wiring in an automobile) or incorporating copper goods into the built infrastructure (transition from stage 3 to 4, e.g., copper tubing for heating systems) "dilute" copper as well. In general, the degree of dilution of copper in this stage is not well known. Information about location, concentration, and specification is a *sine qua non* condition for future management and optimization of copper. For a

---

* For the reduction of the RSE from stage 1 to stage 2, external energy (crushing ores, smelting concentrate, etc.) is required. The impact on the RSE induced by this energy supply is not considered within the system, since the energy supply is outside the system boundary. Whether or not the exclusion of the energy source has an impact on the RSE development depends on the kind of energy source (coal, oil, hydropower) used. However, in this chapter the system boundaries are drawn as described in Spatari et al.[33]

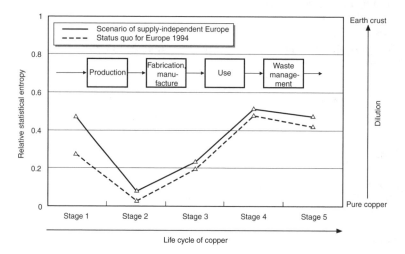

FIGURE 3.23 Change of the relative statistical entropy along the life cycle of copper for the status quo in Europe in 1994 (open system) and for a virtual, supply-independent Europe (closed or autonomous system). The shapes of the trends are identical, but the overall performances (differences between stages 1 and 5) of the systems are different. (From Rechberg, H. and Graedel, T.E., *Ecol. Econ.*, 42, 59, 2002. With permission.)

first hypothesis, it is sufficient to assume that the mean concentration of copper in the stock is the same as the mean concentration in the residues that leave the stock. This concentration level can be determined from copper concentrations and relevant waste-generation rates such as municipal solid waste, construction and demolition debris, scrap metal, electrical and electronic wastes, end-of-life vehicles, etc.[38] During the transition from stage 4 to 5, the entropy decreases, since waste collection and treatment separate copper from the waste stream and concentrate it for recycling purposes. The "V-shape" of the entropy trend — the result of entropy reduction in the production (refining) process and entropy increase in the consumption process (see Figure 3.23) — was described qualitatively, e.g., by O'Rourke et al.,[39] Ayres and Nair,[40] Stumm and Davis,[41] and Georgescu-Roegen.[42]

The differences in the entropy trends between the status quo and the supply-independent system are noteworthy. First of all, the status quo system starts at a lower entropy level, since concentrated copper is imported in goods. The differences in stage 2 are due to the increased ore production in the supply-independent system, resulting in larger amounts of production residues, which are accounted for in stage 2. In stages 3 and 4, the difference between the status quo and the supply-independent system remains rather constant, since the metabolism for both scenarios does not differ significantly in these stages. The effectiveness of waste management is lower in the supply-independent system, as there is no old scrap imported and the recycling rate is therefore lower. In the following, only the supply-independent system and some variations of it are discussed, since it comprises all processes and flows relevant for European copper management and includes external effects within Europe's hinterland.

The overall performance of a system can be quantified by the difference between the RSEs for the first and the final stages. In this case,

Case Studies

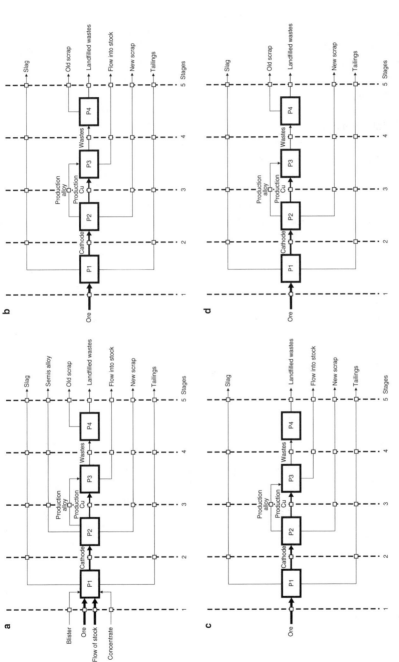

FIGURE 3.24 Assignment of material flows to stages. (a) Status quo, (b) supply-independent system, (c) supply-independent system without recycling, (d) supply-independent system in steady state. (From Rechberg, H. and Graedel, T.E., *Ecol. Econ.*, 42, 59, 2002. With permission.)

$$\Delta RSE_{total} = \Delta RSE_{15} \equiv [(RSE_5 - RSE_1)/RSE_1] \times 100 \qquad (3.8)$$

where $\Delta RSE_{total} > 0$ means that the investigated substance is diluted and/or dissipated during its transit through the system. From a resource conservation and environmental protection point of view, such an increase is a drawback. If maintained indefinitely, such management practice will result in long-term problems. In contrast, scenarios with high recycling rates, advanced waste management, and nondissipative metals use show decreasing RSE trends ($\Delta RSE_{total} \leq 0\%$). Low entropy values at the end of the life cycle mean that (1) only small amounts of the resource have been converted to low concentrations of copper in products (e.g., as an additive in paint) or dissipated (in the case where emissions are considered) and (2) large parts of the resource appear in concentrated (e.g., copper in brass) or even pure form (e.g., copper pipes). Wastes that are disposed of in landfills should preferably have Earth-crust characteristics[43] or should be transformed into such quality before landfilling. Earth-crustlike materials are in equilibrium with the environment, and their exergy approaches zero.[44–46] Thus, waste-management systems must produce (1) highly concentrated products with high exergy that are not in equilibrium with the surrounding environment and (2) residues with Earth-crustlike quality. Low- or zero-exergy wastes can easily be produced by dilution, e.g., by emitting large amounts of off-gases with small concentrations in high stacks, or by mixing hazardous wastes with cement, thus impeding future recycling of the resource. A low RSE value for a stage thus means that both highly concentrated (high exergy) and low-contamination (low exergy) products are generated.

3.2.2.2.2 Recycling in Supply-Independent Europe

The relevance of recycling on the entropy trend is investigated using Figure 3.25. Numbers in parentheses show the supply-independent system without any recycling of old and new scrap. Compared with the supply-independent scenario displayed in

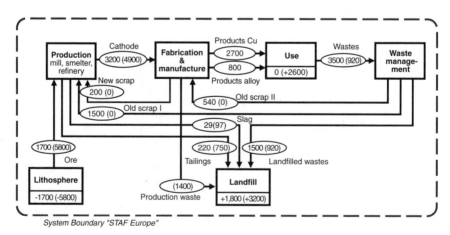

FIGURE 3.25 Copper flows and stocks of a supply-independent Europe with no accumulation of copper in the process "use," kt/year (steady-state scenario). Values in parentheses stand for a scenario without copper recycling. (From Rechberg, H. and Graedel, T.E., *Ecol. Econ.*, 42, 59, 2002. With permission.)

Case Studies

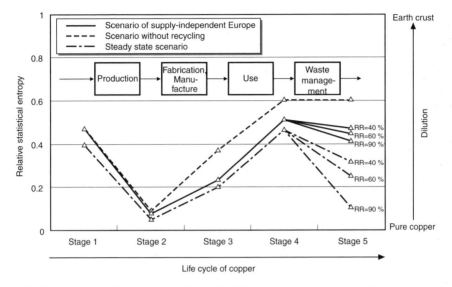

FIGURE 3.26 Comparison of the effect of different scenarios on the relative statistical entropy along the life cycle of copper: scenario supply-independent Europe vs. scenario of no recycling and scenario of steady-state producing no stocks. The assessment shows that waste management and recycling can play a crucial role in future resource use. (From Rechberg, H. and Graedel, T.E., *Ecol. Econ.*, 42, 59, 2002. With permission.)

Figure 3.22, this results in a higher demand for ore (+63%) and larger requirements for landfills for production and consumption wastes (+220%).

The entropy trend for the nonrecycling scenario is given in Figure 3.26. All RSEs are higher, showing the effects of not recycling production residues (new scrap) in stages 2 and 3 and the zero contribution of waste management in stage 5. The resulting $\Delta RSE_{15} = +28\%$ indicates a bad management strategy. At present, the overall recycling rate for old scrap is about 40%. Some countries within the European Union achieve rates up to 60%.[38] Assuming that in the future all countries will achieve this high rate, ΔRSE_{15} would be reduced from –1 to –4% (recycling rate of 90%: $\Delta RSE_{15} = -11\%$) for the supply-independent system. This shows that the impact of today's waste management on the overall performance of the system is limited. The reason is that the copper flow entering waste management is comparatively small.

3.2.2.2.3 Supply-Independent Europe in Steady State

Figure 3.25 also gives the flows for a steady-state scenario in which the demand for consumer goods is still the same as in the status quo, but the output equals the input of the stock in the process "use." This scenario may occur in the future when, due to the limited lifetime, large amounts of materials turn into wastes.[47] Assuming a recycling rate of 60% results in $\Delta RSE_{15} = -47\%$. This shows that in the future, waste management will be decisive for the overall management of copper. A recycling rate of 90% will result in $\Delta RSE_{15} = -77\%$. Such a high recycling rate cannot be achieved with today's design of goods and systems. Also, better information bases on the whereabouts of copper flows and, especially, stocks are needed. If the design process is improved, if necessary information is provided, and if advanced waste management technology is employed, future management of copper can result in declining RSE rates, contributing to sustainable metals management.

3.2.2.2.4 Uncertainty and Sensitivity

The uncertainty of the data (material flow rates $\dot{m}_i$ and substance concentrations c_i) and the accuracy of the results are fundamental pieces of information for the evaluation process. In most cases, data availability constrains the application of statistical tools to describe material management systems. Statistics on material flows do not customarily provide information on reliability and uncertainty, such as a standard deviation or confidence interval. Sometimes, substance concentration ranges can be determined by a literature survey. In Figure 3.27, upper and lower limits of the RSE are presented for the supply-independent scenario. These limits are calculated using the estimated ranges for copper concentrations as given in Table 3.22. Thus, the limits are not statistically derived but estimated. Since the ranges in Table 3.22 have been chosen deliberately to be broad, the possibility that the actual RSE trend lies within these limits is high. This is despite the fact that the uncertainty in the material flow rates is not considered. The range for ΔRSE_{15} lies between –23% and 28% (mean –1%), sufficient for a first assessment. The uncertainties for the different stages vary considerably. The range for stage 1 is due to the range of the copper content in ores (0.5 to 2%). The range for stage 2 is quite small, meaning that the RSE for this stage is determined with good accuracy. The largest uncertainty is found for stage 3, since the average copper concentrations of many goods are poorly known. The uncertainties for stage 4 and 5 are lower, with a range similar to that for stage 1.

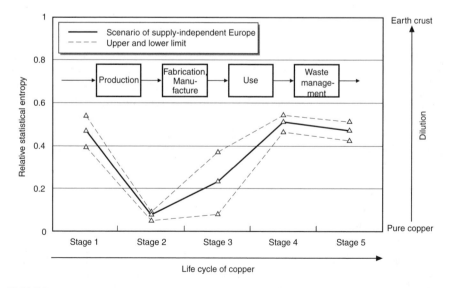

FIGURE 3.27 Variance of relative statistical entropy based on estimated ranges of basic data. (From Rechberg, H. and Graedel, T.E., *Ecol. Econ.*, 42, 59, 2002. With permission.)

Case Studies

The result emphasizes the hypothesis that the stock in use has the potential to serve as a future resource for copper. Both stages 4 and 5 show the same entropy level for one ton of copper. When calculating the RSE, the stock is characterized by the estimated average concentration of copper in the stock, meaning that the copper is evenly distributed and maximally diluted in the stock. This can be regarded as a "worst-case" assumption. Having more information about the actual distribution of copper in the stock would result in lower RSE values for stage 4. Provided that this information can be used for the design and optimization of waste management, the high recycling rates necessary to achieve $\Delta RSE_{15} < -70\%$ should be feasible.

3.2.2.3 Conclusions

Contemporary copper management is characterized by changes in the distribution pattern of copper, covering about 50% of the range between complete dilution and complete concentration. Copper flows and stocks through the (extended) European economy are more or less "balanced" due to recycling of new and old scrap and the small fraction of dissipative use of copper in goods. It is confirmed that the stock of copper currently in use has the potential for a future secondary resource. This can be even further improved by appropriate design for recycling of copper-containing goods. Provided that waste management is adapted to recycle and treat the large amounts of residues resulting from the aging stock, copper can be managed in a nearly sustainable way. Thus, this case study exemplifies how nonrenewable resources can be managed in order to conserve resources and protect the environment.

3.2.3 CASE STUDY 7: CONSTRUCTION WASTES MANAGEMENT

Construction materials are important materials for the anthropogenic metabolism. They are the matrix materials for the structure of buildings, roads, and networks and represent the largest anthropogenic turnover of solid materials (see Table 3.23). They have a long residence time in the anthroposphere and thus are a legacy for future generations. On one hand, they are a resource for future use; on the other hand, they can be a source of future emissions and environmental loadings. An example of reuse would be recycling of road surface materials, which is widely practiced in many countries. Examples of emissions are PCBs (polychlorinated biphenyls) in joint fillers and paints, and CFCs (chlorinated and fluorinated carbohydrates) in insulation materials and foams. Hence, construction materials have to be managed with care in view of both resource conservation and environmental protection. A main future task will be to design constructions in a way that allows the separation of construction materials after the lifetime of a building, with the main fraction being reused for new construction, leaving only a small fraction for disposal via incineration in landfills. (Incineration will be necessary to mineralize and concentrate hazardous substances that are required to ensure long residence times of, e.g., plastic materials.)

In this case study, construction materials are discussed in view of resource conservation. Both "volume" and "mass" are considered as resources. The purpose is twofold. First, it is shown that MFA can be used to address "volume related"

TABLE 3.23
Per Capita Use of Construction Materials in Vienna from 1880 to 2000[100]

Period, Decade	Per Capita Use of Construction Materials, m³/capita/year
1880–1890	0.8
1890–1900	0.4
1900–1910	0.1
1910–1920	0.1
1920–1930	0.1
1930–1940	1.4
1940–1950	0.1
1950–1960	0.1
1960–1970	0.1
1970–1980	2.4
1980–1990	3.3
1990–2000	4.3

resource problems, too. Also, some of the difficulties of bringing construction wastes back into a consumption cycle are explained. Second, two technologies for producing recycling materials from construction wastes are compared by means of MFA.

3.2.3.1 The "Hole" Problem

Excavation of construction materials from a quarry or mine usually results in a hole in the ground. Since construction materials are used to create buildings with residence times of several decades, it takes some 30 to 50 years before these holes can be filled up with construction debris. In a growing economy, the input of construction materials into the anthroposphere at a given time is much larger than the output. Thus, as long as the building stock of a city expands, the volume of holes in the vicinity of the city expands as well.

In Figure 3.28 and Figure 3.29, the total and per capita use of construction materials in Vienna is given for the time span from 1880 to 2000. The extraction of construction materials varies much from decade to decade. The effect of economic crisis, such as the Great Depression of the 1930s and the postwar periods, on construction activities is evident. If accumulated over the time period of 120 years, a total "hole" of 207 million m³ results (Figure 3.29). This corresponds to about 140 m³ per capita for today's population (1.5 million inhabitants).

It is interesting to note that the holes created by the needs of a prosperous, growing city of the 1990s are much larger than the volume of all wastes available for landfilling. In Figure 3.30, Lahner[92] presents a construction-materials balance established for Austria. The input of construction materials exceeds the output of construction wastes by nearly an order of magnitude! Besides the "hole" problem discussed here, another important implication arises from input >> output: The amount of construction wastes available for recycling is small when compared with

Case Studies

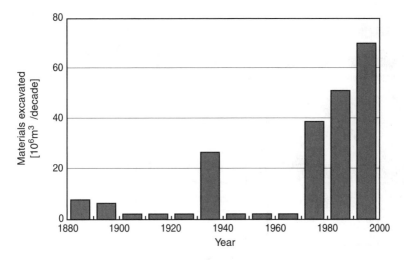

FIGURE 3.28 Construction material input into Vienna from 1880 to 2000, m³ per decade.[100]

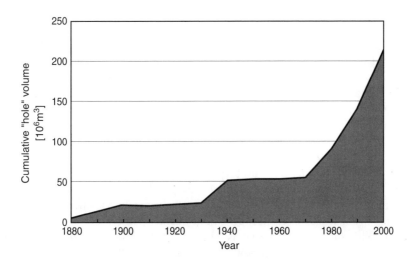

FIGURE 3.29 Cumulative "hole" volume in the vicinity of Vienna due to excavation of construction materials between 1880 and 2000.[100]

the total need for construction materials. Thus, even if all wastes were recycled, they would replace only a small fraction of primary materials. It may be difficult to create a market for a product with such a small market share, especially if there is uncertainty with respect to the quality of the new and as-yet unknown material and if there is only a small advantage in price. For successful introduction of recycling materials, it is necessary to establish technical and environmental standards, to develop technologies that produce sufficiently high quality at a competitive price, and to persuade consumers of the usefulness and advantages of the new product.

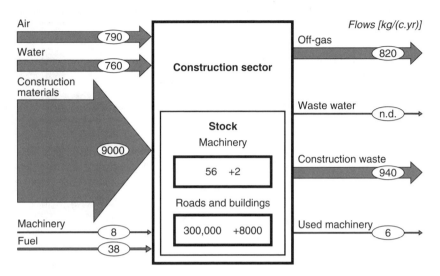

FIGURE 3.30 Materials used for construction in Austria (1995),[92] kg/capita/year. The input of construction materials into a growing economy is much larger than the output of construction wastes. (From Lahner, T., *Müll Magazin*, 7, 9, 1994. With permission.)

In the case of Vienna, the total waste (MSW, construction waste, etc.) generated annually for disposal during the 1990s was about 600,000 tons measuring 800,000 m^3 (or 400 kg/capita at 0.53 m^3/capita). Wastes that are recycled are not included in this figure. This is approximately eight times less than the annual consumption of construction materials (4.3 m^3/capita/year). Thus, it is not possible to fill the holes of Vienna by landfilling all wastes. Note that the actual volume of wastes to be landfilled in Vienna is considerably smaller due to waste incineration, which reduces the volume of municipal wastes by a factor of 10.

Landfilling is usually not a problem from the point of view of quantity (volume or mass); rather, it is an issue of quality (substance concentrations). The wastes that are to be disposed of in landfills do not have the same composition as the original materials taken from these sites. Thus, the interaction of water, air, and microorganisms with the waste material is likely to differ from the original material, resulting in emissions that can pollute groundwater and the vicinity of the landfill. On the other hand, the native material has been interacting with the environment for geological time periods. Except for mining and ore areas, the substance flows from such native sites are usually small ("background flows and concentrations") and not polluting.

The conclusion of the "hole balance" problem is as follows: Growing cities create holes, hence "hole management" is important and necessary. These void spaces can be used for various purposes, such as for recreation or for waste disposal. If they are used as landfill space, qualitative aspects are of prime importance and have to be observed first. Wastes to be filled in such holes need to have "stonelike" properties. They require mineralization (e.g., incineration with aftertreatment), and they should be in equilibrium with water and the environment. The new objective of waste treatment thus becomes the production of immobile stones from waste materials.

3.2.3.2 Use of MFA to Compare Construction Waste-Sorting Technologies

Construction wastes are the largest fraction of all solid wastes. Thus, for resource conservation it is important to collect, treat, and recycle these wastes. There are various technologies available to generate construction materials from construction wastes. Their purpose is to separate materials well suited as building materials from hazardous, polluting, or other materials inappropriate for construction purposes. MFA serves as a tool to evaluate the performance of construction waste-sorting plants with regard to the composition of the products (e.g., production of clean fractions vs. accumulation of pollutants in certain fractions).

In order to design and control construction waste recycling processes, it is necessary to know the composition of the input material that is to be treated in a sorting plant. The composition and quantity of construction wastes depend upon the "deconstruction" process. If a building is broken down by brute force of a bulldozer, the resulting waste is a mixture of all possible substances. If it is selectively dismantled, individual fractions can be collected that represent comparatively uniform materials such as wood, concrete, bricks, plastics, glass, and others. These fractions are better suited for recycling. After crushing, they can be used either for the production of new construction materials or as fuel in industrial boilers, power plants, or cement kilns. Both types of deconstruction yield at least one fraction of mixed construction wastes. While indiscriminate demolition results in mixed construction wastes only, the mixed fraction obtained in selective dismantling is much smaller and comprises mainly nonrecyclables such as plastics, composite materials, and contaminated constituents.

Construction waste-sorting plants are designed to handle mixed fractions. The objectives of sorting are twofold: First, sorting should result in clean, high-quality fractions suited for recycling. Second, sorting should yield nonrecyclables that are ready for treatments such as incineration or landfilling. In Figure 3.31 and Figure 3.32, two technologies for construction waste recycling are presented. They differ in the way they separate materials. Plant A (25 t/h) is a dry process, including handpicking of oversize materials, rotating drum for screening, crusher and pulverizer, zigzag air classifier, and dust filters. In plant B (60 t/h), the construction waste is similarly pretreated before it is divided into several fractions by a wet separator. In order to evaluate and compare the performance of the two processes with regard to resource conservation, both plants are investigated by MFA. The results serve as a base for decisions regarding the choice of technologies for construction-waste sorting.

3.2.3.3 Procedures

Since it is not possible to determine the chemical composition of untreated construction wastes by direct analysis, the input material into both plants is weighed

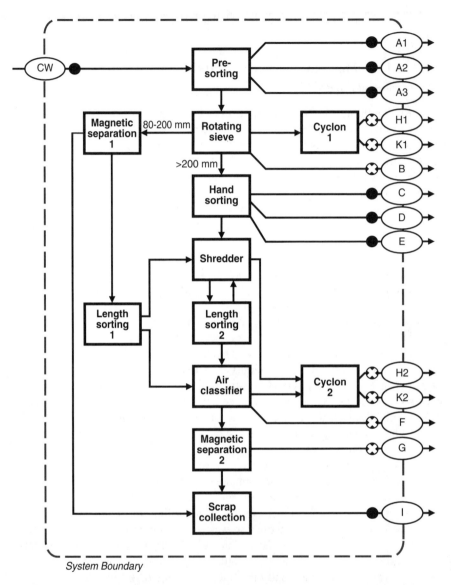

FIGURE 3.31 Construction waste (CW)-sorting plant A, dry process. Fraction A1, large pieces of concrete and stones; A2, metals; A3, oversize combustibles; B, <80 mm; C, concrete and stones; D, metals; E, oversize material; F, light fraction; G, heavy fraction; H1 and H2, dust from cyclones 1 and 2; I, scrap iron; K1, off-gas drum and shredder; K2, off-gas air classifier. (●, measurement of mass flow, m³/h and t/h; X, measurement of substance concentration, mg/kg.)

only and not analyzed. The composition of the incoming waste is established by sampling and analyzing all products of sorting, and by calculating for each substance the sum of all output flows divided by the mass of construction wastes treated within the measuring period. This procedure is chosen because the sorting plants produce

Case Studies

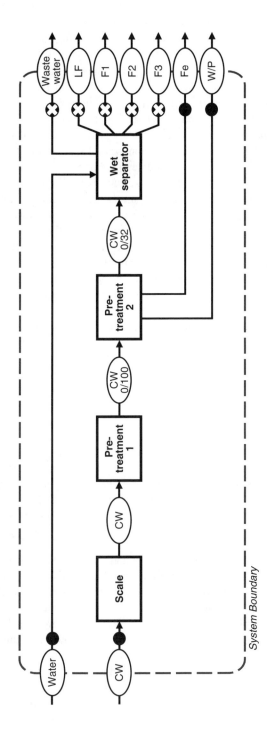

FIGURE 3.32 Construction waste (CW)-sorting plant B, wet process. Pretreatment 1: crusher and sieve, 0 to 100 cm; pretreatment 2: hand sorting, magnetic separator, pulverizer, and sieve, 0 to 32 cm; CW 0/100 and CW 0/32: construction waste crushed, pulverized, and sieved by a mesh size of 100 cm and 32 cm, respectively; wastewater includes settled sludge; LF: light fraction; F1 to F3: construction materials for recycling (F1, 16 to 32 mm; F2, 4 to 16 mm; F3, 0 to 4 mm); Fe, scrap iron; W/P, fraction containing wood and plastics. (•, measurement of mass flows, m³/h and t/h; X, measurement of substance concentrations, mg/kg.)

fractions that are more homogeneous in size and composition, and thus they are easier and less costly to analyze than the original construction waste. The input into both plants is not the same because the two collection systems that supply construction wastes to plant A and B are also different.

The method of investigation is described by Schachmayer et al.[48] and Brunner et al.[49] Mass balances of input and output goods are performed for time periods between 2 to 9 h. The wet process is analyzed in five short campaigns, the dry process in a comprehensive investigation of 9 hours. Samples of all output goods are taken and analyzed for matrix substances (>1 g/kg) and trace substances (<1 g/kg) at hourly intervals. Off-gases and wastewater are sampled according to standard procedures for such materials. The size of solid samples is between 5 and 500 kg. Aliquots of the samples are crushed and pulverized until particles are smaller than 0.2 mm. Metal fractions such as magnetically separated iron are not crushed; their composition is roughly estimated according to the individual components present. Oversize materials of concrete and stones are not analyzed either. For concrete, literature values are taken; composition of stones is assumed to be the same as in other, smaller stone fractions. For fractions that cannot be analyzed due to the lack of pulverized samples, it is tested to see if the overall material balance is sensitive against these assumptions. The fractions not analyzed amount to less than 5% of the total construction waste treated. Since the matrix (bulk) compositions of these fractions are known (e.g., the magnetically separated fraction contains <80% iron), errors in the assumptions proved not to be decisive for the overall mass balance and the transfer coefficients.

3.2.3.4 Results

3.2.3.4.1 Composition of construction wastes

As expected, the composition of the construction wastes treated in plant A is not the same as in plant B (see Table 3.24). The material treated in plant A contains more sulfur (gypsum), organic carbon, and iron than the input into plant B, and the concentration of trace metals is about one order of magnitude higher. Construction wastes treated in plant A are more contaminated and contain less inorganic materials than the product for plant B. The reason for this difference has not been investigated. Possible explanations are:

1. Plant A is located in Switzerland and was analyzed in 1988, while the mass balance of plant B, operating in Austria, was conducted in 1996. During the time period of 8 years, construction waste management experienced swift development. In the 1980s, mixed construction wastes were treated in separation plants, while the 1990s saw a shift toward selective deconstruction and dismantling, resulting in "cleaner" and more uniform input fractions for such plants.
2. At the time of analysis, Switzerland and Austria had distinctly different legislation and practices in construction waste management. In Switzerland, no legislative framework had been established at the time of analysis. The MFA of the sorting plant is a first investigation into the power of

Case Studies

TABLE 3.24
Composition of Construction Wastes Treated in Dry-Separation Plant A and Wet-Separation Plant B Compared with the Average Composition of the Earth's Crust

Substance	Construction Waste Plant A (mixed construction wastes)	Construction Waste Plant B[a] (presorted construction wastes)	Earth's Crust
		Matrix Substances, g/kg	
S	5.8	1.1–2.9	0.3
TC[b]	93	47–79	0.2
TIC[c]	33	35–69	
TOC[d]	60	2–21	
Si	121	100–150	280
Ca	150	120–200	41
Al	9.5	8–15	81
Fe	40	7–20	54
		Trace Elements, mg/kg	
Zn	790	24–66	70
Pb	630	3–103	13
Cr	150	13–32	100
Cu	670	8–23	50
Cd	1.0	0.10–0.22	0.1
Hg	0.2	0.05–0.55	0.02

[a] Data for plant B are the result of four sampling campaigns with different input materials; thus ranges are given.
[b] TC = total carbon.
[c] TIC = total inorganic carbon.
[d] TOC = total organic carbon.

such plants to produce appropriate secondary construction materials. The results are used to establish a new strategy, giving preference to selective deconstruction (see results below). Eight years later in Austria, it was mandatory to separately collect uniform fractions such as wood, metals, plastic, concrete, etc. when a certain mass flow per construction site is exceeded. The cleaner input into plant B indicates that the decision taken in Switzerland (selective deconstruction) is appropriate.

3. Most construction waste stems from demolition and not from new construction sites. Due to different economic cycles (Austria was on a low level of economic development after World War II and was slow in recovering), buildings demolished in Switzerland and Austria are of different time periods. Some of the Swiss construction waste resulted from comparatively new buildings that had been constructed only 20 to 40 years ago, while in Austria, the buildings demolished in the 1990s were older.

Thus, the composition of construction wastes in plant A may resemble the construction materials of the 1950s and 1960s, while for plant B the input most likely stems from prewar periods (1930 to 1940) and hence has a different composition in trace substances.

Note that the three reasons stated here have not been investigated in detail; they are merely given as possible explanations for different compositions of construction materials. In order to derive significant results about differences in the composition of construction wastes, the analysis would have to be planned from a statistical point of view, which was not intended when the mass balance was conducted in plant A.

In summary, at the time of investigation, plant A was fed by mixed construction wastes as received when indiscriminately demolishing a building. Plant B received construction debris that resulted from more-or-less controlled dismantling and represented a fraction that looked well suited for recycling, where much of the unsuitable material had already been removed at the construction site.

3.2.3.4.2 Mass Flow of Products of Separation

The balance of plant A is given in Table 3.22. Dry separation generated 14 different products. Four products are wastes and have no further use (dust from cyclones 1 and 2 and off-gases from the drum, shredder, and air classifier). Some of the remaining 10 fractions are, in part, quite similar. Therefore, they have been rearranged into the five fractions I, II, III, metals, and rest seen at the bottom of Table 3.25.

- Major fractions
  Fraction I, pieces <80 mm
  Fraction II, light materials
  Fraction III, heavy materials
- Minor fractions
  Scrap iron
  Rest, consisting of useless residues (filter dust and off-gas)

The rationale for this new grouping will become apparent when the chemical compositions of the individual fractions are discussed.

The balance of goods for plant B is given in Table 3.26. A priori, this plant produces fewer fractions. Only two of the seven fractions generated are of major importance. The light fraction only amounts to 5.1 g/100 g, indicating again that the input into plant B contains less organic waste (plastic, paper, light wood, and the like) than plant A. In contrast to plant A, plant B produces a significant amount of fine-grain material of <4-mm particle size. The operator of plant B finds a good market for this material, while plant A's customers are asking for more coarse materials. Note that due to waste separation on the construction site, the percentage of the scrap iron fraction is 20 times smaller for plant B than A.

3.2.3.4.3 Composition of Products of Separation

The compositions of the products of the two construction waste-recycling plants are presented in Table 3.27. In both plants, fractions rich in carbonates and silicates and

TABLE 3.25
Mass Flow through Construction Waste-Sorting Plant A

Material	Consisting of	Mass Flow, 10^3 kg/day	Fraction, g/100 g CW
Total input	Construction wastes	225.3	100
Fraction			
A1	Concrete, stones	8.5	3.8
A2	Metals	3.08	1.3
A3	Oversize combustibles	3.75	1.7
B	<80 mm	102	45.3
C	Concrete, stones	4.14	1.8
D	Metals	2.36	1.0
E	Oversize material	0.43	0.2
F	Light fraction	51.4	22.8
G	Heavy fraction	47.7	21.2
H1	Dust cyclone 1	0.16	0.06
H2	Dust cyclone 2	0.10	0.04
I	Iron metals	1.73	0.8
K1	Off-gas drum/shredder	n.d.	n.d.
K2	Off-gas air classifier	n.d.	n.d.
New fractions	I + II + III + metals + rest	225.3	100
I (=B)	<80 mm	102	**45.3**
II (= F + E + A3)	Light fraction	55.6	**24.7**
III (= G + C + A1)	Heavy fraction	60.3	**26.8**
Metals (= A2 + D + E)	Iron	7.13	3.1
Rest (= H + K)	Dust and off-gases	0.26	0.1

Note: n.d. = not determined.

poor in organic carbon are produced. Also, both plants produce light fractions containing approximately 20% of total organic carbon (TOC) and scrap-iron fractions. The difference in chemical composition of the products obtained in the two plants is mainly due to the difference of input materials.

Because of the given input, all fractions of dry separation in plant A exceed concentrations of heavy metals in the Earth's crust. Since construction waste treated in plant B is considerably cleaner, the compositions of the wet products come closer to Earth-crust quality. Nevertheless, concentrations of lead and mercury are above that of the Earth's crust for all fractions analyzed in plant B, too. The fraction most polluted is light fraction II from dry separation. The material is similar to MSW and exhibits a high content of organic carbon (20%). Thus, this fraction is not suited for recycling as a construction material. Instead, it can be utilized to recover energy from waste in an incinerator equipped with sophisticated air-pollution devices to remove acid gases, particulates, and volatile heavy metals like mercury and cadmium.

The light fraction from plant B is similar to the one from plant A. The main differences are that trace-metal concentrations are smaller in B, and the amount of

TABLE 3.26
Mass Flow through Construction Waste (CW)-Sorting Plant B
(Presorted Construction Waste)

Material	Consisting of	Mass Flow, 10^3 kg/h	Fraction, g/100 g CW
Total input		370–380	≈500
Presorted CW		75	100
Water		300	400
Total Output		200–270	260–360
Wastewater		130–190	170–250
(Wastewater sediment)[a]	(Wastewater sludge from pond)	(2.5–3.7)	(3.3–4.9)
LF	Light fraction	3.8	5.1
F1	Sorting fraction 16–32 mm	15	20
F2	Sorting fraction 4–16 mm	27	36
F3	Sorting fraction 0–4 mm	25	33
Fe	Scrap iron	0.13	0.17
W/P	Wood and plastic fraction	0.05	0.07

Note: The difference between input and output is due to the loss of water when the drenched fractions leave the wet process and are stored and dewatered on site without measuring water losses. It is not possible to quantify this difference.

[a] Wastewater sediment is included in wastewater and is generated in a process outside the system's boundary (sedimentation in a wastewater sludge pond).

light fraction that the wet plant B produced per unit of construction waste (5.1 g/100 g CW) is about five times smaller than for the dry plant A (24.7 g/100 g CW). Both differences are due to differences in the input materials for the two plants. Plant B produces a large amount of wastewater containing suspended solids. Most of this wastewater is treated in a sedimentation pond, where a sludge (sediment) is formed and deposited. Contaminant concentration of this sludge is higher than in any other product of plant B, confirming the hypothesis that a lot of heavy metals are present on small particles that are removed and transferred to the water phase during wet separation. A significant amount of less contaminated wastewater is not controlled and is "lost" on the site (the plant stands on a river bank).

3.2.3.4.4 Partitioning of Metals and Transfer Coefficients

The main purpose of construction-waste sorting is to produce "clean" secondary construction materials. In chemical terms, sorting must direct hazardous substances contained in construction wastes to those fractions that are not intended for reuse. Preferably, the resulting substance concentration in recycling fractions is close to the concentration of materials used for the primary production of construction materials such as limestone, granite, and gypsum. A second goal is to maximize mass flows of useful and clean fractions. A third goal is to produce separation wastes that are well suited for disposal, either by landfilling or incineration. All of these

TABLE 3.27
Composition of Products from Dry (A) and Wet (B) Construction Waste Separation

	Products of Plant A						Products of Plant B				
Substance	I	II	III	G	Iron Metals	F1	F2	F3	LF	WW Sludge	Earth's Crust
Matrix Elements, g/kg											
Si	160	n.d.	180	170	n.d	170 ± 10	170 ± 16	190 ± 13	170 ± 8	170	280
Ca	180	91	160	160	n.d.	160 ± 9	160 ± 19	140 ± 18	100 ± 17	160 ± 18	41
Fe	12	16	20	22	800	15 ± 5	16 ± 6	16 ± 5	20 ± 5	20 ± 3	54
TC	62	210	48	47	n.d.	54 ± 4	59 ± 6	59 ± 6	210 ± 90	98 ± 23	0.2
TIC	41	17	38	34	n.d.	53 ± 5	52 ± 10	47 ± 8	22 ± 8	47 ± 6	—
TOC	21	190	9.9	12	n.d.	1.8 ± 1	7 ± 6	11 ± 3	190 ± 95	51 ± 25	—
Al	8.8	8.3	12	12	8.1	15 ± 4	15 ± 5	11 ± 3	21 ± 6	20 ± 3	8.1
S	7.3	5.7	3.9	4.3	n.d.	1.6 ± 0.54	1.3 ± 0.2	1.4 ± 0.2	3.8 ± 0.4	2.4 ± 0.5	0.3
Trace Elements, mg/kg											
Zn	540	1400	170	200	4900	35 ± 8	34 ± 8	48 ± 5	65 ± 9	200 ± 91	70
Cu	47	420	330	410	11,500	16 ± 3	21 ± 6	22 ± 6	30 ± 7	45 ± 4	50
Pb	200	940	930	1200	1800	30 ± 54	16 ± 15	25 ± 10	46 ± 37	75 ± 11	13
Cr	160	90	130	140	760	24 ± 3	25 ± 9	25 ± 10	110 ± 22	41 ± 7	100
Cd	0.7	2.3	0.5	0.6	n.d.	0.12 ± 0.01	0.11 ± 0.005	0.13 ± 0.01	0.2 ± 0.07	0.31 ± 0.08	0.1
Hg	0.2	0.3	0.1	0.1	n.d.	0.11 ± 0.07	0.17 ± 0.08	0.47 ± 0.31	0.7 ± 0.03	3.1 ± 1.7	0.02

Note: n.d. = not determined.

TABLE 3.28
Transfer Coefficients k of Selected Substances in Construction Waste-Sorting Plants A and B, ×10^{-2}

Substance	Plant A				Plant B [a]				
	I	II	III	Metals	F1	F2	F3	LF	Sludge
Mass	45	25	27	3	20	36	33	5.1	≈4
Si	60	n.d.	40	n.d.	21	38	33	3.9	4.6
Ca	56	15	29	n.d.	23	41	28	2.7	5.0
Fe	14	10	13	63	18	34	27	4.3	5.5
TOC	16	80	4	n.d.	2.2	15	21	46	15
Al	42	21	34	3	23	40	25	6.0	6.8
S	57	24	18	n.d.	20	29	24	9.1	6.7
Zn	31	44	5	20	16	28	31	5.5	21
Cu	3	15	13	69	16	38	31	5.5	10
Pb	14	37	40	9	25	24	30	7.2	14
Cd	29	57	14	n.d.	19	33	30	5.9	12
Hg	43	36	12	n.d.	5.7	16	35	6.8	37

Note: n.d. = not determined.

[a] Transfer coefficient k_{Fe} for scrap metals in plant B is 0.11. Transfer coefficient k_S for wastewater in plant B is 0.11. All other k_is for wastewater are <0.003.

goals can be achieved if mechanical sorting succeeds in controlling the flow of hazardous substances to certain fractions of sorting. Hence, it is of first importance to know the partitioning of heavy metals among the sorted products.

Table 3.28 lists the transfer coefficients (partitioning coefficients) for the two plants A and B. The results show that neither the dry nor the wet processes achieve the goal of directing the whole array of hazardous substances from recycling fractions to disposal fractions. Transfer coefficients for mass and substances are quite similar for most fractions, showing that "true" enrichment or depletion does not take place. It becomes clear that the superior qualities of the products of plant B are due to the clean input material and not because of a better separation by the wet process. MFA reveals the potential of the two technologies, and the transfer coefficients allow comparison of the separation efficiencies.

Transfer coefficients display the partitioning of elements only; they do not yet allow direct comparison of the enrichment or depletion of substances. In Figure 3.33, the quotients "substance concentrations in main fractions over concentration in construction waste" are presented for plant A on a log scale. These quotients are chosen to measure accumulation and depletion. In plant A, the most enriched elements are iron, copper, zinc and chromium in the metal fraction. Dry sorting successfully concentrates these metals in the metal fraction. Organic carbon, cadmium, mercury, and lead are enriched in the light (combustible) fraction II. Fractions I and III are similar. In both, the matrix substances Si, Ca, and inorganic carbon are slightly enriched, while organic carbon and some heavy metals are modestly depleted. Except

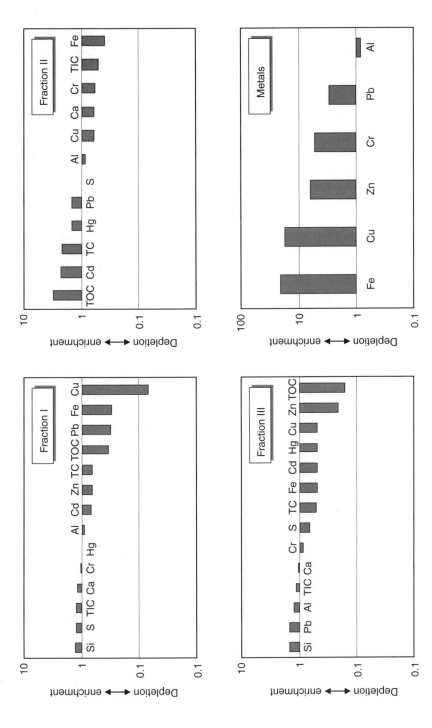

FIGURE 3.33 Enrichment, [concentration of X in fraction i]/[concentration of X in CW], of selected substances in main fractions of CW sorting plant A.

for copper in fraction I, all substances are depleted by less than an order of magnitude in fractions I to III. For mixed construction wastes, this order of magnitude is necessary if the process is to produce materials that are similar to the composition of the Earth's crust or to primary construction materials (see Table 3.24). There are no mechanical means yet to appropriately control the flow of all hazardous substances in sorting of mixed construction wastes.

3.2.3.5 Conclusions

Dry separation in plant A successfully concentrates combustible materials in the light fraction and constructionlike materials in two other fractions. The processing yields about 70% of potentially useful construction products in two fractions and about 3% of metals for recycling. The remaining fraction of 25% is not suited for recycling or landfilling; it has to be incinerated. Plant A is not capable of reducing the contaminant level of any fraction significantly. The main disadvantage of all products is the high trace-metal concentration. When the recycling products from plant A are being used for constructions, the buildings will contain heavy metals that are significantly above Earth crust concentrations. When the light fraction is incinerated, sophisticated and expensive air pollution control is required. Thus, it is most important that contaminants be removed by selective dismantling before entering the construction waste-recycling plant.

Due to a cleaner input, wet separation in plant B results mainly in two comparatively clean fractions well suited for recycling. Although a few of the heavy metal concentrations are elevated compared with the Earth's crust, they are (because of the cleaner input) generally of much lower concentration than in plant A. The overall performance of the wet process is similar to that of the dry process. While it is possible to produce a fraction rich in TOC and combustibles, significant accumulation or depletion of hazardous metals in any of the fractions is not observed. As for plant A, the light fraction contains much organic carbon, too, with the content of TOC reaching nearly 20%. Landfilling of a material with such a high TOC requires long aftercare periods. Thus, it seems appropriate to utilize the light fraction as a fuel. However, due to the presence of heavy metals such as Hg (see Table 3.27), boilers designed to utilize the light fraction must be equipped with efficient air-pollution control devices for atmophilic metals.

Despite the differences between the inputs into the two separation processes, MFA and transfer coefficients allow a comparison of the performances of the two plants. From a recycling point of view, the main differences are the products, with plant A producing gravel substitutes and plant B producing sand and gravel. The regional market situation determines whether sand or gravel is to be preferred. From an environmental point of view, there are no important differences. Because neither plant can sufficiently enrich or deplete hazardous materials, the substance concentrations of the main product fractions are similar to the concentrations of the incoming construction wastes.

The results of the MFA of the two plants support the strategy of selective deconstruction. Neither of the two processes is able to accumulate or deplete significantly (factor 10) hazardous materials in any of the resulting fractions. Once

Case Studies

again, it becomes evident that at today's stage of development, mechanical processes are of limited use for the chemical separation of waste materials. Thus, wastes from indiscriminate demolishing of buildings are not well suited to produce recycling materials in construction-waste sorting plants. For optimum resource conservation, it is important to separately recover materials during the deconstruction process and to recycle uniform fractions such as bricks, concrete, wood, and metals individually. In most cases, the remaining fraction can be mechanically sorted to recover a combustible fraction. Due to the composition of this fraction, containing plastic materials, paints, tubings, and cables, it is mandatory that energy recovery take place in incinerators equipped with state-of-the-art air pollution-control devices suited to remove heavy metals such as mercury.

3.2.4 Case Study 8: Plastic Waste Management

Plastic materials were introduced in the 1930s. Ever since, polymers such as polyvinyl chloride (PVC), polyethylene (PE), polypropylene (PP), and polyamide (e.g., Nylon) have shown large growth rates. Today they are among the most important man-made materials for many activities. At present, most plastics are made from fossil fuels that represent nonrenewable carbon sources. The production of plastics accounts for about 5% of the total fossil fuel consumption. They are used in cars, constructions, furniture, clothes, packaging materials, and many other applications. Often, they contain additives to improve their properties. In particular, long-living plastic materials such as window frames, floor liners, and car fenders have to be protected from degradation and weathering by ultraviolet light, aggressive chemicals, temperature changes, and the like. Hence, plastic materials are usually mixtures of polymers with stabilizers, softeners, pigments, and fillers.

Plastics make up between 10 to 15% of the total MSW flow. In addition, industrial and construction wastes are important sources of plastic wastes. Some plastic wastes (in particular from plastic manufacturing) are relatively clean and homogeneous and thus suitable for recycling. Others are mixtures of several goods and substances and hence cannot be recycled. Most plastic materials have a high energy content, and turned to waste, they can be used as a fuel. Due to stabilizers that contain heavy metals (lead, tin, zinc, cadmium, and others) and the chlorine content of some polymers (PVC, polyvinylidene chloride), thus yielding dioxins during incineration, incinerators for plastic wastes generally must be equipped with advanced air-pollution equipment.

As shown in Table 3.29, packaging materials are comparatively "clean" and may be used as a secondary resource. On the other hand, the stock of long-living plastics contains large amounts of hazardous substances that will have to be dealt with in the future. Hence, plastic recycling and waste management needs tailor-made solutions that are appropriate for the individual material and its ingredients.

Figure 3.34 shows the plastic flows and stocks through Austria.[101] The figure was prepared using data from plastic manufacturers, waste management, and other sources. In the following discussion, the focus is on plastic-waste management, emphasizing plastic wastes as energy resources and as sources of hazardous materials. In 1992, about 8 million Austrian consumers bought roughly 1.1 million tons

TABLE 3.29
Additives in Plastic Materials in Austria

Material	Total Consumption (1992), 1000 t/year	Packing Material Consumption (1992), 1000 t/year	Total Stock (1994), 1000 t
Plastics	1000	250	6700
Softener	14	3	180
Ba/Cd stabilizers	0.25	0.0002	4
Pb stabilizers	1.6	0.002	27
Flame retardants	2	0	34

Note: Plastics with short residence times such as packaging materials are comparatively clean. The long-lasting stock in construction, cars, and other applications contains large amounts of hazardous materials such as cadmium, lead, and organo–tin compounds.[101]

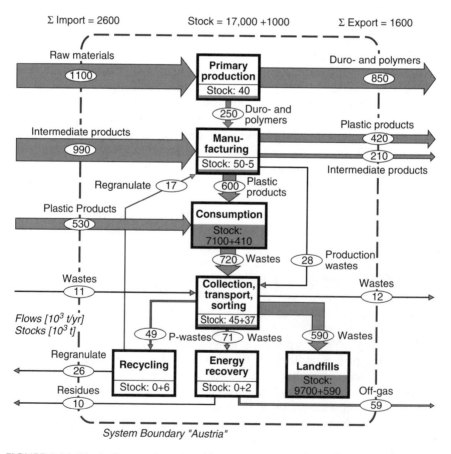

FIGURE 3.34 Plastic flows and stocks in Austria.[101] (From Fehringer, R. and Brunner, P.H., Kunststoffflüsse und die Möglichkeiten der Verwertung von Kunststoffen in Österreich, UBA Monographien Band 80, Umweltbundesamt, Vienna, 1996. With permission.)

of plastic materials. A large portion is used in goods with long residence times (floor liners, window frames, car parts, etc.) and thus is incorporated into the "anthropogenic stock." In Figure 3.34, this stock is assigned to the process "consumption." The rest of the plastic is used for products with short residence times such as packaging materials and other consumer goods. The net flow (input minus output) into the stock of the process "consumption" amounts to 410 kt/year. Of the 720 kt/year of plastic wastes that leave the process "consumption," 590 kt/year are landfilled, and the rest is either incinerated or recycled. It is interesting to note that the packaging ordinance that was instated in Austria in 1992 does not change much of this situation. Only about 7% (49 kt/year out of 759 kt/year) of all plastic wastes are controlled by the packaging ordinance and are directed toward material recycling. About 71 kt/year are being incinerated together with MSW in MSW incinerators. By far the largest amount of plastic wastes (590 kt/year) is still disposed of in landfills. This implies a large waste of energy, since 1 t of plastics corresponds roughly to 1 t of fossil fuels. The landfilling of plastic wastes is not only a waste of resources, it also offends the Austrian Waste Management Act (AWG, 1990[50]). The goals of this law are directed toward the conservation of resources such as energy and materials, and the law explicitly calls for the minimization of landfill space. Neither of these requirements are observed by present plastic-waste management practices.

In Figure 3.35, the advantage of an integrated MFA approach is visualized: If only MSW is considered ("MSW view"), 200 kt/year of plastic wastes are observed, with 80% being landfilled and 20% being incinerated. When public attention is drawn to packaging wastes, leading to legislation such as the Dual System in Germany or the Packaging Ordinance in Austria, a certain amount of plastic wastes (70 kt/year) is separately collected and thus not landfilled (–60 kt/year) or incinerated (–10 kt/year) any more ("Packaging view"). Due to inferior quality, not all separately collected plastic wastes can be recycled as polymers. Hence, a certain percentage is used as an alternative fuel, e.g., in cement kilns, leaving 50 kt/year for substance recycling.

If all plastic wastes are included in the assessment, a much larger amount of landfilled wastes is observed (590 kt/year) ("Total work management view"). It is important to note that without an investigation into the total national flows and stocks of plastic, it is not likely that the large amount of landfilled plastics can be identified. Only a balance of the process "consumption," with estimates of the mean residence time of various plastic materials, allows a reliable assessment of wastes that are leaving consumption. It is a much more difficult, if not impossible, task to directly identify the amount of plastics in the many wastes landfilled. Figure 3.35 shows clearly that rational decisions regarding plastic wastes have to be based on a complete set of flows and stocks of wastes in a national economy ("Resource management view"). The sole focus on a single waste category such as packaging wastes results in solutions that are not optimized regarding resource and waste management.

The benefit of an MFA approach in resource management as discussed in this case study is as follows: a total plastic balance on a countrywide level shows the important flows and stocks of plastics and helps in setting the right priorities in resource management. First, the large and useful stock of plastics (and thus energy) in consumption and landfills is recognized. Second, potential hazards due to toxic

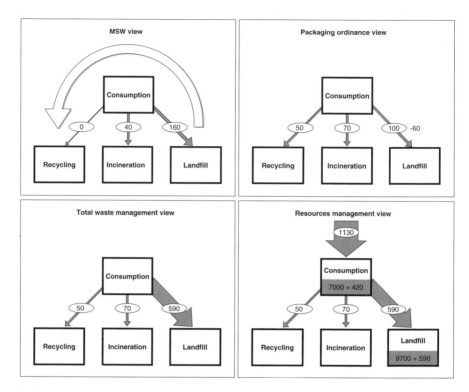

FIGURE 3.35 MFA as a decision-support tool for management of plastic waste.

constituents of plastic materials are identified in both stocks "consumption" and landfill; the toxics will have to be treated with care in the future.

3.2.5 Problems — Chapter 3.2

Problem 3.5: Assume that the production of food by traditional agriculture can be replaced by "hydrocultures" that do not require soil for plant production. What will be the major change regarding total nutrient (N, P) requirements and losses from the activity "to nourish"? Use Figure 3.16 and Figure 3.17 for your discussion.

Problem 3.6: Use the following information to complete the four exercises listed below.

In 1996 about 8.1 million t/year of zinc (Zn) ores and 2.9 million t/year of Zn scrap are processed in order to produce 9.6 million t/year of Zn. Ore processing resulted in approximately 230 million t/year of tailings from milling with a content of about 0.3% Zn and 14 million t/year of slag from smelting with about 5% of Zn, each material flow representing a Zn flow of ca. 0.7 million t/year. Mining wastes are not considered. Zn is further manufactured into products that can be roughly grouped into five categories: galvanized products (3.3 million t/year), die-castings

(1.3 million t/year), brass (1.5 million t/year), Zn sheet and other semi-products (0.6 million t/year), as well as chemicals and other uses (1.4 million t/year).

Galvanizing here stands for all kinds of technologies producing a coating of Zn on iron or steel in order to avoid corrosion. Die-casting is a process to produce strong accurate parts in large quantities by forcing molten Zn alloy under pressure into a steel die (mainly used in the automotive industry). Brass is an alloy based on copper (Cu) and Zn. The Zn content ranges up to ca. 40%. Brass is used as sheet, wire, tube, extrusions, and so on. Zn sheet is produced from Zn or Zn alloy rolled into thin sheets suitable for forming into roofing and cladding and other applications. The last category comprises mainly the "dissipative" uses, where Zn occurs as a trace metal, for example in paints, automotive tires, brake linings, pesticides, animal feed and food additives, pharmaceuticals, cosmetics, etc. Manufacturing also results in production waste (Zn content ca. 1.5 million t/year), mainly in the form of brass and galvanizing residues.

The total amount of Zn in products entering the use phase is 8.1 million t/year. The amount of Zn discarded is estimated at about 2.2 million t/year. Waste management separates 1.4 million t/year of Zn from the waste stream (Zn scrap). The remainder, which has a mean concentration of about 0.1% and comprises waste categories such as municipal solid waste, construction and demolition debris, wastes from electrical and electronic equipment, automotive shredder residues, hazardous wastes, industrial wastes, and sewage sludge is landfilled (0.8 million t/year). This latter figure can only be regarded as a rough estimate. Mass flows of goods, their Zn content, and the resulting Zn flows are given in Table 3.30.

(a) Establish the flow diagram of the described Zn system. (b) Assign the material flows of the flowchart to stages and draw a diagram according to Figure 3.24 in Section 3.2.2. (c) Calculate the statistical entropy trend for the system. Is it a sustainable trend? (d) Calculate what happens if 15% of consumed/used Zn is neither transformed to waste nor remains in the stock, but escapes to the environment. (Assume that this Zn flow is evenly dispersed in the soil.)

Problem 3.7: Consider the following quantitative flowchart for the fluxes and stocks of construction materials within a fictitious region (see Figure 3.36). (a) Which stock will be most important for sand and gravel after 100 years (constant materials management assumed)? (b) Which conditions are required in order that recycling of construction materials can make a substantial contribution to the supply of construction materials (both buildings and underground)? (c) Which differences in materials quality do you expect in the four stocks (which is the fourth stock)?

The solutions to the problems are given on the Web site, www.iwa.tuwien.ac.at/MFA-handbook.

TABLE 3.30
Flows of Zn-Containing Materials, Their Zn Content, and Related Zn Flows for the World Economy

Goods	Mass Flow, million t/year	Zn Content, %	Zn Flow, million t/year
Zn ore	160	5	8.1
Tailings	230	0.3	0.7
Slag	14	5	0.7
Metal	9.6	100	9.6
Production waste	3.0	50	1.5
Zn scrap	17	11	1.4
Products	1500	0.54	8.1
Galvanized products	83	4	3.3
Die castings	1.3	99	1.3
Brass	4.3	35	1.5
Zn sheet and semiproducts	0.6	99	0.6
Chemicals and other	1400	0.1	1.4
Flow into stock	2200	0.27	5.9
(dissipative loss)	(1700)	(0.007)	(1.3)
Wastes	810	0.27	2.2
Landfilled wastes	800	0.1	0.8

Note: Values rounded to two significant digits.

3.3 WASTE MANAGEMENT

MFA is an excellent tool to support decisions regarding waste management because:

1. In waste management, waste amounts and waste compositions are often not well known. MFA allows calculating the amount and composition of wastes by balancing the process of waste generation or the process of waste treatment. Thus, MFA is a well-suited tool for cost-efficient and comparatively accurate waste analysis.
2. As mentioned in the first paragraph of Chapter 1, inputs and outputs of waste-treatment processes can be linked by MFA. Thus, if transfer coefficients are known, one can assess whether a given treatment plant achieves its objectives for a given input. Often, transfer coefficients are not known in waste management, but they can be determined by MFA even if some inputs or outputs are not known.

The following case studies are presented to exemplify how MFA can be used to analyze wastes and to support decisions regarding waste management.

Case Studies

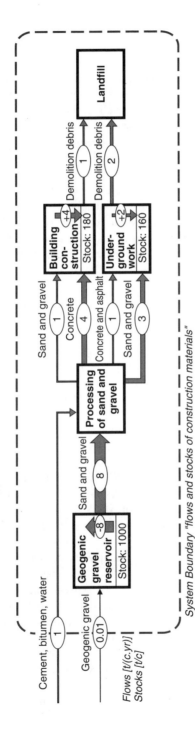

FIGURE 3.36 Fluxes and stocks of construction materials.

3.3.1 USE OF MFA FOR WASTE ANALYSIS

Reliable information on waste composition and waste generation rate is crucial for the following objectives:

1. To identify potentials for recycling (biowaste, paper, metals, plastics, etc.)
2. To the design and maintenance of waste-treatment technologies, including air- and water-pollution control (recycling, incineration, landfilling)
3. To predict emissions from waste-treatment and -disposal facilities
4. To examine the effects of legislative, logistic and technical measures on the waste stream

Because the composition and the generation rate of wastes are changing constantly, it is necessary to analyze them periodically. This is especially true when many new consumer goods are being introduced to the market. Thus, routine, cost-effective determination of waste composition and of time trends is essential for waste management. In this chapter, selected methods of characterizing MSW are presented and discussed. These approaches were originally presented in a paper by Brunner and Ernst.[51]

The parameters that are used to characterize waste materials can be divided into three groups:

1. Materials or fractions of MSW (e.g., paper, glass, metals)
2. Physical, chemical, or biochemical parameters (e.g., density, heating value, biodegradability)
3. Substance concentrations (e.g., carbon, mercury, hexachlorobenzene)

To solve a particular problem of waste management, it is usually not necessary to analyze all parameters. For example, for recycling studies, information on the content of certain fractions in MSW such as paper or glass is required. To predict emissions, the elemental composition of MSW needs to be known.

Generally, there are three main methods for solid-waste analysis (see Figure 3.37). The first involves direct analysis of MSW, while the second and third are indirect methods based on MFA and the mass–balance principle.

1. Direct analysis, also known as the "sample and sort" method. A specified, statistically planned amount of MSW is collected. Samples are taken, screened, analyzed for waste goods, dried, pulverized, and finally analyzed for substances. The sample that is analyzed is usually small compared with the total MSW generated. This method has been used in many waste-characterization studies in the U.S., Europe, and elsewhere.[52–54] Several manuals have been published describing how to conduct such analysis.[55]
2. Indirect analysis of MSW composition by market-product analysis. This approach requires information about the production of goods and about the fate of these goods during use and consumption. Data collected from industrial sources such as key corporations, professional organizations, or

Case Studies

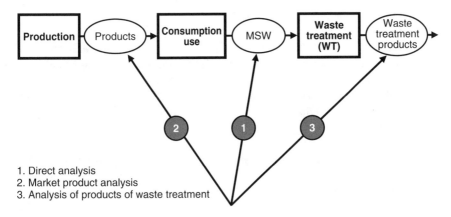

1. Direct analysis
2. Market product analysis
3. Analysis of products of waste treatment

FIGURE 3.37 Methods for MSW analysis.[51] (From Brunner, P.H. and Ernst, W.R., *Waste Manage. Res.*, 4, 147, 1986. With permission.)

from government agencies are used to estimate flows of goods that are produced and consumed. The generation of MSW is calculated by measuring or assuming average life spans for these goods. Various adjustments are made for imports, exports, and stocks in each product category. This method was developed in the early 1970s. Since then, data collection has improved, and databases have evolved. Results are compared with information about wastes that are landfilled, combusted, or recycled and with direct waste-analysis studies. The U.S. Environmental Protection Agency (EPA) applies this approach to estimate MSW generation.[56]

3. Indirect analysis using information about the products of waste treatment to calculate MSW composition. The advantage of this method is that the outputs of waste treatment are usually less heterogeneous than the input waste.

Especially for long-term monitoring, it is more cost-effective and accurate to determine waste composition by indirect methods.[67]

3.3.1.1 Direct Analysis

Direct waste analysis was the first approach used to determine waste composition. Waste samples are collected from different communities or regions based on statistical evaluations. The sample size usually varies between a minimum of 50 kg up to several tons. Samples are classified by hand into a selected number of fractions (paper, glass, etc.). Mechanical equipment is commonly used to separate magnetic metals and to sieve the remaining unidentified material into several additional fractions of different particle size. In order to determine the chemical and physical parameters of each fraction, representative samples are drawn from each material. These samples are further prepared (dried, pulverized, and sieved) for laboratory analysis.

The direct method is useful for:

1. Measuring the concentration of most materials in MSW
2. Determining energy and water content of MSW and its fractions
3. Investigating the influence of geographic, demographic, and seasonal factors on the concentration of materials and some parameters in MSW
4. Assessing changes of waste composition with time
5. Evaluating the impact of separate collection measures on waste composition such as content of paper or glass or the impact of different collection systems (e.g., size of waste containers)

However, the direct method of waste analysis also has a number of limitations and disadvantages. First, it is labor intensive and requires expensive equipment. Provided that adequate technical equipment and sufficient personnel are available, the analysis of one truckload takes at least half a day. A monitoring study on annual changes of MSW is estimated to consume 15 person months of unpleasant and unhealthy labor. Second, the residue of separation that is not assigned to defined fractions such as glass, paper, etc. is usually quite large, often making up as much as 40 to 50% of the total MSW analyzed. As long as the composition of this fraction remains unknown, the value of the other results can be questioned, for example, the assessment of recycling potentials. Third, the determination of trace-element concentrations is problematic. If, for example, mercury batteries and their contribution to heavy metals in MSW are analyzed, one may find a few small batteries in one ton of MSW. This results in an average sample concentration of one to a few milligrams of mercury per kilogram of MSW. However, if only one or two MSW samples of 2 to 20 kg are collected, there is either a great chance of finding no mercury from batteries at all or of finding a high concentration if a single battery turns up in one of the randomly selected samples (Hg content up to 30% for Zn/Hg batteries).

This problem is illustrated in Figure 3.38. If the chosen sample size is too small, the result of the analysis will probably be too low. The possible range of results increases with smaller sample sizes. Sufficiently large samples are needed to achieve results that reflect the actual content of unknown substances. Fourth, for technical and economical reasons, the metal fraction is often excluded from the chemical-physical analysis. However, this fraction may contain considerable amounts of heavy metals. Therefore, results of direct waste analysis may represent minimal values. A fifth problem is erosion and contamination from grinding and pulverization equipment.

These problems highlight some of the distinct limitations of the direct analysis of MSW with regard to the determination of chemical parameters, particularly of trace substances. They indicate that direct chemical analysis of waste materials represents the actual field concentrations only when large sampling campaigns are undertaken, resulting in extremely high costs. The direct approach is well suited for the determination of materials in MSW, but it seems of limited value in analyzing the elemental composition of MSW.

3.3.1.2 Indirect Analysis

The aforementioned problems and limitations of direct waste analysis led to the development of complementary methods, which yield more accurate results with

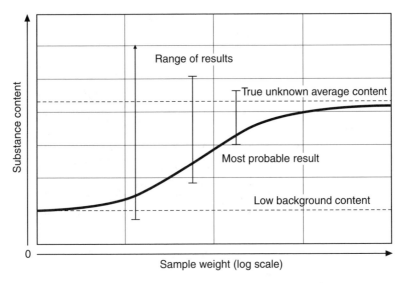

FIGURE 3.38 Illustration showing the drift of the most probable result as the sample size becomes too small.[93] (From Pitard, F.F., *Pierre Gy's Sampling Theory and Sampling Practice*, Vol. II, *Sampling Correctness and Sampling Practice*, CRC Press, Boca Raton, FL, 1989, p. 159. With permission.)

less effort in terms of manpower and costs. Two case studies are presented to illustrate the use of MFA in indirect analysis.

3.3.1.2.1 Case Study 9: Waste Analysis by Market Analysis

Goods are produced and consumed. After use, they are either recycled or discarded as wastes. Since most industrial branches have accurate figures about their production, and since the pathways of many goods are well known, it is often possible to calculate the composition of MSW without field analysis and with high accuracy. This procedure, which can be used to analyze both the contents of materials and the elemental composition, is illustrated by the following examples for paper, glass, and chlorine content in MSW.

3.3.1.2.1.1 Paper

The most abundant single substance in MSW is cellulose, the main constituent of paper. For paper recycling as well as waste treatment, it is of considerable interest to know the amount of paper in MSW. Figure 3.39 shows the flux of paper through the Austrian economy. Data are collected from pulp and paper manufacturers and checked against other available information. The amount of paper in MSW (48 kg/capita/year) is calculated as the difference between total paper consumption (179 kg/capita/year) and separately collected and recycled wastepaper (131 kg/capita/year). The Austrian population in 1996 was around 8.1 million inhabitants, and MSW generation amounted to 1.3 million t/year, which translates to 160 kg/capita/year of MSW for each resident. Based on these figures, a paper content of 30% (48 kg of wastepaper in 160 kg of MSW) can be calculated for average Austrian MSW. This figure has been confirmed by direct analysis.

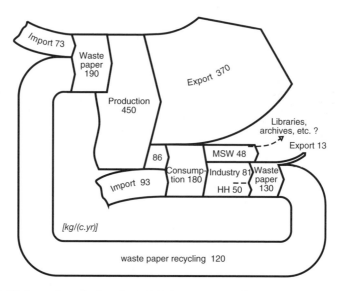

FIGURE 3.39 Paper flows in Austria (1996), kg/capita/year.[94]

3.3.1.2.1.2 Glass

A simple balance for the per capita glass flux in Switzerland in 2000 is given in Figure 3.40.[57] Only packaging glass (bottles, beverage containers, etc.) is considered. The amount of glass in Swiss MSW (2.8 kg/capita/year) is calculated as the difference between glass consumed (46.6 kg/capita/year) and glass recycled (43.8 kg/capita/year). Glass with residence time greater than 1 year is not considered. Accumulation in the stock "household" is assumed to be less than 1% of consumption. With a Swiss population of 7.2 million inhabitants and 2.54 million t of MSW generated

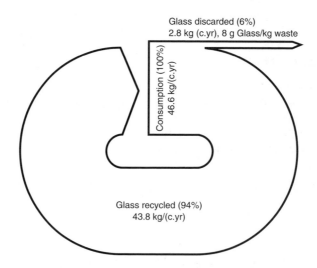

FIGURE 3.40 Recycling of packaging glass in Switzerland in 2000, kg/capita/year.

Case Studies

annually, the per capita allocation of MSW is 350 kg/capita/year. Based on these data, an average concentration of packaging glass in MSW of 8 g glass per kg MSW is calculated.

Paper and glass products have a short lifetime of less than one year. Therefore, it is reasonable to assume that inputs equal outputs over the balancing period. For other products with longer or even unknown residence times (e.g., wood in building materials), attempts to balance are more difficult. Yet, the U.S. EPA studies on MSW generation show that this method is successful. Kampel used this approach to determine differences in waste-glass management among Australia, Austria, and Switzerland.[57]

3.3.1.2.1.3 Chlorine
The main sources of chlorine in MSW are assumed to be PVC and sodium chloride (NaCl). Minor amounts of Cl are contained in plant materials, other plastic materials, and other products. Thus, the content of Cl in MSW can be roughly estimated by the figures on consumption of PVC and table salt and by assumptions on the fate of these products during and after consumption and use. Data about goods such as NaCl and PVC are usually published in annual reports of the specific industrial branch, e.g., salt mine operators and plastics manufacturers. Sodium chloride in private households is utilized for dietary purposes mainly. It is assumed that not more than 10% of the NaCl purchased is discarded with MSW. Most salt is either eaten or discarded with wastewater while preparing food; in both cases, chloride leaves the household via sewage. Residence times of goods containing PVC are difficult to assess. It is estimated that $50 \pm 20\%$ of PVC is used in long-life products, and the other part is used for short-residence-time packaging material and consumer goods. Note that there is not yet a steady state for PVC flows. There is a large yearly growth rate on the input side. Because of the long residence time of some products, PVC is accumulated in the anthroposphere. Therefore, the amount of PVC-derived chlorine in MSW is calculated according to varying percentages of PVC. Despite the fact that chlorine estimates are based on several assumptions, the order of magnitude (5 to 10 g Cl per kg MSW) in Table 3.31 compares well with values resulting from product analysis of 7 to 12 g Cl per kg MSW.

3.3.1.2.1.4 Advantages and Drawbacks
The main advantage of the analysis of MSW by a material balance of market products is the fact that no measurements are needed. MSW composition can be assessed quickly with little effort. In most cases, such rough estimates can give good results on a nationwide level. However, the method is not well suited to identify regional differences. It is usually more important to have reliable figures on the production/consumption side of a product than to have exact estimates of the proportion that enters the waste cycle. Another advantage of this method is the potential to predict trends in waste composition. Because today's products determine tomorrow's waste composition, this method is the only one that can be used to predict future waste composition.

Drawbacks of the method are (1) the dependency on production/consumption data, which are usually known on a national level only and (2) that data are available only for a limited amount of materials and elements. It is not yet possible to

TABLE 3.31
Determination of Chlorine in Swiss Municipal Solid Waste by Market Analysis

	NaCl	PVC min.	PVC	PVC max.
Consumption/use, kg/capita/year	5	8	8	8
Fraction discarded, %	10	30	50	70
Mass in MSW, kg/capita/year	0.5	2.4	4	5.6
Cl content, g/kg	610	580	580	580
Mass of Cl, kg Cl/capita/year	0.31	1.4	2.3	3.2
Contribution to Cl in MSW, g Cl/kg MSW	0.9	3.8	6.3	8.8
Total Cl in MSW (market analysis), g Cl/kg MSW		5–10		
Direct analysis, g Cl/kg MSW		3.4–4.2		
Product analysis, g Cl/kg MSW		7–12		

characterize MSW from a physical point of view by this method (e.g., density and particle size).

3.3.1.2.2 Case Study 10: Analysis of Products of Waste Treatment

The analysis of the products of different waste-treatment processes is a powerful tool to characterize MSW.[51] The main advantage is the homogenizing effect of treatment processes. This is particularly true if incineration is chosen for analysis. The incinerator acts as a large "thermal digester," separating substances from each other and releasing products that are of more uniform composition than the initial MSW. If all residues of the incinerator are analyzed and the total input and output mass flows are determined over a given period of time, the composition of the input into the plant can be calculated. This makes it possible to determine the flows of selected elements through a MSW incinerator and calculate the chemical composition of the waste input. The method has been successfully applied to several incinerators.[58–65]

3.3.1.2.2.1 Procedure

The procedure employed in a full-scale incinerator is as follows. The total mass flows of all input and output goods are determined during a given measuring period. Typical measurement campaigns last from several hours to several days. A (crane) balance measures the weight of the waste material fed to the incinerator. Consumption of water and chemicals is continuously recorded by incinerator control devices. The volume of air used for combustion is calculated based on a final mass balance and data about the energy consumption of the air blower. Solid incineration products are collected separately and weighed as received. Wastewater and off-gas are measured routinely by on-line flowmeters and translated into mass flows. To determine the chemical compositions, samples of bottom ash, filter cake, purified wastewater, and fly ash from the electrostatic precipitator (ESP ash) are taken and prepared for analysis. The bottom ash is the most heterogeneous product and requires extensive processing before analysis. First, it is separated from large pieces of iron, crushed,

Case Studies

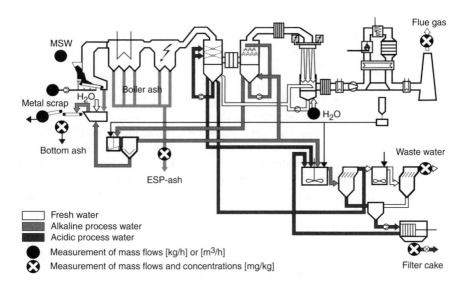

FIGURE 3.41 Locations of sampling and mass–flow metering for indirect waste analysis in an MSW incinerator.

and sieved. The oversize material is weighed but usually not analyzed. It is assumed to consist mainly of iron (an assumption that is not justified for every incinerator). From the pretreated bottom ash, several composite samples are dried (at 105°C for ca. 24 h until a constant weight is achieved) and pulverized in a mill. Again, the oversize material is assumed to be of iron. For balance calculations, all fractions of the bottom ash are taken into account. Composite samples of fly ash are taken as close as possible to the filter device (to avoid time lag) and pulverized to laboratory sample size. Wastewater and filter cake, two rather homogeneous products, are sampled, too. In coordination with the sampling of solid and liquid incineration products, off-gas samples are taken to determine the flows of substances that are not measured continuously (mainly heavy metals). Morf et al.[64,65] provide detailed descriptions about effective sampling plans, procedures, the preparation of samples, and methods of analysis. Figure 3.41 gives an example of appropriate locations for sampling and measurements in an MSW incineration plant.

The element balances are calculated by multiplying the mass flow of goods by the respective concentrations of the elements for every period of the campaign. As mentioned before, input composition is not measured but is calculated indirectly by summarizing the mass flows of each element in the incineration products (minus substance inputs in other input goods such as water, air, or chemicals) and dividing by the mass flow of the waste input (see Equation 3.9).

$$c_{MSW,j} = \frac{\sum_{i=1}^{k} c_{ij} \cdot \dot{m}_i}{\dot{m}_{MSW}} \quad (3.9)$$

where
k = number of incineration products
j = substance

Concentrations of C, Cl, F, S, and several heavy metals in MSW have been determined by this method. Table 3.32 lists the results from six studies of five incinerators in Austria and Switzerland.

When analyzing wastes, it is highly important to consider uncertainty and to assess aspects of quality control. Bauer[66] developed a method for quantifying the statistical uncertainty of such indirect waste analysis. Thus it is possible to determine the effort that is necessary to obtain a given confidence interval for the waste composition. Higher efforts (more samples per time, larger sampling sizes) yield more reliable results (smaller uncertainties). A relationship between cost and accuracy can be established. Results with sufficiently small intervals (below ±20%) with a confidence of 95% can be obtained at reasonable costs. Morf and Brunner[67] extended this approach. Based on MFA and transfer coefficients, they developed a method that allows routine measurement of MSW composition by analyzing only a single product of incineration per substance. They present procedures and examples of how to select the appropriate incineration residue to be analyzed, how to determine the minimum frequency for analyzing the residue, and how to measure the chemical composition of MSW routinely.

3.3.1.2.2.2   Results

Results of such investigations into MSW concentration are shown in Figure 3.42 and Figure 3.43. The monthly mean values of Cl and Hg vary by up to a factor two. The daily flows of the two selected elements Cl and Hg also vary. For Hg, these variations are quite substantial and up to a factor of four within a period of a few days. This emphasizes that random moment investigations are not a sufficient means of determining MSW composition.

The proposed MFA-based method for routine monitoring of waste composition by analyzing single incineration residues has significant advantages regarding data quality compared with the normally applied direct waste analysis. If waste composition were measured in the same way on several MSW incinerators, this would allow comparison of waste compositions in a more cost-effective and objective way than the present practice of direct waste analysis. Future MSW incinerators should be designed for and supplied with hardware and software to apply routine MFA for waste analysis. The additional costs would be small and the return on investment large when compared with the costs and accuracy of traditional approaches.

The main disadvantage of the waste product analysis is that waste fractions cannot be determined, e.g., it is not possible to calculate the contents of paper, plastic, or any other single fraction. This means that, in most cases, the product method is limited to the analysis of elemental composition and parameters like energy content, water content, and the content of total inorganic and organic matter.

3.3.1.2.2.3   Conclusion

It is highly important to choose the method of analysis that is most appropriate to solve a particular problem of waste management. In general, direct waste analysis

TABLE 3.32
Results of Different Indirect Waste-Analysis Campaigns, g/kg[65]

	Biel (CH) 1981	Müllheim (CH) 1984	St. Gallen (CH) 1991	Vienna (A) 1993	Wels (A) 1996	Wels (A) 1996
C	275 ± 55	n.d.	370 ± 40	190 ± 10	252 ± 25	265 ± 28
Cl	6.9 ± 1.7	n.d.	6.9 ± 1.0	6.4	12.2 ± 1.8	10.3 ± 1.2
S	2.7 ± 0.5	n.d.	1.3 ± 0.2	2.9 ± 0.2	4.2 ± 0.14	4.1 ± 0.17
F	0.14 ± 0.06	n.d.	0.19 ± 0.03	1.2 ± 0.1	0.054 ± 0.007	0.060 ± 0.002
Fe	67 ± 35	n.d.	29 ± 5	42 ± 1	37 ± 0.25	43 ± 0.2
Pb	0.43 ± 0.13	0.57 ± 0.43	0.70 ± 0.10	0.60 ± 0.10	0.40 ± 0.079	0.49 ± 0.088
Zn	2.01 ± 1.51	1.1 ± 0.5	1.4 ± 0.2	0.83 ± 0.07	1.2 ± 0.069	1.3 ± 0.14
Cu	0.27 ± 0.07	0.46 ± 0.19	0.70 ± 0.20	0.36 ± 0.03	0.59 ± 0.13	0.52 ± 0.076
Cd	0.0087 ± 0.0019	0.012 ± 0.0056	0.011 ± 0.002	0.008 ± 0.001	0.0107 ± 0.0028	0.0084 ± 0.0026
Hg	0.00083 ± 0.00081	0.002	0.003 ± 0.001	0.0013 ± 0.0002	0.0019 ± 0.00039	n.d.

Note: n.d. = not determined.

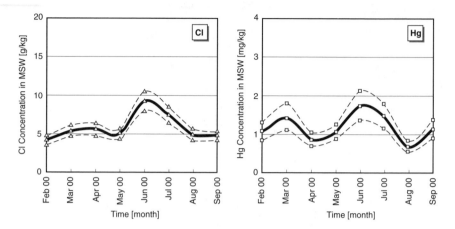

FIGURE 3.42 Time trends for monthly mean MSW concentrations of chlorine and mercury as determined for an incinerator (Spittelau) in Vienna, Austria, between February 1 and September 30, 2000. The figure shows means as well as the lower and upper limits for an approximately 95% confidence interval.[95] (From Brunner, P.H., Morf, l., and Rechberger, H., in *Solid Waste: Assessment, Monitoring, and Remediation*, Twardowsky, I., Allen, H.E., Kettrup, A.A.F., and Lacy, W.J., Eds., Elsevier, Amsterdam, 2003. With permission.)

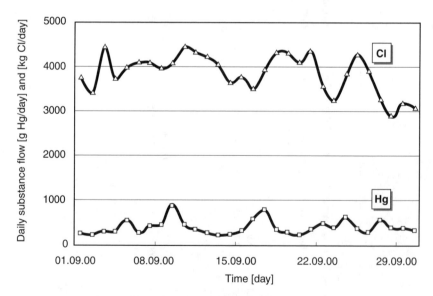

FIGURE 3.43 Time trends for daily flows of Cl (kg/day) and Hg (g/day) through the MSW incinerator (Spittelau) in Vienna, Austria, between September 1 and September 30, 2000.[95] (From Brunner, P.H., Morf, l., and Rechberger, H., in *Solid Waste: Assessment, Monitoring, and Remediation*, Twardowsky, I., Allen, H.E., Kettrup, A.A.F., and Lacy, W.J., Eds., Elsevier, Amsterdam, 2003. With permission.)

yields good results on some fractions in MSW, but it is expensive and labor intensive to determine reliably elemental concentrations by this method. Market–product analysis combined with MFA is an inexpensive and quick method to determine with sufficient accuracy the fraction-based and elemental composition of MSW. In many cases, this method of analysis can be applied in favor of direct waste analysis. However, the method is limited to those materials where information from the producing industries is available and where residence times in stocks are more or less known. MFA-based waste product analysis is well suited for determining element concentrations in MSW, but it does not allow analysis of material composition. It is the superior, cost-efficient method for determining time trends in elemental analysis of wastes.

3.3.2 MFA TO SUPPORT DECISIONS IN WASTE MANAGEMENT

3.3.2.1 Case Study 11: ASTRA

In the case study ASTRA,* selected scenarios for the treatment of combustible wastes are compared in view of reaching the waste-management goals of "environmental protection," "resource conservation," and "aftercare-free landfills."[68] The incentive for this Austrian study is a new federal ordinance on landfilling that became effective in 1996.[69] The ordinance mandates that beginning in 2004, only wastes with a TOC <2 to 5**% may be landfilled. The reason for banning organic carbon in landfills is that organic carbon is transformed by microorganisms. The metabolic products are organic compounds that may be transferred to landfill leachates, and carbon dioxide and methane in landfill gas that contribute to global warming if not collected and treated properly. In addition, organic acids are produced that may mobilize heavy metals. Therefore, leachates of such reactor-type landfills are contaminated with a broad array of organic and inorganic pollutants. This requires treatment of the leachate over long periods (>100 years) and contradicts one of the Austrian objectives of waste management, which is to avoid shifting waste-related problems to future generations ("aftercare-free landfills").

Because of the limit for organic carbon, treatment before landfilling is mandatory for most wastes such as MSW, sewage sludge, and construction debris. Combustion is an efficient means of transforming organic carbon to carbon dioxide. In order to ensure free choice of waste-treatment technologies, the landfill legislation allows an exemption from the TOC limit for the output material of mechanical-biological treatment facilities. These plants produce two fractions: (1) a combustible fraction that is mechanically separated and appropriate for further energy recovery and (2) a product derived from biological digestion. The biological degradation process cannot provide a residue with a TOC <5% because persistent organic compounds such as plastics and lignin cannot be decomposed within months by microorganisms. Thus, an exception is stipulated for this fraction, it may be landfilled if the heating value is below 6000 kJ/kg. In contrast to the limit for TOC, which minimizes

---

* ASTRA is a German acronym for "evaluation of different scenarios for waste treatment in Austria."
** TOC: Total organic carbon; the exact percentage depends on the type of landfill (e.g., monofill, landfill for construction waste, etc.).

reactions in the landfill body and thus supports the objectives of waste management, the limitation of the heating value does not improve landfilling practice or reduce the need for aftercare. Rather, the exemption is based on political decisions. Both limits prevent direct landfilling of untreated MSW after 2004.

Some industry branches are eager to use combustible wastes as a substitute for fossil fuels. This helps to reduce costs of production, since wastes are usually cheaper than fuel. If the wastes are contaminated (e.g., with PCBs) or otherwise difficult to dispose of, they may even create revenue. Also, wastes made up of biogenic carbon are attractive fuels because they do not contribute to global warming.

The following treatment options are available for combustible wastes: incineration, cocombustion in industrial furnaces, mechanical sorting, and biological digestion. All of these options have different environmental impacts and different contributions to the goals of waste management as stated in the Austrian Waste Management Act.[50] In the case study ASTRA, various scenarios for the management of combustible wastes are developed and compared in view of the requirements of the new Landfill Ordinance and of the goals of the Waste Management Act.

3.3.2.1.1 Procedures

The ASTRA project consists of the following steps:

1. Selection of waste treatment processes and defining waste management systems and scenarios
2. Selection of substances
3. Selection of wastes
4. Establishment of mass balances (see Figure 3.44) and substance balances for the actual system
5. Development and selection of criteria to evaluate the scenarios
6. Development of an optimized scenario for improved management of combustible wastes (optimum assignment of wastes to treatment processes)
7. Establishment of total mass balance as well as substance balances for the optimized scenario
8. Comparison between actual system and optimized scenario

For brevity, not all steps of the comprehensive ASTRA study are presented here in detail. The only steps that are discussed are those relevant to the understanding of the results and implications of the case study.

*3.3.2.1.1.1   Selection and Development of Criteria to
              Evaluate Balances*

The starting point is the goals of waste management as listed in the Waste Management Act:

1. To protect human health and the environment
2. To conserve energy, resources, and landfill space
3. To treat landfilled wastes so that they do not pose a risk to future generations

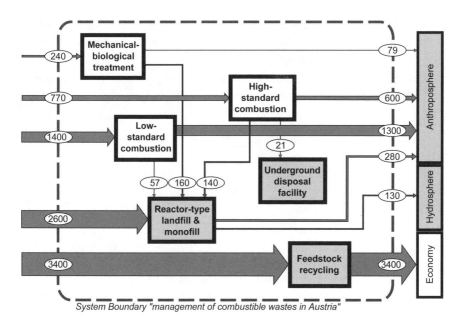

FIGURE 3.44 Mass flows of combustible wastes through the system "waste management" in Austria (1995), 1000 t/year.[68]

The latter goal is part of the precautionary principle. Since the long-term behavior of landfills is not known, future emissions have to be prevented by today's waste treatment and immobilization. In general, these goals are quite abstract and therefore require focus: What are the indicators to decide if human health and the environment are protected? Table 3.33 lists the criteria that have been applied in ASTRA. The chosen metrics or indicators are not absolute measures, since they cannot quantify the extent to which the goals of waste management have been reached. However,

TABLE 3.33
Goals of Waste Management and Assignment of Assessment Methods

Goals of Waste Management as Stated in the Austrian Waste Management Act	Assessment Methods and Criteria
1. Protection of human health and the environment	1. Critical air volume
2. Conservation of energy and resources	2. Efficiency of utilization of the energy content of wastes
3. Conservation of landfill space	3. Volume reduction through treatment
4. Aftercare-free landfill	4a. Total organic carbon in landfilled wastes
	4b. Fate of substances on their way to "final sinks"

they do allow relative comparison of the actual situation with various scenarios, yielding statements such as: "scenario X is Y% better than the actual situation."

3.3.2.1.1.2   Assessment Methods and Criteria

1. The critical air volume as used in ASTRA is adapted from the Swiss Ecopoints approach. It is defined by the following equations:

$$V_{i,crit} = \frac{E_i}{L_i} \qquad (3.10)$$

$$V_{crit} = \sum_{i=1}^{n} V_{i,crit} \qquad (3.11)$$

where

E_i = emission of substance i into the air
L_i = concentration of substance i in ambient air
$V_{i,crit}$ = hypothetical air volume that is needed to dilute E_i to ambient air concentration for substance I

The substance-specific critical volumes are added up to a final assessment indicator (V_{crit}), which has its optimum at small volumes.

2. The efficiency of utilization of waste energy content is calculated as follows:

$$\text{Efficiency} = \frac{\text{substituted fossil fuel [J / yr]}}{\text{energy content in wastes [J / yr]}} \cdot 100 \qquad (3.12)$$

Fossil fuels can be substituted directly and indirectly by waste combustion. Direct substitution takes place when wastes replace fossil fuels,* e.g., when coal is replaced by plastic waste to fire a cement kiln. Indirect substitution is given when wastes are used in an incinerator to produce electricity and/or heat to feed into a network, conserving fossil fuel that would have been required without the MSW incinerator. An efficiency of 100% means that one energy unit of wastes replaces the equivalent energy amount of fossil fuels.

3. Volume reduction by waste treatment is expressed as the difference in landfill space required for the various scenarios.
4a. TOC of final wastes is assessed based on mass and substance balances.
4b. "Fate of substances" means that each substance will finally be transferred to intermediate A and final B sinks. These sinks are:
   1. (A) Recycling products or other new secondary products (e.g., cement, bricks)

---

* It is assumed that wastes replace energy-equivalent units of fossil fuels. Strictly speaking, this is only true when the heating values of wastes and fossil fuels are similar (difference smaller than 20%).

Case Studies   273

   2. (A) The atmosphere
   3. (A) The hydrosphere
   4. (B) The lithosphere as an underground disposal facility
   5. (A+B) The lithosphere as a landfill (see gray boxes in Figure 3.44)

   Sinks 1, 2, and 3 are intermediate sinks for most substances; sink 4 is designed as a final sink; and sink 5 is a sink that is leaching over very long periods of time. For each substance, a suitable sink must be defined. For example, only very minor amounts of cadmium should reach the atmosphere and the hydrosphere. Also, the transfer into recycled goods or cement is not desirable, since cadmium is not required in these products and has to be disposed of at the end of all life cycles. Cadmium in landfills poses a small long-term risk. The disposal in specially designed underground storage facilities that have not been in contact with the hydrosphere for millions of years (e.g., salt mines) represents a long-term solution with an extremely low risk of environmental pollution. Hence, from the point of view of finding an appropriate final sink for cadmium, the underground storage is the most preferred solution.

   For nitrogen, recycling as a nutrient and emission into the air as molecular nitrogen (not nitrogen oxide) are positive pathways and sinks. Most other fates such as nitrate in groundwater or NO_x in air are considered negative paths. For chloride, transport in river systems to large water bodies such as oceans are acceptable solutions as long as the ratio of anthropogenic to geogenic concentrations and flows is small (e.g., <1%). The criterion for fate of substances is expressed as the percentage of a substance that is transferred into appropriate compartments. The reference (100%) is the total flow of the substance in combustible wastes.

3.3.2.1.1.3   Total Mass Balance for the Actual Situation (1995)
The generation and actual flows of combustible wastes in Austria are assessed by analyzing statistics and studies that were commissioned by the authorities responsible for waste-management issues in Austria. As a working hypothesis, combustible wastes are defined as wastes having a heating value >5000 kJ/kg dry substance. This is the range where autarkic combustion is possible. The result is given in Table 3.34. The total amount of wastes is 39 million t/year. It is dominated by construction and demolition debris, including soil excavation. But only a small fraction of this category is combustible (2%). Altogether about 22% or 8.5 million t/year are combustible wastes. The most relevant fractions are waste wood (41%) and wastes from private households and similar institutions (26%). Waste wood comprises bark, sawdust, chips of wood, and other minor fractions. Wastes from water purification and wastewater treatment mainly consist of municipal and industrial sludge and screenings from the sewer and wastewater treatment plants. Other nonhazardous wastes comprise all sorts of industrial wastes. The composition of this fraction is comparatively unknown. Better statistics are available for hazardous wastes. Generally, one can say that the average Austrian produces about one metric ton of combustible wastes per year. Three-quarters of this amount accrue elsewhere (industry, infrastructure) and are not directly visible for the consumer.

TABLE 3.34
Total Waste Generation in Austria[a] and Combustible Fractions

			Combustible Portion of Wastes	
Waste Category	Total Wastes, t/year	%	t/year	% of Total Combustible Fractions
Wastes from private households and similar sources	2,500,000	87	2,170,000	26
Construction and demolition debris including soil excavation	22,000,000	2	500,000	6
Residues from wastewater treatment	2,300,000	41	940,000	11
Waste wood	3,500,000	100	3,500,000	41
Other nonhazardous wastes	7,800,000	14	1,130,000	13
Hazardous wastes	1,000,000	22	220,000	3
Total	39,100,000	22	8,500,000	100

[a] 8.1 million inhabitants.

Figure 3.44 shows the flows of combustible wastes in Austria in 1995. Approximately 40% (3400 kt/year) is used for feedstock recycling. This can be sawdust for chipboard production or wastepaper recycling. About 30% (2600 kt/year) is directly landfilled. After 2004, the disposal of this latter amount will not comply with the landfill ordinance of 1996. New methods of treatment and disposal are needed. Simple incinerators and boilers without advanced air-pollution standards utilize about 17% (1400 kt/year) of the combustible wastes for energy recovery. These plants are equipped with settling and baffle chambers, multicyclones, electrostatic precipitators, or baghouse filters. The standard fuel is oil, coal, or biomass, and emission limits are not as stringent as for MSW incinerators. About 9% (770 kt/year) is incinerated in high-standard facilities. These plants are equipped with advanced air-pollution control systems and easily surpass the most stringent emission regulations. Finally, some 3% (240 kt/year) is treated in mechanical-biological facilities.

3.3.2.1.1.4 Selection of Substances and Characterization of Wastes
The following substances are selected as indicators: carbon, nitrogen, chlorine, sulfur, cadmium, mercury, lead, and zinc. Carbon is selected because of the TOC limit that will apply beginning in 2004. Nitrogen is relevant as a potential nutrient and for the cement industry. Cement kilns are single sources (2.5%) of national NO_x emissions, along with traffic (62%), other industries (17%), and home heating (10%).[70,71] Wastes rich in nitrogen may increase this emission load. Compounds of chlorine, sulfur, and heavy metals are major air pollutants. The heavy metals are also of interest for their potential as resources. An overview of the content of the selected substances in combustible wastes is given in Table 3.35. The ranges are broad and show that there are both extremes of wastes: "clean" wastes that have lower contamination than fuel oil and wastes that show a significantly higher level of pollution than MSW.

Case Studies

TABLE 3.35
Substance Concentrations of Combustible Wastes and Comparison with Other Fuels, mg/kg dry matter

	C	N	S	Cl	Cd	Hg	Pb	Zn
Mean	450,000	9100	2300	4300	5.7	0.8	230	520
Minimum	100,000	200	60	10	0.01	0.001	<1	1
Maximum	900,000	670,000	17,000	480,000	500	10	4000	16,000
MSW	240,000	7000	4000	8700	11	2	810	1100
Coal	850,000	12,000	10,000	1500	1	0.5	80	85
Fuel oil	850,000	3000	15,000	10	<1	0.01	10	20

3.3.2.1.1.5 Criteria for Optimized Assignment of Wastes to Treatment Processes

The variance in chemical composition requires tailor-made assignment of combustible wastes to treatment processes. Not every facility is qualified to treat any waste if the goals of waste management are to be reached. The following criteria are developed to decide upon waste treatment.

First, wastes having lower contaminant concentrations than average coal are appropriate for production processes such as cement kilns or brickworks. Concentrations are not determined per mass but per energy content of the fuel (e.g., mg/kJ). The reason is that 1 ton of waste does not necessarily replace 1 ton of coal. Rather, equivalent energy amounts are substituted by waste utilization. The rationale for this criterion is that it prevents products from becoming a sink for, e.g., heavy metals. It is not clear whether elevated concentrations in concrete, bricks, asphalt, etc. have an impact on the environment. Thus, the precautionary principle is applied by this criterion, and contamination of products is banned. A second argument is that once substances are transferred into such products, they cannot be recovered again. The second criterion considers the impact on air quality by waste combustion. Emissions of state-of-the-art MSW incinerators are smaller, for some substances orders of magnitude smaller, than modern air-pollution regulation demands. Thus, MSW incineration has proved to be environmentally compatible and serves as a reference. The criterion says that emissions from any waste-combustion facility must not exceed typical emissions from state-of-the-art MSW incineration. This can be expressed by the following equation:

$$c_{max} = \frac{TC_I}{TC_{CP}} \cdot c_{MSW} \qquad (3.13)$$

Transfer coefficients of state-of-the-art incineration (TC_I) for relevant substances into the air are known from several investigations. Also, average concentrations in MSW (c_{MSW}) are common. Typical values are given in Table 3.38. Transfer coefficients of the specific combustion process (TC_{CP}) have to be determined by an MFA. The maximum allowable concentration of a substance in a waste to be burned in a specific combustion process is then c_{max}.

TABLE 3.36
Assignment of Combustible Wastes in an Optimized Scenario and Changes Compared with the Actual Situation

	Optimized Scenario	Compared with Actual Situation
MSW incineration	1,500,000	+1,000,000
High-standard industrial combustion	2,000,000	+1,800,000
Hazardous waste combustion	70,000	±0
Wood industry	585,000	−30,000
Biomass cogeneration power station	110,000	+10,000
Pulp and paper industry	550,000	+31,000
Cement industry	170,000	+77,000
Total	5,000,000	+2,900,000

3.3.2.1.2 Results of the Optimized Scenario and Comparison with Actual Situation

Applying these criteria to the actual situation yields a new optimized scenario. In Table 3.36, the optimized assignment of wastes to combustion processes is listed. The capacity for combustion has to be more then doubled from 2.1 to 5.0 million t/year. Some of the required plants already exist (cement, pulp and paper, etc.). Most of them could manage the assigned wastes with little or no process adaptation. These changes can be carried out comparatively quickly. On the other hand, new incineration plants with advanced air-pollution control technology and a total capacity of 2.8 million t/year have to be erected. This may take up to 5 years, including the permitting process, financing, planning, and engineering.

The improvement between actual and optimized situation can be seen when the aforementioned assessment criteria are applied to the materials balances (Figure 3.45).

1. The critical air volume calculated for NO_x, SO_2, HCl, Cd, Hg, Pb, and Zn is reduced by 43%. This is surprising because the quantity of combusted wastes is increased by 140%. The reason for this paradox is that in the actual situation, a comparatively small quantity of wastes is combusted in simple furnaces that lack adequate air-pollution control. The new scenario assigns all wastes to appropriate plants. Noncontaminated wastes are utilized in furnaces that have a lower (but sufficient) standard in flue–gas cleaning. "Dirty" wastes are treated in well-equipped combustion plants.
2. The efficiency of utilization of the energy content of wastes is improved by 150%. In the optimized scenario, one energy unit of waste replaces the energy-equivalent amount of fossil fuel almost completely. The main reason for this progress is that wastes are no longer landfilled without energy recovery; no waste is processed in a mechanical–biological treatment plant anymore.

Case Studies

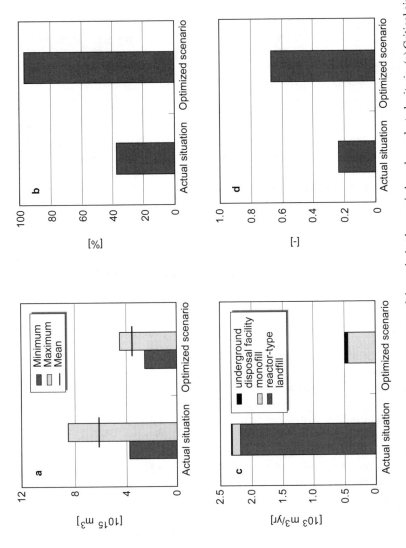

FIGURE 3.45 Comparison between status quo waste management and the optimized scenario based on selected criteria. (a) Critical air volume, (b) energy efficiency, (c) volume reduction, (d) substance management and "final sink."

3. Consumption of landfill space is reduced by 80%. Again, the main reason for this improvement is the ban of direct landfilling of wastes and the abandonment of mechanical–biological waste treatment. Combustion reduces the volume of wastes between 80 to 98%, depending on the ash content of the specific waste.
4a. The TOC of all residues landfilled is below 3%. This is an important step away from reactor-type landfills to "final storage" landfills that require no aftercare.
4b. The percentage of substances that are transferred into appropriate final sinks is increased by 180%. This indicates a relevant improvement in substance management.

3.3.2.1.3 Conclusions

This case study shows that MFA facilitates goal-oriented waste management. General goals on a high hierarchic level, as stated in various waste-management acts,* can be translated into well-defined, concrete assessment procedures with appropriate criteria. ASTRA outlines a way for this to be achieved. Note that a single goal of waste management may require two or more assessment methods for comprehensive evaluation. MFA is used at several levels in the study:

1. To describe the actual situation of the system "management of combustible wastes"
2. To reveal deficits and develop criteria for waste assignment
3. To compile the optimized scenario
4. To demonstrate the differences between the actual and the optimized situation

The findings of the study reveal the capacities for new plants and may serve as a basis for planning and engineering. A next step is to assess costs (including uncertainties) for the scenarios. In fact, this is also done in ASTRA. The result is that the optimized scenario can be realized without significantly raising total costs for disposal (collection, separation, treatment, and landfilling). The main drawback of the optimized scenario, and also the main reason why this scenario will take a long time to be accomplished, is the following: A large proportion of the wastes that are landfilled at present will be incinerated in the future. For landfill owners and operators, this may cause a severe economic situation because the landfills will lose business. Considering that landfills are long-term investments with filling times between 25 to 50 years, it is clear that such a stern change cannot be pushed through in a short time. It is also clear that landfill operators will use every possible legal and economic means to postpone strategic changes endangering landfilling.

---

* Examples are the European Waste Framework Directive, the Swiss Guidelines for Waste Management, and the German "Kreislaufwirtschaftsgesetz."

3.3.2.2 Case Study 12: PRIZMA

In the case study PRIZMA,* the utilization of combustible wastes for energy recovery in cement kilns is investigated for Austria.[72] In this country, production of cement requires about 10 million GJ/year to produce 3 million t/year of clinker that is further processed into cement. This amount of energy corresponds to some 400,000 tons of wastes with an average heating value of 25 MJ/kg. Today, wastes cover about 27% of the energy demand of the cement industry. Energy consumption for clinker production is quite high and represents a significant share of total production costs. For example, the European Cement Association estimates that energy accounts for 30 to 40% of the production cost of cement.[73] Hence, the cement industry strives to minimize energy-specific costs. One possibility is to reduce costs for fuel consumption. Wastes are alternative fuels. Compared with standard fuels such as natural gas, oil, or coal, they are less expensive. Other ways to reduce costs are to improve energy efficiency or to use different raw materials and technologies.

The Austrian cement industry has a long history of experience with energy recovery from wastes. Traditional waste fuels are used tires, waste oil, and solvents. Test runs have been carried out with sewage sludge, mixed plastics, and waste wood as well as meat and bone meal (as a result of the disposal crisis caused by "mad cow disease"**) and others. Sorted fractions of MSW are also considered as a fuel alternative. The Austrian cement industry aims to cover 75% of its energy demand by wastes within a few years. Besides the expected cost relief, this goal contributes to the reduction of carbon dioxide emissions by the branch. Since direct landfilling of organic wastes is forbidden after 2004 in Austria, the cement industry will help to provide the required capacities for treatment. On the other hand, not all wastes are appropriate for the cement process. Wastes with high contamination of heavy metals may lead to environmentally incompatible emissions and a polluted product (cement, finally concrete). Hence, operators as well as authorities want to have clear regulations specifying which kinds of wastes are appropriate for energy recovery. One possibility for establishing such an instrument is to generate a so-called positive list. The positive list specifies and characterizes waste types that are appropriate for recovery in cement kilns. The objective of PRIZMA is to develop criteria to establish such a list.

3.3.2.2.1 The Cement Manufacturing Process

The prevailing technology for cement manufacturing in Austria is the cyclone preheater type. A scheme of such a facility is displayed in Figure 3.46. The heart of each cement factory is a massive steel tube up to 100 m in length and up to 8 m in diameter. It is slightly inclined to the horizontal (3 to 4°) and slowly rotates at about one to four turns per minute. The main raw materials for the feed of the kiln are limestone, chalk, marl, and corrective materials (e.g., ferrous materials). The chemical properties of these materials and the desired properties of the clinker govern the correct mixture. Mixing is an important step in the process to ensure an even

---

* PRIZMA is a German acronym for "positive list for utilizing of residues in the cement industry: methods and approaches."
** The formal name for the disease is bovine spongiform encephalopathy (BSE).

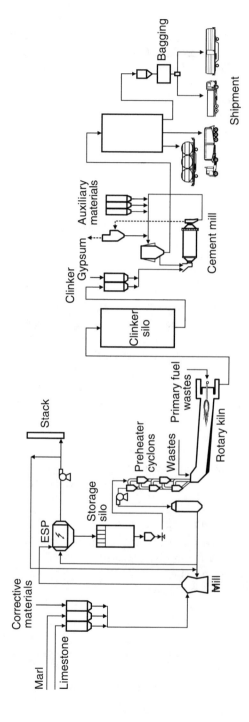

FIGURE 3.46 Scheme of a cement kiln (cyclone preheater type).

Case Studies

distribution of the properly proportioned components of the raw material so that the clinker will be of a uniform quality. The raw material is ground in a mill, from where it is pneumatically boosted into a mechanical predeposition (baffle) and a subsequent electrostatic precipitator (ESP). There, the raw material is collected and conveyed into a storage silo. Afterwards, the raw material runs through four stages: evaporation and preheating, calcining, clinkering, and cooling.

Evaporation and preheating remove moisture and raise the temperature of the raw material. This process takes place in the cyclones, where raw material counter-flows the hot off-gas from the kiln. The raw material enters the kiln at the back (the upper end of the kiln), and gravity and the rotation of the kiln allow the mix to flow down the kiln at a uniform rate through the burning zone. The tube is lined with refractory bricks to avoid heat damage of the kiln. Calcining takes place at 600 to 900°C and breaks the calcium carbonate down into calcium oxide and carbon dioxide. Approximately 40% by weight of the raw material is lost by this process. Clinkering completes the calcination stage and fuses the calcined raw material into hard nodules resembling small gray pebbles. The clinker leaves the kiln at the front (the lower end of the kiln) and falls onto a reciprocating grate, where it is cooled. The primary fuel is introduced and burnt at the same end of the kiln. The flame is drawn up the kiln to the burning zone, where the heat intensity is highest and fusion of chemicals in the raw material takes place. Hot combustion gases continue to flow up the kiln and exit from the back end. Secondary firing at the upper end of the kiln maintains the energy-intensive calcination process. Product temperatures in the burning zone are around 1450°C. The primary flame has a temperature of 2000°C. The cooled clinker is stored in a silo. Cement is produced by grinding of the clinker and blending with gypsum and other materials (e.g., fly ashes from coal combustion) to produce a fine gray powder. The last stage is bagging of cement and preparing the product for transportation. The cooled flue gas (heat-transfer to raw material in the cyclones) is cleaned by the ESP and emitted via a stack.

Generally, for any substance there are only two ways to enter and to leave the process: enter via raw materials or fuels and exit via off-gas or the product. The partitioning for any substance A between raw materials and fuels can be determined by measurement. The result will be that X% of A enters the process via fuels and Y% via raw materials, with X + Y = 100%. For the off-gas and the product, only total amounts of substance A can be determined. It is not possible to say that $\tilde{X}\%$ of A stems from fuels and $\tilde{Y}\%$ stems from raw material in the clinker (again $\tilde{X} + \tilde{Y} = 100\%$), and it cannot be assumed that $\tilde{X}$, etc. The same applies to the off-gas. Only some qualitative information is available about the behavior of substances in a clinker manufacturing process. However, for the given problem, which is waste combustion in cement kilns, it is mandatory to know how fuel-borne* substances behave in the process.

A special characteristic of the process is that two kinds of cycles evolve:

1. The so-called inner cycle arises when a substance i vaporizes in the kiln. It is then transferred to cooler parts in the cyclone, where i may condense

---

* The term *fuel-borne* characterizes a substance that enters the process via fuels.

at the surface of raw material particles. So i travels back to the hot kiln. There i vaporizes again. The cycle is closed and built up until some kind of equilibrium is established (theoretically). Operators try to break such cycles by bypassing the cyclones.
2. The so-called outer cycle develops because cyclones cannot intercept fine particles (<5 µm). But interception in the cyclones is a prerequisite for raw material particles to enter the kiln and leave the process as clinker. Fine particles are carried away with the flue gas to the ESP. There they are removed from the flue gas (efficiency >99%) and reenter the cyclones, where they are not intercepted again. The loop is closed and built up.

These cycles make it difficult to establish closed substance balances for the process and to predict the behavior of substances in the process. However, the following working hypothesis can be put forward: Fuel-borne substances (heavy metals) are predominantly embedded in an organic matrix. These substances are destroyed in the flame, which is an area with temperatures around 2000°C. This means that fuel-borne substances will vaporize to a high extent. Contrary to fuels, heavy metals in the raw material are fixed in a mineral matrix. The raw material is heated up to 1450°C, and it can be assumed that not all metals will vaporize. A portion will remain in the solid phase and contribute in the clinkering process (sintering). In other words, the possibility for a metal to reach the gaseous phase is higher for fuel-borne substances than for substances descending from raw material. When the flue gas is cooled down (in the cyclones), condensation of metals on particle surfaces (raw material, ashes) takes place. This process happens at the same rate for all metals, regardless of their origin (fuel or raw material). Let us assume that the partitioning of a substance χ between fuel and raw material is X:Y. Then the mentioned vaporization and condensation processes imply that there is a different ratio $\tilde{X}:\tilde{Y}$ of χ in particles in the cyclone with $X/Y < \tilde{X}/\tilde{Y}$. Fine particles will pass the cyclones, and a small fraction will also pass the ESP. Evidence of this process is that particulates emitted from cement plants are enriched with heavy metals compared with raw materials (see Table 3.37). The conclusion is that fuel-borne and raw-material-borne substances show different behavior in the process and

TABLE 3.37
Mean Substance Concentrations in Raw Material and Emitted Particulates

	Cl	Cd	Hg	Pb	Zn
Raw material, mg/kg	150	0.15	0.15	15	37
Emission, mg/kg	46,000[a]	8	2000[a]	400	150
Enrichment	300	50	13,000	27	4

[a] Gaseous and solid emissions are related to the total mass of emitted particulates.

Case Studies

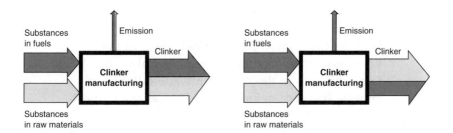

FIGURE 3.47 Assumptions for the fate of fuel-borne heavy metals in the clinker manufacturing process. The left/right assumption yields the lower/upper limit for the transfer coefficient of fuel-borne substances into the atmosphere.

therefore have different transfer coefficients. However, for the problem of waste combustion, it is crucial to know the transfer coefficients for fuel-borne substances.

3.3.2.2.2 Assessment of Transfer Coefficients

The uncertainty concerning transfer coefficients for fuels leads to the following approach: A range for transfer coefficients is established by determining hypothetical extreme values (see Figure 3.47). For the lower limit, it is assumed that the partitioning between fuel-borne and raw-material-borne substances is the same in the off-gas and the product. Based on the aforementioned considerations, this will underestimate the transfer of fuel-borne substances into the off-gas. The upper limit is given by the assumption that the emission only contains fuel-borne substances. This assertion certainly overestimates the influence of fuels for emissions. On the other hand, this is a reliable upper limit, since higher transfer coefficients are not possible. The real transfer coefficient has to be somewhere between these extremes. In cases where the range is large (e.g., one order of magnitude), it is safe to apply the upper limit, i.e., the higher value.

3.3.2.2.3 Criteria for Waste Fuels

Which substances should be incorporated into a "positive list" that defines wastes suited for cement kilns? Chapter 3.3.2.1 gives criteria for the selection of substances with regard to waste combustion. The list can be shortened for cement manufacturing because carbon compounds are most efficiently destroyed in the cement kiln. With the exception of carbon dioxide, emissions of carbon compounds are very small. The emission of nitrogen oxide is a problem of cement manufacturing, but this has little to do with waste recovery. The high temperatures and long residence times of gases in the process, which are essential for efficient mineralization of carbon compounds, are responsible for nitrogen oxide formation. The nitrogen content of wastes plays only a minor role in the formation of nitrogen oxide. Sulfur from wastes is efficiently contained into clinker. Recorded emissions of sulfur dioxide stem from certain kinds of raw material. As in the case of nitrogen, these emissions are not a result of waste combustion. Chlorine poses a limit for the quality of the product and may cause blockage in the cyclones. Therefore, it has to be part of the positive-list. The heavy metals cadmium and mercury are chosen because of their toxicity and volatility. Lead and zinc are also potentially toxic and are important resources.

TABLE 3.38
Data for Criterion A: Transfer Coefficients of the Reference Technology Incineration, Typical Concentrations in the Reference Waste (MSW), and Extreme Transfer Coefficients for Off-Gas of a Cement-Manufacturing Process

	Cl	Cd	Hg	Pb	Zn
TC_I	0.0005	0.0005	0.02	0.0001	0.0002
c_{MSW}	10,000	10	2	500	1000
$TC_{CM,min}$	0.01	0.0002	0.4	0.0002	0.0001
$TC_{CM,max}$	0.02	0.0004	0.8	0.0008	0.0001

Wastes as fuel for clinker manufacturing influence input as well as off-gas and clinker. Therefore, three criteria — A, B, and C — are developed for the positive list:

Criterion A deals with the off-gas and has been described in Section 3.3.2.1. The criterion says that emissions from clinker manufacturing must not exceed typical emissions from state-of-the-art incineration. This can be expressed by the following equation:

$$c_{max} = \frac{TC_I}{TC_{CM}} \cdot c_{MSW} \qquad (3.14)$$

TC_I is the transfer coefficient of state-of-the-art incineration (known). Also, the average concentrations in MSW (c_{MSW}) are commonly known (see Table 3.38). Transfer coefficients for clinker manufacturing (TC_{CM}) have to be determined according to aforementioned considerations about a reliable range for transfer coefficients. The result of criterion A, c_{max}, is the maximal allowable concentration of a substance in a waste.

Criterion B controls the quality of the product clinker. It is based on the principle that anthropogenic material flows must not exceed the natural fluctuations of geogenic flows (see Chapter 2, Section 2.5.1.7). The criterion now considers raw material chemically as a geogenic flow. As any natural material, raw materials show a certain variance in chemical composition. To apply the criterion, the clinker composition is assessed for (a) average and (b) maximal raw material concentrations. Criterion B says that the changes in clinker concentration caused by waste recovery must not exceed the calculated geogenic variance. For both scenarios (a) and (b), clinker is produced with an average fuel mix consisting of coal (52%),* oil (21%), natural gas (3%), used tires (6%), plastics (5%), waste oil (9%), and others. The calculation of criterion B requires the following assumptions: The mass ratio between raw materials (RM) and fuels (F) is ca. 10:1. Transfer coefficients for substances stemming from raw materials and fuels are identical. The loss of carbon dioxide is 40%, and the

---

* Percentages are based on energy equivalents.

Case Studies

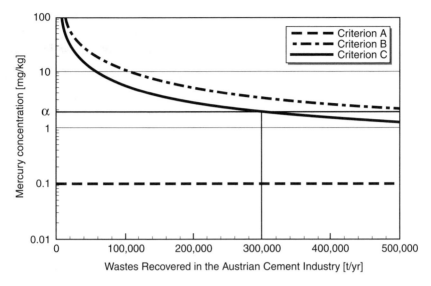

FIGURE 3.48 Results of criteria A, B, and C for Hg serve to support decisions regarding the utilization of wastes in cement kilns.[72] Criterion A: emissions, Criterion B: product quality, Criterion C: dilution of metals.

ash content of the fuel is considered to be negligible (usually ca. 1% of raw material mass). Then the concentration of clinker (Cl) is

$$c_{Cl} = \frac{(c_F \cdot 1 + c_{RM} \cdot 10) \cdot TC_{Cl}}{10 \cdot (1 - 0.4)} \qquad (3.15)$$

The data needed to calculate criterion B are summarized in Table 3.39. The result of criterion B is the maximum load of a substance that can be added to the clinker by waste recovery. The load (e.g., mg of a substance per ton of clinker) gives the dependence between the total mass of recovered wastes and the substance concentration in the waste. The result is a curve (see Figure 3.48).

Criterion C considers the input into cement manufacturing. If the total national consumption is assigned a value of 100%, then combustible wastes contain roughly 40% of cadmium and mercury (see Table 3.40). Hence, combustible wastes are important carriers of some heavy metals. This is one reason why waste management plays such an important role for the total turnover of some heavy metals. The consequential question for the cement industry is: Which share of these substances shall enter the processes and finally end up in the product cement? There is no uniform answer to this question. Some cement manufacturers do not want their products associated with hazardous materials and thus are cautious in using contaminated wastes as a fuel. Other manufacturers see a chance for economic advantage over their competitors by using inexpensive waste-derived fuel and use large amounts of wastes. In Table 3.40, results are presented assuming that cement manufacturers take over 15% of the metals contained in combustible wastes. As for criterion B,

TABLE 3.39
Calculation of Maximum Allowable Load on Clinker through Waste Recovery

	Cl	Cd	Hg	Pb	Zn
Mean concentration in fuel mix, mg/kg	1100	0.9	0.4	60	65
Mean concentration in raw material, mg/kg	150	0.15	0.15	15	37
Maximum concentration in raw material, mg/kg	400	0.5	0.5	42	110
TC into clinker	0.99	0.99	0.6	0.99	0.99
Mean concentration in clinker, mg/kg	430	0.4	0.19	35	72
Maximum concentration in clinker, mg/kg	840	1.0	0.54	79	190
Maximum load through waste recovery, mg/kg	410	0.6	0.35	45	120
Maximum load through waste recovery, t/year	1200	1.7	1.1	130	360

Note: Based on a clinker production of 3 million t/year.

TABLE 3.40
Estimated National Consumption of Selected Substances, Content in Combustible Wastes, and Maximum Flow into Cement Manufacturing

	Cl	Cd	Hg	Pb	Zn
National consumption, t/year	450,000	80	10	32,000	43,000
Combustible wastes, t/year	30,000	36	3.9	1600	3900
In combustible wastes, %	6.6	45	39	5	9
Input into cement, %	15	15	15	15	15
Input into cement, t/year	4500	5.4	0.6	240	590

the result is a curve showing the dependency between total mass of wastes and substance concentration in the wastes (see Figure 3.48).

3.3.2.2.4 Results

Criteria A to C are calculated for the selected substances and provide either maximum concentrations for wastes or maximum amounts of substances that can be transferred into clinker. Required parameters for calculation are transfer coefficients of waste-borne substances into the off-gas and into the clinker. They should be determined for each cement plant separately, as results may vary considerably among different technologies. In Figure 3.48, the result is given for mercury. Consider a waste (α) having a mercury concentration of 2 mg/kg. Criterion B allows waste recovery of 500,000 t/year. This means that, in practice, mercury does not pose a limit for the quality of clinker. Self-restriction of the cement industry (criterion C) still allows 300,000 t/year of waste α. A really severe limit poses criterion A: Only wastes having Hg concentrations lower than 0.1 mg/kg are qualified as a waste fuel. This reduces the amount of potentially available wastes considerably. With respect to mercury, it should be evaluated whether the investment costs for improving air-pollution control technology are paid back within an acceptable time frame by the savings for a cheaper waste fuel.

3.3.2.2.5 Conclusion

Case study PRIZMA exemplifies how MFA can be used to establish environmental regulations. The proposed criteria consider system-specific as well as plant-specific constraints. Criteria A and B guarantee that waste utilization in cement kilns does not pollute the atmosphere (and subsequently the soil) or lower the quality of clinker. The extended systems approach assures that only a limited amount of resources is transferred into cement and concrete (criterion C). Note that the selected substances are not required in cement and are lost for recovery and recycling. A limit for this kind of sink is therefore reasonable. Waste recovery in state-of-the-art cement kilns that fulfill criteria A, B, and C can be considered as environmentally compatible. Thus, the utilization of wastes in cement kilns can be a valuable contribution to goal-oriented waste and resource management.

3.3.2.3 Case Study 13: Recycling of Cadmium by MSW Incineration

In mineral ores of commercial value, cadmium is usually associated with other metals. Greenockite (CdS), the only cadmium mineral of importance, contains zinc, sometimes lead, and complex copper-lead-zinc mixtures. Hence, zinc and lead producers have no choice, and they usually produce cadmium, too. Most cadmium (>80%) is produced as a by-product of beneficiating and refining of zinc metal from sulfide ore concentrates. An estimated 90 to 98% of the cadmium present in zinc ores is recovered in the zinc extraction process. About 3 kg of cadmium are produced for every ton of refined zinc. Small amounts of cadmium, about 10% of consumption, are produced from secondary sources such as baghouse dust from electric arc furnaces (EAF) used in the steelmaking industry and the recycling of cadmium products. Total world production of cadmium in 2000 was about 19,300 tons.[74] The International Cadmium Association has made the following estimates of cadmium consumption for various end uses in 2001: batteries 75%, pigments 12%, coatings and plating 8%, stabilizers for plastics and similar synthetic products 4%, and nonferrous alloys and other uses 1%.[75] Usage (or consumption) of cadmium in developed countries is estimated between 0.005 and 0.016 kg/capita/year.[75–77] Currently, consumption of cadmium shows a declining trend due to increasing regulatory pressure to reduce or even eliminate the use of cadmium. This is a result of the growing awareness of cadmium being potentially toxic to humans and of the risks presented by its accumulation in the environment. Unlike other heavy metals (e.g., zinc, selenium), cadmium is not essential to the biosphere and has no known useful biological functions. It accumulates in the kidneys and liver and affects protein metabolism, causing severe disorder, pain, and even death. In addition, it may cause a variety of severe damages such as lung cancer[78,79] and bone diseases (osteomalacia and osteoporosis).[80]

In MSW, the typical concentration of cadmium is between 8 to 12 mg/kg.[64,81,82] This is about 50 times the average concentration found in the Earth's crust (0.2 mg/kg). When MSW is landfilled, cadmium can be mobilized by organic acids and reach the groundwater. Cadmium and some of its compounds such as chlorides have low boiling temperatures (Cd, 765°C; $CdCl_2$, 970°C). Therefore, cadmium belongs to the group of atmophilic elements (Hg, Tl, Zn, Se). In combustion processes, these substances tend to volatilize and escape via the flue gas from the combustion chamber.

The average generation of MSW in Europe is between 150 and 400 kg/capita/year. In the U.S., currently about 700 kg/capita/year are collected. There are several reasons to explain the differences in quantities. First, the term MSW is defined operationally. MSW usually designates all mixed wastes that are collected at the curbside on a daily, weekly, or biweekly basis. For example, in some areas bulky wastes are collected separately and therefore not included in MSW data. In other areas, waste containers are large and allow collection of MSW together with bulky wastes. MSW includes mixed wastes from private households and may also include wastes from the service sector and small shops and companies. Decisive factors include the kind of wastes the collector accepts, the collection frequency, the size of the bins or containers for collection, and how statistical data are compiled. A second reason for differences is the extent of recycling. Separate collection of paper, biowaste, glass, metals, etc. may reduce the weight of MSW up to 50%. Third and fourth reasons are differences in lifestyle (e.g., small or large family and household size, packed food vs. open food, etc.) and purchasing power of the consumer. At an average MSW generation rate of 250 kg/capita/year, about 2.5 g/capita/year of cadmium are collected via MSW. This is about 25% of the average national per capita consumption of cadmium. The remaining 75% is mostly incorporated in goods with long residence times that will enter waste management in the future. A smaller part of cadmium is expected in other wastes (estimated 20%, mainly contained in industrial wastes and the combustible fraction of construction and demolition waste), and a small quantity is lost to the environment via emissions and fugitive losses.

Modern incinerators are equipped with advanced air-pollution control (APC) devices. As mentioned before, cadmium is transferred to the flue gas during incineration and removed by the APC devices. During off-gas cooling in the heat exchanger, volatile cadmium is condensed on very small particles that offer a large surface area. Consequently, more than 99.9% of cadmium can be removed by particle filters such as electrostatic precipitators or fabric filters if these filters have been designed to capture small particles. Remaining quantities are removed upstream in wet scrubbers with high pressure drops (venturi scrubbers) or adsorption filters. Very small amounts are emitted via the stack (<0.01%). Figure 3.49 shows transfer coefficients for cadmium in a state-of-the-art MSW incinerator.

Typical Cd concentrations in incinerator fly ash range from 200 to 600 mg/kg. In comparison, filter dust from electric arc furnaces that is recycled contains 500 to 1000 mg/kg Cd.[83–87] This shows that incinerator fly ash could be used for cadmium recovery.

Thermal treatment of bottom ash and/or fly ash can improve the potential for recovery. For example, adequate thermal treatment of incinerator bottom ash results in three products:

1. A silicate product, with an average cadmium concentration similar to the Earth's crust, that can be utilized for construction purposes

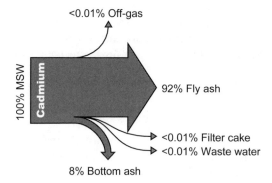

FIGURE 3.49 Transfer coefficients for cadmium in a state-of-the-art MSW incinerator.[60] (From Schachermayer, E. et al., Messung der Güter- und Stoffbilanz einer Müllverbrennungsanlage, Monographien Bd. 56, Bundesministerium für Umwelt, Vienna, Austria, 1995. With permission.)

2. A metal melt containing mainly iron, copper, and other lithophilic metals (metals of high boiling points)
3. A concentrate of atmophilic metals

Applying such technologies to ashes makes cadmium and other metals accessible for efficient recovery. For cadmium, recycling efficiencies from MSW up to 90% are realistic. Another possibility is to use thermal processes not to produce a concentrate for recovery, but to immobilize metals in a ceramic or vitreous matrix (vitrification). Such residues may come close to final storage quality.[43]

Today, most of the obsolete cadmium enters landfills, where it remains a potential hazard for generations. The advantage of recycling is that the consumption of primary cadmium is reduced. Thus, the quantity of cadmium that enters the anthroposphere is reduced, facilitating the management and control of this resource. Technologies to immobilize cadmium are required to safely dispose of the large amounts already

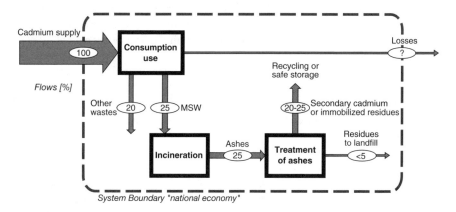

FIGURE 3.50 Recycling of cadmium by MSW incineration.

in the anthroposphere. The new task is to continually collect and transform cadmium that is stored in infrastructure and long-living goods into a form where it can be safely stored within the anthroposphere. Note that for both options, a concentration step is indispensable; thermal processes are proven technologies that can achieve such concentrations.

3.3.3 PROBLEMS — SECTION 3.3

Problem 3.8: Plastic wastes have a high calorific value, which makes them a potential fuel for cement kilns, blast furnaces, and municipal incinerators. Packaging plastics have been successfully incinerated in cement kilns: Production costs are reduced; the quality of cement does not change; and emissions are not altered significantly. Wastes from longer-lasting plastic materials (containing about 10% PVC) have a high chlorine content, rendering these wastes unsuitable as a fuel in cement kilns because they exceed the capacity of the process for chlorides. Using Table 3.29, evaluate whether nonpackaging plastics are better suited for blast furnaces or for MSW incinerators. Take into account environmental and resource considerations only, and do not consider economic (additional investments, fuel savings) or technological (pretreatment, adaptation of feed or furnace, etc.) aspects. Discuss "final sinks" for heavy metals. Use the transfer coefficients in this handbook for MSW incineration, Table 3.38, and look for data about blast furnaces in the library or the World Wide Web.

Problem 3.9: Assess the paper content in MSW of a country of your choice. First, determine the appropriate system (processes, flows, system boundaries). Second, carry out an Internet search for the annual report of the pulp and paper industry of the selected country and determine the flows through and within your system. Third, find out the national MSW generation rate (e.g., contact the EPA Web site) and calculate your result.

Problem 3.10: The combustion of biomass is described by Obernberger et al.[88] Calculate the Cd concentration in cereals based on the information given in the paper. Using the approach described in Section 3.3.1.2, calculate the composition of the input (cereals) from the composition of the output (different ash fractions). Compare with Cd values for cereals you find in the literature.

Problem 3.11: Figure 3.51 gives the Cd balance for the management of combustible wastes in Austria. Discuss the flowchart together with the total waste mass balance for Austria as given in Section 3.3.2.1, Figure 3.44. Consider resource potentials and potentially dangerous environmental loadings.

Problem 3.12: Summarize the reasons why a cement manufacturer association might decide to limit the annual flow of heavy metals into cement kilns with 15 % of the national consumption of heavy metals.

Problem 3.13: Assume that incineration of 1 ton of MSW (copper [Cu] content ca. 0.1%) yields the following solid residues: 250 kg of bottom

Case Studies

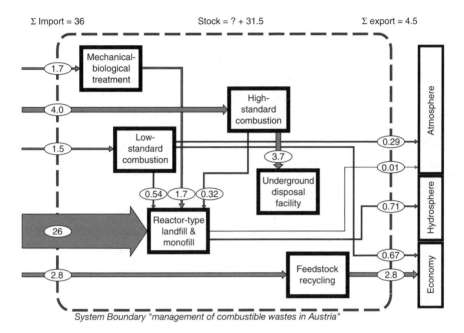

FIGURE 3.51 Flows of cadmium induced by the management of combustible wastes in Austria (1995), t/year.

ash, 25 kg of fly ash, 3 kg of iron scrap, and 3 kg of neutralization sludge from the treatment of scrubber water. About 90% of the Cu leaves incineration via bottom ash and 10% via fly ash. The Cu flow via other residues such as off-gas, iron scrap, etc. is <1% and can be neglected. Investigations show that by mechanical processing of bottom ash, approximately 60% of the Cu can be separated in the small fraction of metals concentrate. The Cu content of this fraction (ca. 50%) can be recovered in a metal mill. (a) What is the recovery efficiency for the combined process "MSW and mechanical processing of bottom ash"? (b) Calculate the substance concentrating efficiency (SCE) for the process chain "incineration, mechanical processing, and metal mill." (c) As a decision maker, would you support such a technology, and why?

The solutions to the problems are given on the Web site, www.iwa.tuwien.ac.at/MFA-handbook.

3.4 REGIONAL MATERIALS MANAGEMENT

The objective of regional materials management is to protect the environment, to conserve resources, and to minimize wastes in one combined effort. Regional materials management is an integrated approach that links all three issues and strives for an optimum solution. Instead of focusing on one topic alone, all three are taken into account at the same time. This comprehensive procedure requires less effort and results in more information than three separate studies. It also ensures that the results from the different fields are compatible and that conclusions regarding all three fields can be drawn. For regional materials management, it is essential to know the main anthropogenic as well as natural sources, conveyor belts (transport paths), stocks, and sinks of materials in a region. Without this information, regional materials management is not possible. In order to achieve the stated goals, a long-term view must be taken. Material flows and stocks have to be balanced over decades to centuries in order to examine whether harmful or beneficial accumulations and depletions of materials are taking place in the region. All materials used within a region must find a safe final sink. If a safe final sink is not available, use of the material should be phased out or controlled tightly and accumulated over long time periods with a clear purpose and economic plan for future reuse.

3.4.1 CASE STUDY 14: REGIONAL LEAD MANAGEMENT

This example of regional lead management is drawn from Case Study 1 (Section 3.1.1), which described systems definition and data collection. The following discussion covers only those results and conclusions that are important for regional materials management.

3.4.1.1 Overall Flows and Stocks

A total of 340 t/year of lead is imported into the region, and 280 t/year is exported (see Figure 3.1). The main import consists of used cars that are crushed in a car shredder. The main exports are filter residues from a steel mill that produces steel for construction from the shredded cars, lead contained in construction steel, and lead in MSW. The difference between imports and exports amounts to 60 t/year, which is accumulated mainly in landfills. The geogenic stock "soil" includes about 400 t of lead (the term "geogenic" is not actually precise here, since a certain fraction of lead in soils is of anthropogenic origin, [see Baccini et al.[89]]). The anthropogenic stock "landfill" is much larger and amounts to >600 t (>10 years of landfilling 60 t/year). Like most materials in urban regions (Chapter 1, Section 1.4.5.4), lead is accumulated in this region. Imports and exports of lead by geogenic conveyor belts (air, water) are marginal and are <1%. From the point of view of resource management, shredder residues and filter dusts are of prime interest. From an environmental point of view, depositions on the soil as well as potential leaching of lead from landfills to surface and groundwater are important.

The advantage of a regional material balance is that with one single balance, present and future hot spots for environmental, resource, *and* waste management can be detected. For example, the potentially large but at present unknown flow of lead from local landfills to the hydrosphere cannot be hypothesized without information about local landfills and their constituents. Quantitative information about landfills in general is not available, but it is known that most of the shredder residue is landfilled within the region. By a simple balance of the car shredder, assuming a certain lead input based on car manufacturers' information and the number of cars

Case Studies

treated in the shredder, and the lead in the metal fraction used and analyzed by the smelter, it is possible to roughly assess the amount of landfilled lead.

3.4.1.2 Lead Stock and Implications

The existing stock of lead in landfills totals >600 t. A doubling time (τ_{2x}) for the lead stock of ≈10 years can be calculated. In other words, if the regional anthroposphere remains the same for the next 100 years, the stock will have increased from 600 t to 7000 t. According to Chapter 1, Section 1.4.5.1, there are no indications yet that waste lead flows will decrease. What makes this case study special is the huge extent of the accumulation. Nearly 20% of the lead imported does not leave the region and stays there probably for 10,000 years until erosion slowly removes the landfill. All lead landfilled and deposited on the soil is of no further use. The concentration is comparatively low, and the heterogeneity of the landfilled materials is much larger than that of lead ores. Hence, economic reuse of this stock is at present not feasible. Emissions from this stock are likely but not known. A conscientious approach to regional materials management would dictate that, in the future, this lead stock be managed in a different way, turning it from a hazard into a positive asset. Means for upgrading and reuse have to be explored (see item 2 in Section 3.4.1.4).

3.4.1.3 Lead Flows and Implications

Lead flows can be divided into flows in products, in wastes, and in emissions. The management goal is to maximize the use of lead in products, to reuse lead in wastes, and to reduce emissions to an acceptable level. MFA shows where the large lead flows are, and thus points out key processes and goods for control and management. For each environmental compartment (water, soil, air), potential sources are identified and sometimes quantified. Thus, priorities can be set when measures for the protection of the environment are taken.

Figure 3.1 shows that lead increases by 1.4 t/year in the river between the point of entry and exit out of the region. While 0.31 t/year are due to leaching from soils, and 0.14 t/year due to treated wastewater, 1 t/year has not been accounted for by MFA. This flow is so large that it is not likely due to an error in measuring soil leaching or effluent from WWTP. The most probable source of this large amount of lead is leachates of shredder residue from landfills. It is not efficient to reduce the comparatively small amount of lead emitted by WWTP effluents. The first step is to investigate the hypothesis that shredder residue landfills are really leaching such a large amount of lead to the surface waters. The second step is to reduce the loadings of the soil, e.g., by banning leaded gasoline (as was done in the late 1980s) or by incinerating sewage sludge and landfilling the immobilized ash.

MFA supports environmental impact assessment and serves as a design tool. Figure 3.1 shows that emissions from landfills (and any other point source) are not relevant if they are in the range between 0.002 to 0.02 t/year (0.1 to 1% of present aquatic export flow). Considering a total stock of ≈1,000 t of lead landfilled, one can calculate that no more than about 2 to 20 ppm (mass) of lead may be mobilized in the landfill if there is to be no significant effect on the river water concentration. This figure can serve as a goal for the design of waste treatment such as immobilization or solidification. Note that this calculation does not consider groundwater pollution. If a local groundwater flow is small and the residence time is high, the lead flows from landfills calculated above may be large enough to exceed drinking water standards. Hence, it is important to take groundwater into consideration, too.

3.4.1.4 Regional Lead Management

For the region, it is more efficient to manage lead in a comprehensive way than to segregate the lead issue into different problem areas. This is exemplified by the following three conclusions:

1. Lead not in use should be accumulated actively and purposefully in safe, intermediary stocks with residence times of several decades. The objective is to build up concentrated stocks of lead and other metals and to reuse these stocks once they have reached a size that makes them viable for economic reuse. In order to concentrate lead as much as possible, the shredder residue should be treated in an incinerator with advanced air-pollution control. Mineralization will increase lead concentration by at least a factor of ten. Many materials are well suited for such accumulation. The region could offer to take back filter residues from MSW incineration and to accumulate these materials together with car shredder residues. Intermediary lead stocks are distinctly different from fluff or MSW in landfills. They are highly concentrated in metals, and the chemical form is such that economic metallurgical reuse is facilitated. Hence, solidification with cement is not recommended. The intermediary stocks are engineered sites that are designed and constructed to last for a predefined period during which they have to be maintained. The period is calculated according to economic considerations. Due to the economy of scale, and depending on the technology applied, there is a minimum size required for economic reuse of materials. This minimum size divided by the waste generation rate yields the time span needed for accumulation of an amount of material for economic recycling.
2. Accumulation and mineralization for reuse ensures that most lead is controlled within the anthroposphere and no longer poses a threat to the environment. Information about anthropogenic lead flows allows identification of those processes that emit lead. Based on the regional flows and depositions and on dispersion models, the acceptable flows and depositions for lead in water and soil can be calculated. "Acceptable" can be defined from a toxicology point of view (limiting value for lead content in water or soil) as well as a precautionary principle point of view (lead input into soil or water equals lead output). In any case, concentrations *and* flows of lead have to be taken into account. Also, potential accumulation of lead in downstream regions needs to be considered. Information

Case Studies   295

about input flows and acceptable output flows of processes is useful in designing transfer coefficients that ensure regional environmental protection over long periods of time.

3. Monitoring based on materials accounting allows one to track accumulation or depletion as well as harmful flows of lead. Efficient monitoring points are:

   a. The products "construction steel" and "filter residue of the smelter." These two goods are routinely analyzed for production and quality control purposes. The results allow the determination of lead in shredded cars and indicate whether a change in landfilled lead is to be expected.

   b. Concentration of lead in gasoline. This figure is supplied by gasoline producers.

   c. Filter residues from MSW incineration. This information combined with known transfer coefficients allows calculation of lead flows in MSW.

   d. Sewage sludge. Routinely sampled and analyzed sewage sludge yields information about the sewage network as a potential source. This analysis is instrumental for identifying new emissions or for confirming that loadings to the sewer have been successfully eliminated.

   e. Surface waters. For water quality assessment, sampling surface water at the outflow of the region yields adequate information about the total load of the hydrosphere, especially if the same information is available from upstream regions. Monitoring of soil samples may be adequate initially to get an overview of lead in soils. However, as mentioned in Section 3.1.1, routine monitoring by soil sampling is expensive and inefficient, and it does not allow early recognition of harmful accumulations or depletions in soils.

3.4.2 PROBLEMS — SECTION 3.4

Problem 3.14: Taking the lead example in Figure 3.1 as a starting point, establish an MFA for cadmium in the same region, assuming no major industrial application of cadmium. Use data given in this handbook, such as Table 3.29, Figure 3.50, and data from the Internet on cadmium in soils, MSW, etc. Assume that, as in problem 3.1, MSW from 280,000 persons is incinerated in the region. (a) What are the major flows and stocks of cadmium in the region with "old" incineration and air-pollution control technology (transfer coefficient to air = 0.10)? (b) How do these flows and stocks change if advanced air-pollution control equipment is applied and the transfer coefficient is changed to 0.00001? (c) Evaluate environmental and resource implications arising from the two technologies in the region.

The solutions to the problems are given on the Web site, www.iwa.tuwien.ac.at/MFA-handbook.

REFERENCES

1. König, A., personal communication, Hong Kong, September 3, 2002.
2. Brunner, P.H., Baccini, P., Deistler, M., Lahner, T., Lohm, U., Obernosterer, R., and Van der Voet, E., Materials Accounting as a Tool for Decision Making in Environmental Policy — MAcTEmPo Summary Report, paper presented at Fourth European Commission Programme for Environment and Climate, Research Area III, Economic and Social Aspects of the Environment, ENV-CT96-0230, Institute for Water Quality and Waste Management, Vienna, 1998.
3. Brunner, P.H., Daxbeck, H., Hämmerli, H., Rist, A., Henseler, G., Gajcy, D., Scheidegger, R., Figi, R., von Steiger, B., Baccini, P., Langmeier, M., Beer, B., Moench, H., Sturzenegger, V., Piepke, G., and Bächler, M., RESUB — Der regionale Stoffhaushalt im Unteren Bünztal: Entwicklung einer Methodik zur Erfassung des regionalen Stoffhaushaltes (Development of a methodology to assess regional material management), Eidgenössische Anstalt für Wasserversorgung, Abwasserreinigung und Gewässerschutz EAWAG, Abteilung Abfallwirtschaft und Stoffhaushalt, CH-8600 Dübendorf, Switzerland, September 2, 1990.
4. Beer, B., Schlussbericht RESUB Luft, Eidgenössische Anstalt für Wasserversorgung, Abwasserreinigung und Gewässerschutz EAWAG, Abteilung Abfallwirtschaft und Stoffhaushalt, CH-8600 Dübendorf, Switzerland, 1990.
5. Primault, B., Du calcul de l'evapotranspiration, *Arch. Met. Geophys. Bioclimatology Ser. B*, 12, 124–150, 1962.
6. Udluft, P., Bilanzierung des Niederschlagseintrags und Grundwasseraustrags von anorganischen Spurenstoffen, *Hydrochemisch-hydrogeologische Mitteilungen*, 4, LOCATION, 1981, pp. 61–80.
7. Von Steiger, B. and Baccini, P., Regionale Stoffbilanzierung landwirtschaftlicher Böden, Nationales Forschungsprogramm Boden NFP 22, Bericht 38, Bern-Liebefeld, Schweiz, 1990.
8. Baccini, P. and Lichtensteiger, Th., Conclusions and outlook, in *The Landfill, Reactor and Final Storage*, Baccini, P., Ed., Springer, Berlin, 1989, pp. 427–431.
9. Scheffer, F., *Lehrbuch der Bodenkunde*, 12th ed., Schachtschabel, P., Blume, H.-P., Brümmer, G., Hartge, K.-H., and Schwertmann, U., Eds., Ferdinand Enke, Stuttgart, Germany, 1989.
10. Monteith, J.L. and Unsworth, M.H., *Principles of Environmental Physics*, 2nd ed., Edward Arnold, London, 1990.
11. Henseler, G., Scheidegger, R., and Brunner, P.H., Determination of material flux through the hydrosphere of a region, *Vom Wasser*, 78, 91–116, 1992.
12. BAS, *Statistisches Jahrbuch der Schweiz 1978/88*, Bundesamt für Statistik, Birkhäuser, Basel, 1987.
13. Lentner, C., *Geigy Scientific Tables*, 8th ed., Vol. 1, Ciba-Geigy Ltd., Basle, Switzerland, 1981.
14. PHARE, Nutrient Balances for Danube Countries, Final Report Project EU/AR/102A/91, Service Contract 95-0614.00, PHARE Environmental Program for the Danube River Basin ZZ 9111/0102, Consortium TU Vienna Institute for Water Quality and Waste Management, and TU Budapest Department of Water and Waste Water Engineering, Vienna, Austria, 1997.
15. Mee, L., The Black Sea in crisis: a need for concentrated international action, *Ambio*, 21, 278–286, 1992.
16. Somlyódy, L., Brunner, P.H., and Uroiss, H., Nutrient balances for Danube countries: a strategic analysis, *Water Sci. Technol.*, 40, 9–16, 1999.

17. Brunner, P.H. and Lampert, Ch., The Flow of Nutrients in the Danube River Basin, *EAWAG News*, 43E, CH-8600 Dübendorf, Switzerland, June 1997, p. 15–17.
18. Zessner, M., Fenz, R., and Kroiß, H., Waste water management in the Danube Basin, *Water Sci. Technol.*, 38, 41, 1998.
19. Canter, L.W., *Environmental Impact Assessment*, McGraw-Hill, New York, 1996.
20. European Commission, Directive on the Assessment of the Effects of Certain Public and Private Projects on the Environment, 85/337/EEC, *Official Journal of the European Communities*, Brussels, Belgium, 1985.
21. Brunner, P.H. and Baccini, P., Regional material management and environmental protection, *Waste Manage. Res.*, 10, 203, 1992.
22. Schachermayer, E., Rechberger, H., Brunner, P.H., and Maderner W., Systemanalyse und Stoffbilanz eines kalorischen Kraftwerkes (SYSTOK), Austrian Environmental Agency, Monographien Bd. 67, Vienna, 1995.
23. Greenberg, R.R., Zoller, W.H., and Gordon, G.E., Composition and size distribution of particles released in refuse incineration, *Environ. Sci. Technol.*, 12, 566, 1978.
24. Jasinski, S.M., The materials flow of mercury in the United States, *Resour. Conserv. Recycling*, 15, 145, 1995.
25. Bergbäck, B., Johansson, K., and Mohlander, U., Urban metal flows — a case study of Stockholm, *Water Air Soil Pollut. Focus*, 1, 3, 2001.
26. Steen, I., Phosphorous availability in the 21st century — management of a nonrenewable resource, *Phosphorous Potassium*, Issue No. 217, 1998.
27. Larsen, T.A., Peters, I., Alder, A., Eggen, R.I., Maurer, M., and Muncke, J., Reengineering the toilet for sustainable wastewater management, *Environ. Sci. Technol.*, 35, 192A–197A, 2001.
28. Metallgesellschaft Aktiengesellschaft, *Metallstatistik (Metal Statistics)*, Frankfurt am Main, Germany, 1993.
29. Baccini, P. and Bader, H.-P., *Regionaler Stoffhaushalt*, Spektrum Akademischer Verlag, Heidelberg, 1996, chap. 1.
30. Meadows, D.H., Meadows, D.L., Randers, J., and Behrens, W.W., III, *The Limits to Growth*, Earth Island, London, 1972.
31. Kesler, S., *Mineral Resources, Economics, and the Environment*, Macmillan, New York, 1994.
32. Becker-Boost, E. and Fiala, E., *Wachstum ohne Grenzen — Globaler Wohlstand durch nachhaltiges Wirtschaften*, Springer, New York, 2001.
33. Spatari, S., Bertram, M., Fuge, D., Fuse, K., Graedel, T.E., and Rechberger, H., The contemporary European copper cycle: 1 year stocks and flows, *Ecol. Econ.*, 42, 27, 2002.
34. Rechberger, H. and Graedel, T.E., The contemporary European copper cycle: statistical entropy analysis, *Ecol. Econ.*, 42, 59, 2002.
35. Rechberger, H., Entwicklung einer Methode zur Bewertung von Stoffbilanzen in der Abfallwirtschaft, Ph.D. thesis, Vienna University of Technology, Vienna, 1999.
36. Krauskopf, B., *Introduction to Geochemistry*, McGraw Hill, New York, 1967.
37. Graedel, T.E., Bertram, M., Fuse, K., Gordon, R.B., Rechberger, H., and Spatari, S., The construction of technological copper cycles, *Ecol. Econ.*, 42, 9, 2002.
38. Bertram, M., Rechberger, H., Spatari, S., and Graedel, T.E., The contemporary European copper cycle: waste management subsystem, *Ecol. Econ.*, 42, 43, 2002.
39. O'Rourke, D., Connelly, L., and Koshland, C.P., Industrial ecology: a critical review, *Int. J. Environ. Pollut.*, 6, 89, 1996.
40. Ayres, R.U. and Nair, I., Thermodynamics and economics, *Phys. Today*, 37, 62, 1984.

41. Stumm, W. and Davis, J., Kann Recycling die Umweltbeeinträchtigung vermindern? in *Recycling: Lösung der Umweltkrise?*, Gottlieb Duttweiler-Institut für wirtschaftliche und soziale Studien, Zurich, 1974.
42. Georgescu-Roegen, N., *The Entropy Law and the Economic Process*, Harvard University Press, Cambridge, MS, 1971.
43. Baccini, P., The landfill-reactor and final storage, *Lecture Notes in Earth Sciences*, Vol. 20, Springer, Berlin, 1989.
44. Ayres, R.U., Eco-thermodynamics: economics and the second law, *Ecol. Econ.*, 26, 189, 1998.
45. Ruth, M., Thermodynamic implications for natural resource extraction and technical change in U.S. copper mining, *Environ. Resour. Econ.*, 6, 187, 1995.
46. Ayres, R.U. and Martinás, K., Waste Potential Entropy: The Ultimate Ecotoxic, 94/05/EPS, Centre for the Management of Environmental Resources, INSEAD, Fontainbleau, France, 1994.
47. Brunner, P.H. and Rechberger, H., Anthropogenic metabolism and environmental legacies, in *Encyclopedia of Global Environmental Change: Causes and Consequences of Global Environmental Change*, Vol. 3, Douglas, I., Ed, Wiley, London, 2001, p. 54.
48. Schachermayer, E., Lahner, T., and Brunner, P.H., Assessment of two different separation techniques for building wastes, *Waste Manage. Res.*, 18, 16–24, 2000.
49. Brunner, P.H. and Stämpfli, D.M., Material balance of a construction waste sorting plant, *Waste Manage. Res.*, 11, 27–48, 1993.
50. Anon., Austrian Waste Management Act, Austrian Ministry for Family Affairs, Youth and the Environment, Vienna, 1990.
51. Brunner, P.H. and Ernst, W.R., Alternative methods for the analysis of municipal solid waste, *Waste Manage. Res.*, 4, 147, 1986.
52. Barghoorn, M., Dobberstein, J., Eder, G., Fuchs, J., and Goessele, P., Bundesweite Müllanalyse 1979–1980, Forschungsbericht 103 03 503, Technical University of Berlin, Berlin, 1980.
53. BUWAL, Abfallerhebung, Schriftenreihe Umweltschutz No. 27, Bundesamt für Umweltschutz, Bern, 1984.
54. Maystre, L.Y. and Viret, F., A goal-oriented characterization of urban waste, *Waste Manage. Res.*, 13, 207, 1995.
55. Yu, Ch.-Ch. and Maclaren, V., A comparison of two waste stream quantification and characterization methodologies, *Waste Manage. Res.*, 13, 343, 1995.
56. U.S. EPA, Municipal solid waste in the United States: 2000 facts and figures, EPA 530-R-02-001, Office of Solid Waste and Emergency Response, Washington, DC, June 2002.
57. Kampel, E., Packaging Glass and Chlorine Fluxes of Australia, Austria and Switzerland and their Assessment with Different Software Tools, Master's thesis, Vienna University of Technology, Vienna, 2002.
58. Brunner, P.H. and Mönch, H., The flux of metals through municipal solid waste incinerators, *Waste Manage. Res.*, 4, 105, 1986.
59. Reimann, D.O., Heavy metals in domestic refuse and their distribution in incinerator residues, *Waste Manage. Res.*, 7, 57, 1989.
60. Schachermayer, E., Bauer, G., Ritter, E., and Brunner, P.H., Messung der Güter- und Stoffbilanz einer Müllverbrennungsanlage, Monographien Bd. 56, Bundesministerium für Umwelt, Vienna, Austria, 1995.
61. Vehlow, J., Heavy metals in waste incineration, paper presented at DAKOFA Conference, Copenhagen, Denmark, 1993.

62. Belevi, H., Dank Spurenstoffen ein besseres Prozessverständnis in der Kehrrichtverbrennung, *EAWAG News*, 40D, EAWAG, Dübendorf, Switzerland, 1995.
63. Belevi, H. and Mönch, H., Factors determining the element behavior in municipal solid waste incinerators, part 1: field studies, *Environ. Sci. Technol.*, 34, 2501, 2000.
64. Morf, L.S., Brunner, P.H., and Spaun, S., Effect of operating conditions and input variations on the partitioning of metals in a municipal solid waste incinerator, *Waste Manage. Res.*, 18, 4, 2000.
65. Morf, L., Ritter, E., and Brunner, P.H., Güter- und Stoffbilanz der MVA Wels (Project Mape III), Vienna University of Technology, Institute for Water Quality and Waste Management, Vienna, 1997.
66. Bauer, G., Die Stoffflussanalyse von Prozessen der Abfallwirtschaft unter Berücksichtigung der Unsicherheit, Ph.D. thesis, Vienna University of Technology, Vienna, 1995.
67. Morf, L.S. and Brunner, P.H., The MSW incinerator as a monitoring tool for waste management, *Environ. Sci. Technol.*, 32, 1825, 1998.
68. Fehringer, R., Rechberger, H., Pesonen, H.-L., and Brunner, P.H., Auswirkungen unterschiedlicher Szenarien der thermischen Verwertung von Abfällen in Österreich (Project ASTRA), Vienna University of Technology, Institute for Water Quality and Waste Management, Vienna, 1997.
69. Austrian Landfill Ordinance, DVO, BGBl, No. 164/96, Austrian Ministry for Agriculture, Forestry, Water and Environment, Vienna, 1996.
70. Hackl, A. and Mauschitz, G., Emissionen aus Anlagen der österreichischen Zementindustrie II, Zement + Beton Handels- und Werbeges.m.b.H, Vienna, 1997.
71. Gangl, M., Gugele, B., Lichtblau, G., and Ritter, M., Luftschadstoff-Trends in Österreich 1980–2000, Federal Environmental Agency Ltd., Vienna, 2002.
72. Fehringer, R., Rechberger, H., and Brunner, P.H., Positivlisten in der Zementindustrie: Methoden und Ansätze (Project PRIZMA), Vereinigung der österreichischen Zementwerke (VÖZ), Vienna, 1999.
73. European Cement Association (CEMBUREAU), Alternative Fuels in Cement Manufacture: Technical and Environmental Review, CEMBUREAU, Brussels, 1999.
74. U.S. Geological Survey, Mineral Commodity Summaries, Cadmium 2001, http://minerals.usgs.gov/minerals/pubs/commodity/cadmium/.
75. U.S. Geological Survey, Minerals Yearbook, Cadmium 2001, http://minerals.usgs.gov/minerals/pubs/commodity/cadmium/.
76. Llewellyn, T.O., Cadmium (Materials Flow), Information Circular 9380, U.S. Bureau of Mines, Washington, DC, 1994.
77. Bergbäck, B., Johansson, K., and Mohlander, U., Urban metal flows — a case study of Stockholm, *Water Air Soil Pollut. Focus*, 1, 3, 2001.
78. Waalkes, M.P., Cadmium carcinogenesis in review, *J. Inorg. Biochem.*, 79, 241, 2000.
79. Heinrich, W., Carcinogenicity of cadmium — overview of experimental and epidemiological results and their influence on recommendations for maximum concentrations in the occupational area, in *Cadmium*, Stoeppler, M. and Piscator, M., Eds., Springer, Berlin, 1988.
80. Verougstraete, V., Lison, D., and Hotz, Ph., A systematic review of cytogenetic studies conducted in human populations exposed to cadmium compounds, *Mutation Res.*, 511, 15, 2002.
81. Brunner, P.H. and Mönch, H., The flux of metals through municipal solid waste incinerators, *Waste Manage. Res.*, 4, 105, 1986.
82. Schachermayer, E., Bauer, G., Ritter, E., and Brunner, P.H., Messung der Güter und Stoffbilanz einer Müllverbrennungsanlage (Projekt MAPE I), Monographie No. 56, Umweltbundesamt Wien, Vienna, Austria, 1995.

83. Stegemann, J.A., Caldwell, R.J., and Shi, C., Variability of field solidified waste, *J. Hazard. Mater.*, 52, 335, 1997.
84. Donald, J.R. and Pickles, C.A., Reduction of electric arc furnace dust with solid iron powder, *Can. Metall. Q.*, 35, 255, 1996.
85. Youcai, Z. and Stanforth, R., Integrated hydrometallurgical process for production of zinc from electric arc furnace dust in alkaline medium, *J. Hazard. Mater.*, B80, 223, 2000.
86. Jarupisitthorn, C., Pimtong, T., and Lothongkum, G., Investigation on kinetics of zinc leaching from electric arc furnace dust by sodium hydroxide, *Mater. Chem. Phys.*, 77, 531, 2002.
87. Xia, D.K. and Pickles, C.A., Microwave caustic leaching of electric arc furnace dust, *Miner. Eng.*, 13, 79, 2000.
88. Obernberger, I., Biedermann, F., Widmann, W., and Riedl, R., Concentrations of inorganic elements in biomass fuels and recovery in the different ash fractions, *Biomass Bioenergy*, 12, 211, 1997.
89. Baccini, P., von Steiger, B., and Piepke, G., Bodenbelastungen durch Stoffflüsse aus der Anthroposphäre, in *Die Nutzung des Bodens in der Schweiz*, Brassel, K.E. and Rotach, M.C., Eds., Zürcher Hochschulforum 11, Zurich, 1988.
90. Baccini, P. and Brunner, P.H., *Metabolism of the Anthroposphere*, Springer, New York, 1991.
91. Kelly, T., Buckingham, D., DiFrancesco, C., Porter, K., Goonan, T., Sznopek, J., Berry, C., and Crane, M., Historical Statistics for Mineral Commodities in the United States, U.S. Geological Survey Open-File Report 01-006, Version 5.9, http://minerals.usgs.gov/minerals/pubs/of01-006/.
92. Lahner, T., Steine und Erden, *Müll Magazin*, 7, 9, 1994.
93. Pitard, F.F., *Pierre Gy's Sampling Theory and Sampling Practice*, Vol. II, *Sampling Correctness and Sampling Practice*, CRC Press, Boca Raton, FL, 1989, p. 159.
94. Austrian Paper Industry, personal communication, 1996.
95. Brunner, P.H., Morf, l., and Rechberger, H., Thermal waste treatment — a necessary element for sustainable waste management, in *Solid Waste: Assessment, Monitoring, and Remediation*, Twardowsky, I., Allen, H.E., Kettrup, A.A.F., and Lacy, W.J., Eds., Elsevier, Amsterdam, 2003.
96. Scheffer, F., *Lehrbuch der Bodenkunde*, 12th ed., Schachtschabel, P., Blume, H.-P., Brümmer, G., Hartge, K.-H., and Schwertmann, U., Eds., Ferdinand Enke, Stuttgart, Germany, 1989, p. 305.
97. DKI Deutsches Kupferinstitut Kupfer, *Vorkommen, Gewinnung, Eigenschaften, Verarbeitung, Verwendung*, Informationsdruck Deutsches Kupferinstitut, Duesseldorf, 1997.
98. Zeltner, C., Bader, H.P., Scheidegger, R., and Baccini, P., Sustainable metal management exemplified by copper in the USA, *Reg. Environ. Change*, 1, 31, 1999.
99. Gordon, R.B., Production residues in copper technological cycles, *Resour. Conserv. Recycling*, 36, 87, 2002.
100. Fischer, T., Zur Untersuchung verschiedener methodischer Ansätze zur Bestimmung entnommener mineralischer Baurohstoffmengen am Beispiel des Aufbaus von Wien, M.S. thesis, Vienna University of Technology, Vienna, 1999.
101. Fehringer, R. and Brunner, P.H., Kunststoffflüsse und die Möglichkeiten der Verwertung von Kunststoffen in Österreich, UBA Monographien Band 80, Umweltbundesamt, A-1090 Wien, 1996.

4 Outlook: Where to Go?

As indicated by the case studies presented in the previous chapters, the systematic investigation into material flows and stocks of anthropogenic systems allows a new view of the anthroposphere. MFA is a key discipline for this new view because it links anthropogenic activities with resource consumption and environmental loadings. If the important flows and stocks of materials used by humans are uniformly analyzed by MFA as described in this handbook, and if the information obtained through many individual MFAs is linked to large databases on material sources, pathways, intermediate stocks, and final sinks, then a more efficient use of resources will become possible. Eventually, if MFA is applied (1) on different cultural systems and economies with different metabolisms and (2) on various levels such as primary and secondary production sectors, the service sector, consumers, and governments, then completely new schemes of utilizing resources may become feasible. This chapter points out some of the fascinating future potentials associated with MFA.

4.1 VISION OF MFA

The development of MFA is driven by two ideas:[1]

1. That materials accounting will become a standard analytical procedure for enterprises, regions, and nations. The goals are (1) to save resources by conserving materials and energy and (2) to produce less waste and minimize environmental loadings. These objectives are achieved by analyzing, evaluating, and controlling the flows and stocks of goods and substances within the anthroposphere. As in financial accounting, stakeholders who account for materials have a comparative advantage over those who do not use MFA and materials accounting. Hence, as the actors within modern societies progressively adopt and apply MFA, resource conservation and environmental protection will be accomplished in an efficient way.
2. That MFA will become a major design tool to optimize the metabolism of the anthroposphere. MFA is instrumental in explaining the metabolic processes of complex cultural systems that have developed over time. Furthermore, this tool will support the development of new urban systems by informing decisions concerning the design of more efficient processes that optimize the consumption of energy and materials. MFA will be essential for the design of new goods and the choice of new materials, and it will be employed as a tool for early recognition of human impacts

on the environment. Thus, MFA will be instrumental for the continuous improvement of anthropogenic systems by supporting the choice and design of the set of processes and materials that are necessary to sustain human activities.

How far are these visions from today's reality? Much methodological development has been undertaken, particularly in modeling flows and stocks, and in the field of evaluation of MFA results. MFA is applied in diverse areas such as industrial ecology, environmental management and protection, resource management, and waste management. It is a basis for life-cycle assessments, ecobalancing, environmental impact statements, and waste-management concepts. Several groups around the world routinely apply MFA for these tasks. Much data have been collected in studies ranging from polyvinyl chloride accounting, to heavy metal inventories, to nutrient balances of large watersheds.

By linking mining, production, consumption, and waste management, these first studies have demonstrated the utility of MFA in making the transition from waste management to resource management. Recent MFA works point to waste-management measures that are inefficient and have helped in identifying those measures within the total economy that are more efficient and goal oriented. Thus, MFA will play an important role in reaching a new, advanced level in waste management and in defining the limits of waste management, beyond which the measures in other economic sectors will become more effective and powerful.

However, MFA has yet to achieve a major breakthrough on the level of production, manufacturing, trade, commerce, or at the institutional level. It is not used as a standard analytical tool in everyday decisions on materials management. While the chemical industry has always practiced mass balancing of chemical reactants as a standard procedure, most other enterprises so far do not see an advantage in using this tool. Nevertheless, MFA is being applied in some other sectors, proving that this tool is also well suited for environmental and resource management on the corporate level. However, because the accounting of substances is a new, laborious, and costly task, so far most companies do not see an economic advantage in applying MFA to their enterprises. Hence, there is still a long way to go until the various sectors of society come close to the routine use of MFA as an analytical tool, as envisioned above.

At the beginning of the 21st century, it appears that the motivation to use technologies and systems that conserve resources is still comparatively small. Economic signals that indicate immediate resource scarcities do not exist. The warnings presented by The Club of Rome in *The Limits to Growth*[2] seem to be falsified by reality; instead of running out of resources at the turn of the 20th century, resources have never been as cheap. New studies even suggest "growth without limits."[3] On the other hand, many MFA studies (see Chapter 3) point to severe deficiencies of the current industrial economy in terms of environmental impact and inefficient use of resources. The metabolism of the anthroposphere relies on fossil fuels that will become scarce in the distant future; the specific energy demand is high. Many materials are dispersed after use as off-products, and some of them are overstraining the carrying capacity of water, air, and soil. Detrimental impacts have been observed on the regional as well as on the global level. The hypothesis that the limits to growth appear first at the back end of the anthropogenic system has not been falsified yet. On the contrary, problems like global accumulation of dichlorodiphenyltrichloroethane (DDT) and polychlorinated biphenyls (PCB), ozone-layer deterioration due to chlorofluorocarbons (CFC), and climate change (in part due to anthropogenic carbon emissions) all support this hypothesis. Natural reservoirs such as soils, lakes, and the atmosphere, which serve as sinks, have limited carrying capacities. They have to be regarded as finite natural resources that must be conserved. Applying the materials-balance principle to the anthroposphere, it becomes clear that safe final sinks are needed for all materials used for industrial and consumption activities. It is a major task of the coming decades to bring sources and sinks together; the materials exploited from the Earth's crust must match the capacity of the final sinks. A common concept is needed:

1. To control material flows in a way that leads materials to appropriate and proven final sinks
2. To monitor the grade of exploitation of the carrying capacity of final sinks
3. To gradually phase out those substances that have no such sink

In the past as well as today, the metabolism of the anthroposphere is constantly improved and adapted to changing boundary conditions by many stakeholders of various disciplines. The community of research groups and organizations working for the increase in resource productivity, for environmental management, for the decrease of ecological footprints, for sustainable development, and for a suitable definition of progress is constantly growing. The result may be a significantly different use of resources, namely energy, materials, and information. The last is an important resource that has not yet been specifically addressed in this book. Information includes all knowledge of a society, be it in science, technology, art, education, warfare, or administration. Without information, the resource "material" is a resource without value. Societies as we know them today depend highly on their stock of information and its continuous expansion. The rise and fall of cultures such as the Roman Empire in Europe, the Egyptian dynasties in Africa, or the Mayas in the Americas was always accompanied by a tremendous development and subsequent loss in the information resource. Today, looked at from a global perspective, this resource stands at a very high level. The so-called global village embraces all cultures and thus is the greatest pool of information that ever existed on this globe. All our complex systems (supply of energy, water, food, and construction materials; transport of goods, persons, and information; etc.) are large assemblies of information and know-how, and they could not function without a comprehensive information base.

MFA is an important means of identifying, collecting, and supplying information about the "material world." MFA links the resource "information" with the resource "materials" and is instrumental in better understanding, designing, and controlling materials within the anthroposphere. At present, it is often difficult to find good information and data, in particular about regional issues. It is necessary to develop a new set of tools for data acquisition about historical and present flows and stocks of materials, such as data reconstruction, back channeling, and the use of redundancy to investigate regional systems.[4] In the future, MFA should also become an important tool for the advancement of information resources.

What can be done to foster a broader use of MFA and thus come closer to achieving the objectives mentioned above? A whole array of measures can promote future application of MFA. The first step is to foster a general awareness of the importance of materials as resources, as environmental loadings, and as constituents of final sinks. Integration of MFA into the academic curriculum and research agenda will promote broader application and further development of MFA. A further requirement is standardization of MFA methods and the development of user-friendly software. Linkage of MFA to other disciplines is also necessary, e.g., the combined analysis and illustration of materials and financial flows, known as economically extended MFA (EE-MFA).[5,6] This will stimulate further development of MFA methodologies. And finally, there is a need to incorporate MFA into selected parts of legislation related to the environment, wastes, and resources.

4.2 STANDARDIZATION

Standardization is seen as one requirement to establish MFA as an ordinary tool in materials-management decisions. An ultimate goal could be the development of a publicly available common materials database that (1) includes information about the most important anthropogenic processes, flows, and stocks of goods and substances and (2) is constantly updated by newly available information. If such a database is standardized, individual interpretative material flow accounting tools and their evaluation methods (ecological footprints, sustainable process index [SPI], life-cycle assessments [LCA], etc.) can be attached, too.

The prerequisites for successful standardization are as follows:

1. National and international bodies must define terms and definitions in a uniform way.
2. Students and professionals must be taught and trained in standardized MFA methodology (terms and procedures), thus creating a new generation of professionals familiar with MFA. This group would comprise civil engineers, process engineers, architects, urban planners, landscape planners, environmental engineers, resource managers, waste-management experts, product designers, and others. These experts will design the goods, buildings, and infrastructure of the future anthroposphere.
3. Tailor-made software (including statistical tools) must be developed to support the application of MFA and to facilitate modeling of metabolic systems. This will advance active application in academia, in consulting, in engineering, in industry, and by government authorities and agencies. The software should include interfaces for links to MFA databases.
4. Professionally developed and maintained MFA databases must be created. These databases should be available worldwide and contain information about specific generation rates and compositions of materials, transfer coefficients of technologies, and other relevant data. It is indispensable that the data be collected, processed, organized, and stored uniformly so

that it can be easily linked to commercial MFA software and utilized to solve problems in environmental, resource, and waste management. These databases should be accessible to professionals as well as the public at affordable costs. Currently, there are no such databases and no concepts for how, where, and by whom such databases can be established.

A drawback of standardization might be that progress and development are slowed down. Once a method is standardized, most experts use the same methodology; the variety of approaches decreases rapidly; and users are less likely to have new and creative ideas. The history of science and technology, as well as of fine arts and music, shows that standardization and rules are crucial for all disciplines. At the same time, standardization has never prevented talented people from further developing their fields by extending or breaking rules and routines. It seems timely to standardize the application of MFA, even if some negative effects cannot be ruled out.

4.3 MFA AND LEGISLATION

If legislation in the fields of environmental management, resource management, and waste management is based on MFA, decisions and measures are likely to be more cost effective. Especially for goal-oriented legislation, MFA in combination with social science methods such as cost-benefit or cost-efficiency analysis can serve as a powerful tool to analyze whether the objectives of regulations are met. Such combined approaches have been used in the past, e.g., to analyze quantitatively how far the objectives of waste management can be reached by several management scenarios.[7]

Banning hazardous substances by legislation requires MFA-based design and control. When the stocks and flows of a specific substance are known, the effects of the ban can be predicted by scenario analysis. The modeled development for the substance can then be compared with actual measurements. If there is a relevant difference, decision makers have the option of interfering and correcting again. For example, a stop in the production of a substance may not provide instant relief for the environment because of stocks that serve as reservoirs for continuing emissions. If the presence of those stocks is not accounted for in the model, the effectiveness of actions may be overrated. One example for such a case is CFCs, where foams used in construction represent a stock that is responsible for ongoing CFC emissions despite the total ban on production.[8] Besides predicting the decline of substance flows and stocks, a complete MFA also helps to identify appropriate points for monitoring the trend of a banned substance. Monitoring points should facilitate analysis of the relevant flows and guarantee that the information processed is significant. Such an approach maximizes the accuracy and cost-effectiveness of the monitoring process. The same approach also applies to other legal regulations such as taxes on energy from fossil fuels, nonrenewable material resources (e.g., copper, gravel), or hazardous substances (e.g., cadmium).

Regional and national waste-management plans can be improved when based on MFA. Today, these plans focus on the treatment and disposal of single waste streams that are defined by the way they are collected (e.g., separate collection of paper or industrial wastes, mixed municipal solid waste [MSW] collection) or by the functions of the goods that have become wastes (packaging function). In future, the collection and management of wastes should focus more on the substances contained in individual wastes; i.e., it is more efficient to collect polyethylene wastes or cellulose wastes than to collect packaging plastics or packaging paper.

Investigations have also shown that considerable amounts of both valuable and hazardous substances (e.g., zinc, copper, mercury, cadmium) are collected in waste streams. With the volume of waste streams growing steadily — particularly with construction and demolition debris, as well as wastes from electronic equipment — this potential will become increasingly important in the future. MFA supplies substance-based information for waste-management decisions. This is important because the main objectives of waste management — protection of human health and the environment as well as resource conservation — are mainly substance-oriented and not goods-oriented objectives. Waste-management strategies should focus primarily on the levels of substances and only secondarily on goods. This approach facilitates the identification of appropriate recycling and treatment technologies and appropriate final sinks for residues. Information on substances is also useful in preventing the mixing of contaminated and clean wastes and in deciding which wastes and substances are to be collected separately.

Such MFA-based developments are beginning to emerge at the political and administrative level. The following two examples from Germany and Austria are the first encouraging signs of how MFA and materials accounting can become routine instruments for planning of effective waste management.

Since 1999, the German province of North Rhine Westphalia has required an MFA before permitting the operation of new waste-treatment facilities. A local ordinance stipulates that a license for the construction and operation of new waste-treatment plants is contingent upon a complete material flow analysis of the total plant. The MFA must account for all input and output goods and concentrations of relevant substances in these goods. Finally, the prospective facility must establish a material balance with transfer coefficients for selected substances. To support the implementation of this ordinance, government authorities have prepared a database containing waste compositions, emissions data, and transfer coefficients for selected waste-treatment processes.[9] The purpose of this ordinance is to ensure that decision makers have access to the same level of information for all of the processes that are taken into account in waste management. So far, information about state-of-the-art MSW incinerators has been much more abundant than on any other thermal, mechanical, and biochemical process, making it an uneven base for decisions regarding the selection of alternative technologies.

Another example of new advances at the administrative level is the 2001 edition of Austria's federal waste-management plan,[10] which contains several models showing how waste management can be improved by the use of MFA. The plan also identifies strategies for gradually implementing MFA as a tool to optimize goal-oriented waste management. MFA is included in the waste-management plan because MFA-based studies on packaging materials, plastic waste management,[11] and scenario evaluation[6] have considerably improved the Austrian packaging ordinance and the design of new waste-management concepts.

4.4 RESOURCE-ORIENTED METABOLISM OF THE ANTHROPOSPHERE

Due to innovation in technology and efforts to reduce costs, the substances, goods, processes, and systems necessary for the metabolism of the anthroposphere are constantly changing. From a materials point of view, this transformation, at present, happens more or less at random. There are no clear objectives yet for the choice and use of materials. If long-term efficient utilization of resources is set as a goal, MFA can serve as an important tool to support decisions regarding the reconstruction, design, and maintenance of the anthroposphere. It should be kept in mind that anthropogenic systems for infrastructure, transportation, communications, etc. have lifetimes of 25 to 100 years. Hence, within this time range, new systems have to replace old ones. Sections 4.4.1 through 4.4.3 outline a three-step procedure showing how MFA can be used in a comprehensive way to plan and design the anthroposphere of tomorrow. The result can be new ways of managing materials. The purpose of this section is not so much to predict the future (which is impossible), but to stimulate readers to develop their own thoughts about the potential of MFA to create new strategies for resource management.

4.4.1 MFA AS A TOOL FOR THE DESIGN OF PRODUCTS AND PROCESSES

First, MFA should become a standard tool for the design of products and processes. It can support industry, consumers, government, and public-interest groups as they strive for better goods and services. MFA should be used for environmental-impact assessments, life-cycle assessments, and other tools for environmental management. Decisions about the choice of waste-management scenarios, about nutrient management, and about management of metals and other resources should also be based on MFA. This will improve decisions by engineers, designers, communities, and government authorities. The goal is to create a "critical mass" of users and applied case studies, and to show the advantages of using MFA for the design process. An increase in the number of users, along with communication among users, will stimulate and promote application of MFA in this and neighboring fields. Readily available, user-friendly, and well-supported MFA software will further advance the use of MFA.

4.4.2 MFA AS A TOOL TO DESIGN ANTHROPOGENIC SYSTEMS

In a second step, knowledge and information gained by these individual applications of MFA are collected and evaluated using a systemic approach. Based on this new know-how, future design takes into consideration the whole system that is influenced by a good, a process, or a service. Material balances become a requirement for goal-oriented design and management of larger systems, such as regional transport, supply, and disposal systems. A uniform census and processing of data facilitate the establishment of these larger and more complex systems. Materials accounting is introduced on regional and national levels. National material flow and stock-accounting indicators are used for regular reporting, and the results provide a basis for developing policy. For selected substances, materials accounting is also implemented at the industrial level. Companies collect data and use the information to optimize their performance. Public as well as private databases are established. All of these developments abet the transition from a system that relies mainly on end-of-pipe measures to a resource-oriented materials-management system. The control of material flows to the environment by end-of-pipe measures is then supplemented by a more comprehensive approach focusing on the optimization of the total anthropogenic resource flows.

4.4.3 MFA as a Tool to Design the Anthropogenic Metabolism

The third level is the design of the total anthropogenic metabolism based on MFA. The objective is to control the total material flows and stocks in pursuit of resource conservation and long-term environmental protection. Data are collected, organized, processed, and evaluated in a consistent way. The puzzle of the anthropogenic metabolism is systematically put together step by step. Comparative metabolic studies are performed in an organized way, and comprehensive databases are set up for complex systems and for many materials.

On the third level, primary ore mining is replaced by secondary "urban mining."[12] The amounts of some resources in the urban stocks are of the same order of magnitude as those found in natural deposits. Hence, materials supply for urban systems must be shifted from primary resources, which are extracted from natural deposits, to secondary resources, so-called urban ores, recovered from the urban stock itself. New methodologies for mining the urban stock need to be developed, used, and maintained with an eye toward low consumption of energy and primary materials. First approaches for restructuring urban systems have been developed by Baccini, Oswald, and colleagues.[13–15] In their study, they link architectural and social structures and metabolic processes of urban regions and develop new strategies for resource-oriented cities. The result of their interdisciplinary combination of MFA with architecture, planning, and socioeconomic issues is a product called "network city," a restructured urban region with a more sustainable metabolism.

The demand of urban systems for new virgin resources has to be covered mainly by renewable materials. Dissipation of materials while being processed, used, recycled, or disposed of is controlled and reduced to acceptable long-term levels. Acceptable refers to environmental *and* resource issues. Materials are used in a cyclic manner if resulting environmental loadings are smaller, energy requirements are low, no further substance needs arise, and economic benefits exist or can be created. For all losses from cycles as well as linear use, appropriate final sinks must be found or designed. If they are not available, the losses must be minimized further to meet the capacity of the available sinks, or the substances have to be phased out. Optimized materials management prevents long-term, harmful accumulation of hazardous substances in environmental compartments and averts depletion of finite resources within a few generations.

Many substances will no longer be sold; instead, they will be leased to manufacturers and end users, ensuring minimum dissipation due to the strong interest of the substance owner not to lose his stock. This will stimulate the development of new products with better resource performance (product life cycles, logistic systems). New objectives for the design process arise: minimize waste, emissions, and wear during use; and enable easy disassembly for reuse of components or substances after consumption, thus facilitating refurbishment and reuse. These goals are achieved by standardizing materials, avoiding complex composite materials, minimizing the use of hazardous materials, reducing the number of substances incorporated in a product, and using mainly recyclable or abundant substances.

Action is possible at all levels and scales, with new products, new processes, new systems, and new structures applied at regional, national, and international scales. The regional level is especially attractive because it is well suited for trying out new strategies and concepts to gain experience. Restructuring of the anthroposphere has an impact on lifestyle. Vice versa, lifestyle has a controlling effect on material flows and stocks. Future development of consumer lifestyle and expansion of a customer base that is motivated to choose a resource-conserving lifestyle will be crucial factors in future materials management. As Steindl-Rast and Lebell[16] suggest in their introduction, there are several ways to deal with goods: they can be taken in and given out in balance, leading to wealth and happiness; or they can be accumulated only, leading to hostile and deteriorating conditions. The old question of what is important — to be or to have — deserves new answers in view of resource management. New concepts are wanted that show how to improve the quality of life by consuming less energy and fewer resources. Eventually, the focus is shifted from material growth to nonmaterial welfare and development. MFA is a necessary element for such a change in paradigm. Nevertheless, it is only a tool and not a driving force. It is up to the user to apply MFA in a beneficial way.

REFERENCES

1. Brunner, P.H., Materials flow analysis — vision and reality, *J. Ind. Ecol.*, 5, 3, 2001.
2. Meadows, D.H., Meadows, D.L., Randers, J., and Behrens, III, W.W., *The Limits to Growth*, Earth Island, London, 1972.
3. Becker-Boost, E. and Fiala, E., *Wachstum ohne Grenzen*, Springer, New York, 2001.
4. Eklund, M., Reconstruction of historical metal emissions and their dispersion in the environment, Ph.D. thesis, Linköping University, Linköping Studies in Arts and Science, 127, 1995.
5. Kytzia, S., Wie kann man Stoffhaushaltssysteme mit ökonomischen Daten verknüpfen? in *Ressourcen im Bau*, Lichtensteiger, Th., Ed., vdf Hochschulverlag AG, ETH, Zurich, Switzerland, 1998, chap. 5.
6. Kytzia, S. and Faist, M., Joining economic and engineering perspectives in an integrated assessment of economic systems, in Proceedings of the International Conference on Input-Output Techniques, Montreal, Oct. 2002.
7. Brunner, P.H., Döberl, G., Eder, M., Frühwirth, W., Huber, R., Hutterer, H., Pierrard, R., Schönbäck, W., and Wöginger, H., Bewertung abfallwirtschaftlicher Maßnahmen mit dem Ziel der nachsorgefreien Deponie (Projekt BEWEND), Monographien des UBA Band 149, Umweltbundesamt GmbH, Vienna, 2001.
8. Obernosterer, R., Flüchtige Halogenkohlenwasserstoffe FCKW, CKW, Halone — Stoffflußanalyse Österreich. Masters thesis, Vienna University of Technology, Vienna, 1994.

9. Alwast, H., Koepp, M., Thörner, T., and Marton, C., Abfallverwertung in Industrieanlagen, Ministerium für Umwelt und Naturschutz, Landwirtschaft und Verbraucherschutz des Landes Nordrhein-Westfalen, Berichte für die Umwelt, Bereich Abfall, Band 7, Düsseldorf, Germany, 2001.
10. Bundesministerium für Land- und Forstwirtschaft, Umwelt und Wasserwirtschaft, Bundes-Abfallwirtschaftsplan — Bundesabfallbericht 2001, Vienna, 2001.
11. Fehringer, R., Flows of plastics and their possible reuse in Austria, in *From Paradigm to Practice of Sustainability*, Proceedings of the ConAccount Workshop, Leiden, The Netherlands, Jan. 1997.
12. Obernosterer, R. and Brunner, P.H., Urban metal management: the example of lead, *Water Air Soil Pollut. Focus*, 1, 241, 2001.
13. Baccini, P., Kytzia, S., and Oswald, F., Restructuring urban systems, in *Future Cities: Dynamics and Sustainability*, Moavenzadeh, F., Hanaki, K., and Baccini, P., Eds., Kluwer Academic Publishers, Dordrecht, The Netherlands, 2002, chap. 1.
14. Baccini, P. and Oswald, F., *Netzstadt: Transdiziplinäre Methoden zum Umbau urbaner Systeme*, 2nd ed., vdf Hochschulverlag AG, ETH, Zurich, Switzerland, 1999.
15. Oswald, F. and Baccini, P., *Netzstadt — Designing the Urban*, Birkhäuser, Basel, Switzerland, 2003.
16. Steindl-Rast, D. and Lebell, S., *Music of Silence*, Seastone, Berkeley, CA, 2002.

Index

A

Accounting, materials, 51
Accuracy, software, 82
Activities, 4, 44–48
Actual vs. optimized scenarios, 276–278
Agriculture
    consumer lifestyle and nutrient loss, 219
    as source of lead, 174–175, 176, 182
    as source of nutrient pollution, 197
    as source of phosphorus, 191
Air pollution control devices, 288
Anthropogenic metabolism, 19–28
    consumption vs. production emissions, 25–27
    growth and, 19–22
    linear urban material flows, 23–24
    MFA in design of, 308–309
    stocks accumulation, 25
    surpassing geogenic, 22–23
    waste quantity and composition, 28
Anthropogenic nitrogen solution, 220
Anthropogenic systems, 3
    MFA in design of, 307–308
Anthropogenic vs. geogenic flows, 147–149
Anthroposphere, 48–49
    compartments of, 48
    outlook for, 307–309
Application, 13–28. *See also* Case studies; *subtopics*
    anthropogenic metabolism, 19–28
    environmental management and engineering, 14
    industrial ecology, 14–16
    resource management, 16–17
    waste management, 17–19
Area calculations, 138–139
Ash, cadmium recycling from, 287–290
Ash landfills, 203–204
ASTRA case study, 269–278
Atmophile vs. lithophile elements, 56
Austria. *see also* Construction waste management
    ASTRA case study, 269–278
    construction materials used in, 238
    Graz University of Technology sustainable process index, 137–139
    paper flows, 262
    plastic waste management, 251–256
    PRIZMA case study, 279–287
    software case study, 83–85
    Vienna Institute of Technology statistical entropy analysis, 149–159
Austrian Waste Management Act, 253, 270, 306
Automatization, software, 82
Avogadro's number, 153

B

Balance
    mass of goods, 50
    material, 49
Biosphere. *see* Anthroposphere
Boundaries
    establishment of, 172–173
    geographic, 58
    spatial, 43
    temporal, 43, 58
Brussels, ecosystem analysis, 9
Building construction, 46. *See also* Construction wastes management

C

Cadmium
    in fly ash, 159
    increase in flow of, 23
    in landfills, 289–290
    recycling, 287–290
    in waste fuels, 283–287
Carbon, 203–204
    coal-fired power plants as source, 207
    landfill management and, 269–270
Case studies
    environmental management, 168–214
        environmental impact assessment (SYSTOK study/coal-fired power plants), 200–213

nutrient pollution of watersheds (Danube Basin), 194–200
regional lead pollution (RESUB/Bunz Valley study), 168–184
regional phosphorus management (RESUB/Bunz Valley study), 184–194
resource conservation, 215–216, 215–256
    construction wastes management, 225–251
    copper management, 220–235
    nutrient management, 215–220
    plastic waste management, 251–256
waste management, 256–292
    cadmium recycling, 287–290
    decision making, 269–290
    regional lead management, 292–295
    waste analysis, 258–269
Cellulose, 261–262
Cement kilns
    ASTRA case study, 269–278
    PRIZMA case study, 279–287
Cement manufacturing process, 279–283
Center for Industrial Ecology (Yale U.), 227
Chemical elements, 36. *See also* Substances
Chlorine
    waste analysis, 263
    in waste fuels, 283–287
Chlorofluorocarbons (CFCs), 303
Chromium, Swedish consumption vs. production emissions, 26
Clean Air Act, 54, 203
Cleaning, 44–46
Clinker manufacturing, 284–285
Closed loops, 15
Club of Rome, 221, 301–302
Coal-fired power plants
    regions of impact, 207–210
    SYSTOK study, 200–213
Collection vs. concentrating processes, 150–151
Colonization, 12
Combustible wastes, 269, 270, 273. *See also* Incineration
Commodity. *See* Goods
Communicating, 47
Compartments, anthroposphere, 48
Compatibility, software, 82
Composting, as source of phosphorus, 193
Compounds, 36. *See also* Substances
Concentrating vs. collection processes, 150–151
Conservative substances, 36
Construction materials, increase in flow of, 21
Construction materials balance ("hole" problem), 236–239
Construction waste management, 225–251

comparison of sorting technologies, 239
composition of separation products, 244–246
composition of wastes, 242–244
"hole" problem, 235–239
mass flow of separation products, 244
partitioning of metals/transfer coefficients, 246–250
procedures, 239–242
selective deconstruction, 250–251
Consumer lifestyle, nutrient loss and, 219
Consumption, consumer product, 21
Consumption vs. production emissions, 25–27
Copper
    as nutrient, 220–235
    recycling of, 232–233
    relative statistical entropy, 228–232
Copper management, 220–235
Cost benefit, of software, 82
Cost–benefit analysis, 145–147
    modified, 146
Cycles, 43

D

Danube Basin, nutrient pollution of watersheds, 194–200
Database development, 304–305
Data uncertainties, 69–80. *See also* Uncertainty
Decision making
    ASTRA case study, 269–278
    PRIZMA case study, 279–287
Deconstruction, selective, 250–251
Dematerialization, 15, 137
De Medicina Statica Aphoroismi (Santorio), 7
Denitrification, 196
Denmark, Kalundborg as industrial ecosystem, 16
Design
    of anthropogenic metabolism, 308–309
    of anthropogenic systems, 307–308
    MFA's role in product and process, 307
Dichlorodiphenyltrichloroethane (DDT), 303
Documentation, software, 80
Dry-separation vs. wet-separation methods, 243, 247
Dupuit, J., 145

E

Early recognition, of environmental hazards, 180–182, 192–193
Ecoefficiency, 137
Economic cycles, construction waste and, 243–244

Index

Ecosystems, industrial, 16
Efficiency, of energy content utilization, 272
Electrical power generation (SYSTOK study), 200–213
Electricity, area for consumption, 138–139
Elements
    atmophile vs. lithophile, 56
    indicator, 56, 57
Emissions. *See also specific components*
    calculation of regional, 210–211
    from coal-fired power plants, 210–213
    consumption vs. production, 25–27
    exergy as measurement, 143–144
    heavy metal, 210–213
    statistical entropy analysis, 222
Energy content utilization, efficiency equation, 272
Entropy, 149–159. *See also* Statistical entropy analysis
Entropy trends, 227–235
    recycling and, 232–233
Environmental engineering, 14
Environmental hazards, early recognition, 180–182, 192–193
Environmental impact assessment (SYSTOK study/coal-fired power plants), 200–213
Environmental management, 14, 168–214
    environmental impact assessment (SYSTOK study/coal-fired power plants), 200–213
    nutrient pollution of watersheds (Danube Basin), 194–200
    regional lead pollution (RESUB/Bunz Valley study), 168–184
    regional phosphorus management (RESUB/Bunz Valley study), 184–194
    resource conservation, 215–216
    software, 83–85
Environmental monitoring. *See* Monitoring
Environmental protection, 192
Environmental Protection Agency (EPA), 259
Environmental stressors, 141–142
Erosion, 196
Evaluation, 133–159
    cost–benefit analysis, 145–147
        modified, 146
    exergy, 142–145
    life-cycle assessment (LCA), 140–141
    material intensity per service-unit, 136–137
    objectivity vs. subjectivity, 133–135
    single vs. multiple systems, 134
    statistical entropy analysis, 149–159. *See also* Statistical entropy analysis

sustainable process index (SPI), 137–139
Swiss ecopoints, 141–142, 272
Exergy, 142–145
Extensive and intensive properties, 143

F

Factor-X debate, 137
Fate of substances, 272–273
Fe, standard chemical energy, 142–143
Feces, as source of N and P, 219–220
Flexibility, software, 82
Flows, 39–40
    anthropogenic vs. geogenic, 147–149
    defined, 4
    examples, 41
    key, 65–66
    lead, 292–294
    of plant production, 191
    specific, 40
    total, 40, 61–63, 63–64
Fluxes, 39–40
    defined, 4
    examples, 41
Fly ash, 288–289
Food. *See* Agriculture; nutrient management
Fossil raw materials, area for, 138
Fuels, waste, 283–286. *See also* ASTRA study; PRIZMA study

G

GaBi software, 113–133
    potential problems, 132
    program description, 113–114
    quickstart, 114–132
Gasoline, lead in, 295
Geochemistry, 56
Geographic boundaries, 58
Gibbs's energy of formation, 143
Glass, waste analysis, 262–263
Global warming, 142, 303
Goods
    defined, 36–37
    element concentration in, 55
    mass balance of, 50
    material vs. immaterial, 4
    transformation, 38
    transport, 38
Goods/EC ratio, 55
Graz University of Technology sustainable process index, 137–139
Greenhouse effect, 142, 303

Greenockite, 287. *See also* Cadmium
Growth, 19–22
 in consumption, 19, 20

H

Hazardous substances, 18, 36. *See also specific substances*
 growth and, 19–22
 policy and legislation issues, 305
 zinc as, 26–27
Heavy metals. *See also specific metals*
 coal-fired power plants as source, 207, 210–213
 partitioning from construction wastes, 246–250
 in power plant ash, 203–204
 regional emissions calculation, 210–211
 RESUB/Bunz valley regional lead pollution study, 168–184
 significance of, 220–221
 sustainability and, 221–222
 in waste fuels, 283–287
History, 5–13
 anthroposphere metabolism, 12–13
 Leontif's economic input–output methodology, 5, 8
 metabolism of cities, 8–9
 regional material balances, 9–12
"Hole" problem, 235–239
Homogeneity, 66
Hong Kong, metabolism, 9
Households. *See also* Municipal solid waste
 food consumed per capita, 216
 as lead pollution sources, 169–171, 172
 nitrogen and phosphorus flow per capita, 219
 as source of phosphorus, 186
Hudson-Raritan Basin, regional study, 10–11
Human metabolism, 5–7
 information sources, 217
Hydrological balance, 175, 185–186, 189

I

Incineration, 150–151
 lead residues, 295
 transfer coefficients, 275
 waste analysis, 264–269
Indicator elements, 56, 57
Industrial ecology, 14–16
Industrial ecosystems, 16
Industrial metabolism, 15
Industry
 nutrient loss through processing, 219
 as source of lead, 177–178, 179
 as source of phosphorus, 190
Information, importance of, 303–304
Inputs defined, 4
Institute for Polymer Testing and Polymer Science (U. Stuttgart), 113–114. *See also* GaBi software
Intensive and extensive properties, 143
International Institute for Applied Systems Analysis (IISA), 11

J

Journal of Industrial Ecology, 15

K

Kalundborg, Denmark, as industrial ecosystem, 16
Key processes, flows, and stocks, 65–66

L

Land conservation analysis (LCA), 17
Landfills, 17, 18
 areas for investigation, 182–183
 cadmium in, 289–290
 carbon content and, 269–270
 "hole" problem and, 238
 plastic wastes in, 253
 for power plant ash, 203–204
 precautionary principle and, 271–272
 recreational use, 238–239
 as source of lead, 178–180
 as source of phosphorus, 190
 water contamination by, 178–180
Lavoisier, Antoine, 5
Lead
 in gasoline, 295
 increased in flow of, 19, 22
 industry as source, 177–178, 179
 private households as sources, 169–171, 172
 problems related to, 294–295
 regional management, 292–295
 RESUB/Bunz valley regional study, 168–184
 in sewage sludge, 295
 sewer systems as source, 175, 177
 soils as source, 174–175
 surface waters as source, 178–180
 urban areas as source, 176
 in waste fuels, 283–287
 wastewater treatment plant as source, 171–173
Lead flows, 292–294

Index

Lead stocks, 292–294
Legislation, 305–306. *See also specific laws and regulations*
Leontief, Wassily W., 8
Leontif's economic input-output methodology, 5, 8
Life-cycle assessment (LCA), 140–141
Life cycle inventory, 133
Limits of Growth, The (Club of Rome), 221, 302
Loops, closed, 15
Los Angeles, regional study, 9–12

M

Market product analysis, in waste management, 258–259
Mass, of heavy metals vs. properties, 220–221
Mass–balance principle, 59, 61–62
Mass flows, 59–60
    from coal-fired power plants, 204–207
Mass input/output balance, 242
Material
    defined, 3–4, 37
    fluxes and flows (inputs), 4
    renewable raw, 138
Material balance, 49
Material flow analysis, defined, 49–51
Material flow analysis (MFA). *See also subtopics*
    application, 13–28
    bulk, 51
    concept, 305
    data uncertainties, 69–80
    defined, 49–51
    evaluation methods for results, 133–159
    history, 5–13
    legislation and policy issues, 305–306
    objectives, 28
    procedures, 53–69
    in product and process design, 307
    scale, independence of, 194
    software, 80–133
    standardization, 304–305
    symbols, 38
    terms, 3–4, 35–53
    vision driving concept, 301–304
Materials accounting, 51, 64–67
    initial MFA, 64–65
    key processes, flows, and stocks, 65–66
    mean substance concentrations, 67
    resource assessment, 88
Material vs. immaterial goods, 4
Mean substance concentration, 67
Merchandise. *See* Goods

Mercury
    coal-fired power plants as source, 205–207, 212–213
    in waste fuels, 283–287
Metabolism, 48–49
    anthropogenic, 19–28. *See also* Anthropogenic metabolism
    of anthrosphere, 12–13
    of cities, 8–9
    human, 5–7, 5–17, 217
    industrial, 15
Material, stock of, 38–39
Minerals, area for, 138
Monitoring, 192–193
    of lead pollution, 183–184
    soil, 193
Municipal solid waste (MSW)
    cadmium recycling from, 287–290
    hazardous substances and, 18
    incineration product analysis, 264–269
    as source of lead, 183
    as source of phosphorus, 187

N

Nitrogen
    in Danube River, 199
    recycling, 273
    in urine vs. feces, 219–220
Nitrogen flow per capita, 218
Nitrogen solution, anthropogenic, 220
North Rhine Westphalia, 305
Nourishing activity, 44
    nitrogen flow per capita, 218
    nutrient management, 215–220
    phosphorus consumption, 218, 219
Nutrient management, 215–220
Nutrient pollution of watersheds (Danube Basin), 194–200

O

Objectives, 28
Objectivity vs. subjectivity, 133–135
Organization for Economic Cooperation and Development, 135
Ozone deterioration, 303

P

Packaging materials, 251, 252. *See also* Plastic-waste management
Paper, waste analysis, 261–262

Particulate removal, from power plant emissions, 212–213
Phosphorus
  agriculture as source, 191
  consumption as nutrient, 218, 219
  in Danube River, 199, 200
  industry as source, 190
  private households as source, 186
  RESUB/Bunz Valley study, 184–194
  sewer systems as source, 190
  soils as source, 191
  surface waters as source, 187–190
  in urine vs. feces, 219–220
Plant production flow, 191
Plastic waste management, 251–256
Policy issues, 305–306
Polyamides, 251–256
Polychlorinated biphenyls (PCBs), 303
Polyethylene, 251–256
Polymer wastes, 251–256
Polypropylene, 251–256
Polyvinyl chlorides (PVCs), 251–256, 263
  Austria software example, 83–85
Power plants, coal-fired (SYSTOK study), 200–213
Precautionary principle, 271–272
Priority setting, 182–183, 193
Private households. *see* Households
PRIZMA case study, 279–287
Procedures, 53–69
  identification of relevant components, 58–59
  mass flow determination, 59–60
  presentation of results, 63–67
  stock determination, 59–60
  substance concentrations, 60–61
  substance selection, 54–56
  time/space definition of system, 56–58
  total flow and stock assessment, 61–63
Processes
  defined, 4, 37–39
  key, 65–66
  relevance, 58–59
Processing, nutrient loss through, 219
Products. *See* Goods
Proxy data, 61

R

Raw material
  area for fossil, 138
  renewable, 138
Recycling, 18
  cadmium, 287–290
  copper, 232–233
  entropy trends and, 232–233
  glass in Switzerland, 262
Reference species substances, 143
Region, as term, 11
Regional material balances, early studies, 9–12
Regional phosphorus management (RESUB/Bunz Valley study), 184–194
Regions of impact
  administratively defined region, 208
  defining, 207–210
  potential environmental impact region, 208–210
  product-related region, 207–208
Relative statistical entropy, 225, 228–232
Relevance, of substances, 54–55
Renewable raw material, 138
Reside and work activity, 46–47
Resource conservation, 215–256
  construction wastes management, 225–251
  copper management, 220–235
  nutrient management, 215–220
  plastic waste management, 251–256
  regional lead management, 292–295
Resource management, 16–17
Resources, kinds of, 16
RESUB/Bunz valley studies
  regional lead pollution, 168–184
  regional phosphorus management, 184–194
Results presentation, materials accounting, 64–67
Rivers
  as source of lead, 178–180
  as source of phosphorus, 187–190

S

Sample and sort waste analysis, 258
Sankey diagrams, 108–109, 129–130
Santorio's analysis of human metabolism, 5–7
Selective deconstruction, 250–251
Selenium, coal-fired power plants as source, 207, 212–213
Sensitivity, 234–235
Service units, 136–137
Sewage sludge
  lead in, 171, 183–184, 295
  monitoring of, 183–184
Sewer systems
  as source of lead, 175, 177
  as source of phosphorus, 190
Society for Environmental Toxicology and Chemistry (SETAC), life-cycle assessment method, 140–141

Index

Software, 80–133
　calculation methods, 85–86
　case study, 83–85
　comparison, 132
　GaBi, 113–133
　general requirements, 80–82
　Microsoft Excel®, 86–88
　need for dedicated, 304
　products considered, 83
　special MFA requirements, 82–83
　trial versions, 132
　Umberto, 89–113
Soils
　agricultural as source of lead, 174–175, 176, 182
　agricultural as source of phosphorus, 191
　urban as source of lead, 182
Sorting, of construction wastes, 239
Sorting plant design, 240
Spatial boundaries, 43
Specific flow, 40
Stability, software, 82
Standardization, 304–305
Statistical entropy, relative, 225, 228–232
Statistical entropy analysis, 149–159
　application, 159
　background, 149–151
　calculation of, 222
　first transformation, 153–154
　information theory, 151–153
　second transformation, 154–156
　third transformation, 156–159
　trends, 227–235
Statistical sensitivity, 234–235
Statistical uncertainty, 234–235
Stocks, 38–39
　defined, 4
　determination, 59–60
　fast-changing, 62
　increased accumulation, 25
　key, 65–66
　lead, 292–294
　relevance, 58–59
　total, 61–63, 63–64
　total mass of, 62
　urban, 25
Stressors, environmental, 141–142
Subjectivity vs. objectivity, 133–135
Substance concentrating efficiency, 157
Substance concentration, 60–61
Substance flow analysis (SFA), 50–51
Substances
　concentration, 60–61
　conservative, 36
　defined, 35–36

　fate of, 272–273
　hazardous. *See* Hazardous substances; Heavy metals *(and specific materials)*
　mean concentrations, 67
　reference species, 143
　relevance, 54–55
　selection, 54–56
Sulfur
　in coal, 204–205
　from waste fuels, 283
Support, software, 82, 132
Surface waters
　as source of lead, 178–180
　as source of phosphorus, 187–190
Sustainability, 148
　heavy metals as lacking, 221–222
Sustainable process index (SPI), 137–139
Sweden, chromium emissions, 26
Swiss Ecopoints, 141–142, 272
Switzerland
　glass recycling, 262
　RESUB/Bunz valley studies
　　regional lead pollution, 168–184
　　regional phosphorus management, 184–194
System boundaries, 43
Systems, 4, 43

T

Technosphere. *see* Anthrosphere
Temporal boundaries, 43, 58
Time/space definition of system, 56–58
Total flow, 40
Total mass of stock, 62
Total material flow, 61–63
Total stocks, 61–63
Toxic Equivalent (TE), 145
Transfer coefficients, 40–42
　for cement kiln fuels, 283
　for construction waste separation, 248–250
Transformation, 38
Transport, 38, 47
Trends, entropy, 227–235

U

Umberto software, 89–113
　potential problems, 111–113
　program description, 89
　quickstart, 89–111
Uncertainty, 234–235
　Gauss's law, 69–72

least-square data fitting, 75–78
Monte Carlo simulation, 72–75
Urban areas, as source of lead, 176, 182
Urban material flows, 23–24
Urine, as source of N and P, 219–220
U.S. Environmental Protection Agency (EPA), 259
U.S. National Environmental Policy Act of 1999, 201
User friendliness, of software, 82

V

Vienna Institute of Technology statistical entropy analysis, 149–159

W

Waste analysis
  chlorine, 263
  direct, 259–260
  glass, 262–263
  indirect, 260–269, 261–262
  methods, 258–259
  paper, 261–262
  products of waste treatment, 264–269
Waste management, 256–292. *See also* Wastes
  actual vs. optimized scenarios, 276–278
  applications in, 17–19
  cadmium recycling, 287–290
  construction, 225–251. *See also* Construction waste management
  decision making, 269–290
    ASTRA case study, 269–278
    PRIZMA case study, 279–287
  entropy analysis and, 149–151

incineration, 150–151
policy and legislation issues, 305–306
regional lead management, 292–295
waste analysis, 258–269. *see also* Waste analysis
Wastes. *See also* Waste management; Waste treatment
  combustible, 269, 270, 273
  composition of construction, 242–244
  efficiency of energy content utilization, 272
  municipal solid. *see* Municipal solid waste
  quantity and composition changes, 28
  sorting of construction, 239
Waste treatment, products of, 264–269
Wastewater treatment plants
  as lead sources, 171–173
  as source of nutrient pollution, 197
Water (hydrological) balance, 175, 185–186, 189
Waters, surface
  as source of lead, 178–180
  as source of phosphorus, 187–190
Wingspread Conference, 147
Wolman, Abel, 8
World Resource Institute, 135

Y

Yale University Center for Industrial Ecology, 227

Z

Zinc
  production flow, 27
  technology and emissions, 26–27
  in waste fuels, 283–287